T0358969

Omega-3 Oils
Applications in Functional Foods

Editors
Ernesto M. Hernandez and Masashi Hosokawa

AOCS PRESS

Urbana, Illinois

Library of Congress Cataloging-in-Publication Data

Omega-3 oils : applications in functional foods / editors, Ernesto Hernandez, Masashi Hosokawa.
   p. cm.
   Includes bibliographical references and index.
   1. Omega-3 fatty acids. 2. Functional foods. I. Hernandez, Ernesto, 1955- II. Hosokawa, Masashi, 1966–

QP752.O44O456 2011
612.3'97--dc22

2010052415

Contents

•● Preface ●•

The growing awareness regarding nutritional properties of edible oils has prompted a great deal of research in the development of new sources of edible oils with more healthful compositions. Besides playing an important role in food processing and nutrition, lipids are essential components of many metabolic functions and this is particularly true for omega-3 fatty acids such as eicosapentaenoic acid (EPA) and (DHA). The role of essential fatty acids in nutritional and disease prevention, particularly, omega-3 fatty acids have been studied in numerous health applications. As the general population grows older, formulation and composition profiles of edible fats and oils have become more specialized and the need for novel sources of omega-3 fats has also increased. The general awareness of the health benefits of omega-3 fatty acids is reflected by the growth in consumption of omega-3 fats either as dietary supplements or in fortified foods. EPA is generally associated with cardiovascular protection, and has been reported to have strong anti-inflammatory, anti-thrombotic, anti-arrhythmic, and anti-atherogenic. DHA on the other hand is generally related with cell structures and has been found to be particularly important in neurologically related metabolism, such as brain and retina development and function, and has become an important nutrient in prenatal and post-natal nutrition for mother and children.

This book addresses the role omega-3 plays in general in health and disease. Some chapters review the latest clinical evidence on the impact of n-3 PUFA consumption on prevalent human diseases such as inflammation related illnesses in general and cardiovascular disease in particular. These reviews emphasize preferentially on selected meta-analyses and some chapters are actually original data to demonstrate the beneficial effects of long chain omega-3s. When processing omega-3 oils, it is important and precautions have to be taken to minimize conditions that can destroy the nutritional properties of the omega-3s and generation of unwanted fishy taste and aroma making the oil and food products unacceptable for consumption. The book also examines the different aspects of processing fish oil and other omega-3 fats taking into account issues such as preservation of the nutritional properties, the essential fatty acids, and how fats interact with other components and nutrients in food products in a focused and coordinated manner. Lastly but not least important, this book also explains different methods to deliver omega-3 to the consumer, either through food fortification, nutritional supplements, and also newly developed pharmaceutical products. Including methods used in the protection of the oil against oxidation and techniques of incorporation into foods.

Ernesto M. Hernandez
Masashi Hosokawa

•1•

Omega-3 Fatty Acids in Health and Disease

Fereidoon Shahidi
*Department of Biochemistry, Memorial University of Newfoundland,
St. John's, Newfoundland, Canada*

Introduction

The omega-3 (n-3) fatty acids belong to the family of polyunsaturated fatty acids (PUFAs) with three or more double bonds with the first unsaturation site occurring on the third carbon from the methyl end group. The location of the first unsaturation site dictates the biological activity of the molecules involved. Other double bonds are positioned in a methylene-interrupted manner with respect to the first and subsequent double bonds. Omega-6 fatty acids constitute another family of PUFAs; the first member of this family is linoleic acid (LA, 18:2n-6).

The first member of the omega-3 family and the parent molecule in this series is alpha-linolenic acid (ALA, 18:3n-3), which is abundant in flaxseed oil and is also present in canola, soybean, and walnut oils, among other commodities. Meanwhile, stearidonic acid (18:4n-3) has been detected in viable amounts in several species of algae, fungi, and animal tissues, as well as seeds of Echium (Boraginaceae). It has also been produced in soybean transgenitically. The other important omega-3 fatty acids, namely eicosapentaenoic acid (EPA, 20:5n-3), docosahexancoic acid (DHA, 22:6n-3), and to a lesser extent, docosapentaenoic acid (DPA, 22:5n-3), are found in high amounts in certain algal species and in marine oils, primarily those from the body of fatty fish species such as herring and mackerel, the liver of lean white fish such as cod and halibut, and the blubber of marine mammals such as whales and seals. These important all *cis* omega-3 oils may also be recovered from processing by-products of wild catch and aquaculture fisheries (Shahidi, 2007; Zhong et al., 2007). While alpha-linolenic acid may be metabolized in the body via a series of desaturation and elongation reactions (Fig. 1.1), the production of long-chain omega-3 PUFA from ALA is limited and may vary from 1 to 5% (Shahidi & Finley, 2001). This is because conversion of ALA to stearidonic acid (SA, 18:4n-3) via the action of DELTA-6 desaturase is the slowest and thus the rate-determining step. This enzyme may be impaired and its activity reduced owing to aging, disease condition, lifestyle

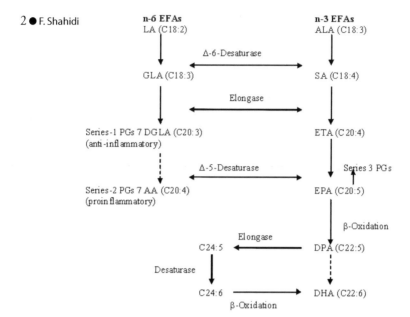

Fig. 1.1. The n-6 and n-3 fatty acids and their metabolites. Abbreviations: EFA, essential fatty acids; LA, linoleic acid; ALA, α-linolenic acid; GLA, γ-linolenic acid; SA, stearidonic acid; PG, prostaglandin; DGLA, dihomo-γ-linolenic acid; ETA, eicosatetraenoic acid; AA, arachidonic acid; EPA, eicosapentaenoic acid; DPA, docosapentaenoic acid; DHA, docosahexaenoic acid.

factors, micronutrient deficiency, or to use of certain drugs. Thus, adequate dietary intake of long-chain PUFA is generally recommended to maintain health and reduce disease risk. The fatty acid composition of selected plant, algal, and animal sources of omega-3 fatty acids are summarized in Table 1.A.

Table 1.A. Fatty Acid Composition of Omega-3 Oils.[a]

Fatty acid	Flax oil	Menhaden	Cod liver oil	Seal blubber	Algal oil
14:0	—	9.4	2.7	3.7	8.7
16:0	5.3	16.0	12.4	6.0	22.2
16:1	—	12.8	5.0	18.0	—
18:0	4.1	2.9	6.1	0.9	0.7
18:1	20.2	10.4	16.3	20.1	0.7
18:2	12.7	2.1	3.0	1.5	0.5
18:3n-3	53.3	1.1	1.0	0.4	0
20:5n-3	—	11.4	13.2	6.4	1.4
22:5n-6	—	0.5	0	4.8	16.3
22:6n-3	—	11.3	15.4	7.6	41.1

[a]Seal blubber oil contains 12.2% of 20:1 and 2.0% of 22:1.

Omega-3 Fatty Acids in Food and Daily Intake Requirements

As it was noted, plant sources of omega-3 oils provide ALA, which may promote health. However, the importance of omega-3 fatty acids is often ascribed to the long-chain PUFAs, especially EPA and DHA. The total content of EPA and DHA in various fish varies and depends on the type of fish and their habitat. The proportion of omega-3 fatty acids in fish muscle is higher in fatty fish such as herring, mackerel, and salmon than in lean fish such as cod, flounder, haddock, perch, and halibut (Table 1.B). Meanwhile, shellfish such as crab (0.3%–0.4%), shrimp (0.2%–0.5%), and lobster (0.3%–0.4%) have low levels of omega-3 fatty acids, and mollusks may also contain low levels of omega-3 fatty acids (e.g., squid, scallops, oysters, and mussels, 0.2%–0.6%) or no omega-3 fatty acids (abalone and some species of clam) (Shahidi, 2006; 2007; Barrow & Shahidi, 2008).

The total recommended daily intake of EPA and DHA is age-dependent. It should be up to 1.5 g for children 1–8 years old, 2.0 g for 9–13 years old, and 2.5 g for 14–18 years old; this value may reach up to 3.0 g/day for adults, depending on the purpose and type of studies conducted. Owing to the oxidizability of such oils, their supplementation with antioxidants such as vitamin E is recommended. However, most suppliers of fish oil capsules generally add vitamin E to their products in order to arrest oxidative processes; this would also address the increasing need of the body on antioxidants because of consumption of highly unsaturated lipids which would otherwise cause oxidative stress. Although manufacturers as well as consumers of dietary supplements may use alpha-tocopherol, mainly *d*-alpha-tocopherol, manufacturers of omega-3 capsules often use mixed tocopherols from soybean oil-refining deodorizer-distillate, which is rich in gamma-tocopherol, as a source of vitamin E for stabilizing the oil because gamma-tocopherol is a much more effective antioxidant than alpha-tocopherol in vitro and hence it better protects the oil from oxidative deterioration.

Table 1.B. Content (g/100g) of Omega-3 Fatty Acids (EPA + DHA)[a] in Selected Fish.

Anchovy	1.4	Halibut	0.2–0.5
Bass	0.3–0.8	Herring	1.3–1.8
Capelin	1.2	Mackerel	0.6–2.2
Catfish	0.3–0.5	Perch	0.2–0.4
Cod	0.2–0.3	Salmon	1.0–1.4
Eel	0.9	Trout	0.6–2.0
Flounder	0.2	Tuna	0.4–1.6
Haddock	0.2		

[a]EPA, eicosapentaenoic acid; DHA, docosahexaenoic acid.

In order to take advantage of omega-3 fatty acids, there has been a surge in the retail market in the appearance of fortified foods as well as a myriad of dietary supplements containing omega-3 fatty acids—occasionally ALA, but mainly EPA and DHA (Shahidi, 2006). Although use of such oils has traditionally been restricted owing to their instability and flavor reversion, availability of stabilized products such as microencapsulated and coacervated marine oils have allowed manufacturers to offer a variety of products. These include cereal-based products such as crackers, bars, bread, and pasta; dairy-based products such as milk, yogurt, and specialty dairy-based drinks; and juices, such as orange juice, as well as other commodities. Table 1.C summarizes the types of products in which omega-3 oils may be used for fortification and/or enrichment purposes. In addition, candies, pastries, and other confectionary products containing omega-3 fatty acids have appeared. Furthermore, meat products and even seafoods have been fortified in test markets and in product development. These formulated commodities either use microencapsulated and coacervated marine oils and do not release their content until it reaches the gastrointestinal tract or employ stabilized omega-3 oils by using sophisticated refining techniques and powerful antioxidants.

For therapeutic purposes, concentrates of omega-3 fatty acids—for example, up to 75% omega-3 content, or even capsules containing pure omega-3 fatty acids—may be used for certain health conditions. The concentrates may be in the ethyl ester form or as triacylglycerols, reconstituted to appear as natural or true-to-nature. Commercially available concentrates may contain equal amounts of EPA and DHA, or predominantly EPA or DHA, depending on the end use. Procedures used for concentration of omega-3 fatty acids are varied and are beyond the mandate of this brief overview (Shahidi & Wanasundara, 1998).

Table 1.C. Different Classes of Food Products Enriched with Omega-3 Fatty Acids.

Class	Items
Dairy	Milk, yogurt, yogurt-based drinks
Grain-based	Bread, cereals, pasta, bars
Confectionary	Sweets, candies, bars
Spreads	Margarine, butter, spreads
Dressings	Salad dressing, mayonnaise, etc.
Juices	Orange juice, other fruit juices, etc.
Muscle foods	Meat, fish, and poultry products
Infant foods	Formula, etc.
Eggs	Hen eggs and products thereof
Others	Specialty products

Omega-3 Fatty Acids in Health and Disease

Marine oils and their omega-3 fatty acids provide the best example of functional food ingredients and nutraceuticals that may serve as a continuum for their perceived and demonstrated health benefits. Thus, they may prevent ailments caused by their deficiency, such as those encountered in the body, and secondly they aid in reducing disease risk, and thirdly they act as therapeutic agents in treating certain diseases or conditions.

A large body of literature provides information about the health benefits of omega-3 fatty acids, mainly those arising from EPA and DHA. EPA and DHA serve as building blocks of cells in vital organs, particularly those with electrical activity such as the brain, the heart, and the eye, and also are important for normal growth and development (Shahidi & Miraliakbari, 2004; 2005). Therefore, adequate intake of omega-3 fatty acids by pregnant and lactating women is encouraged in order to address the need of the fetus or the infant for these essential components that are necessary for the development of the retina and the brain. Whereas most studies have examined the health benefits of EPA and DHA, research on DPA is rather limited because this fatty acid occurs in minor amounts in seafoods and marine oils. However, DPA, which is found in relatively high abundance in blubber of marine mammals such as seal blubber oil, may play a major role in health promotion and disease prevention (Kanayasu-Toyoda et al., 1996). It is worth noting that ALA supplementation increased the concentration of ALA, EPA, and DPA (but not DHA) in breast milk lipids (Francois et al., 2003).

As mentioned earlier, DHA is an important component of the brain, and studies have clearly indicated that infant formulas devoid of omega-3 fatty acids may adversely affect the intelligence quotient, verbal skill, and general performance of the infants when compared with mother's milk that is rich in omega-3 fatty acids. It has also been shown that omega-3 fatty acids are important in addressing attention deficit/hyperactivity disorder, dyslexia, mental health issues, depression, schizophrenia, and memory problems as seen in Alzheimer's disease and dementia, as well as in certain respiratory disorders.

With respect to heart health and associated benefits, omega-3 fatty acids are important in providing a myriad of benefits. These benefits are ascribed to their effect in reducing the risk of stroke and heart disease; reducing blood thickness and blood pressure; reducing the risk of blood clotting and hardening of the arteries; providing relaxation to blood vessels; reducing the level of triacylglycerols; and finally protecting against arrhythmia, ventricular tachycardia, and fibrillation. The latter two effects have been fully documented in the existing literature.

In addition, omega-3 fatty acids may have a positive effect on patients with type-2 diabetes and those with inflammatory disorders, including arthritis and bowel problems, such as Crohn's disease, among others. Skin disorders are also often alleviated, and the shine of the hair (or fur in animals) is improved by consumption of omega-3 fatty acids. Fish oils may also increase low-density lipoprotein (LDL) cholesterol concentration in the body. It is therefore beneficial to use garlic and/or garlic extracts or phytosterols together with fish oils in multicomponent products in order to take advantage of the cumulative effect of such mixed products. Omega-3 oils may also be useful in the treatment of certain types of cancer and mental disorders and could affect gene expression. A cursory account of these health benefits and influence on gene expression is provided in the following section.

Omega-3 Fatty Acids and Cardiovascular Disease

Cardiovascular disease is the common term for all diseases that affect the heart and the circulatory system, including ischemic heart disease, non-ischemic myocardial heart disease, hypertensive heart disease, and valvular heart disease. It is the leading cause of death in the western societies (de Lorgeril et al., 2002) and has been linked to the high fat intake, particularly saturated fat, common in western diets (Dolocek & Granditis, 1991). The hallmark of cardiovascular disease is cardiac dysfunction, which in most cases is caused by hypertension due to the narrowing of large arteries with atheromatous plaques, or the total occlusion of coronary arteries (thrombus) caused by atheromatous blockages leading to myocardial tissue necrosis. Both conditions reduce the heart's ability to pump blood and can result in either chronic or sudden heart failure.

It is well known that regular consumption of fatty fish or fish oils containing n-3 long-chain PUFA lowers the rate of incidence and death from cardiovascular heart disease (Albert et al., 2002; Hu et al., 2002). The cardioprotective effects of marine oils were first postulated in the 1950s based on cross-cultural studies done on Greenland Inuits and Danish settlers of Greenland (Sinclair, 1956). These studies demonstrated that the Greenland Inuits had a significantly lower incidence of heart disease compared to the Danish settlers, despite their comparable fat intakes (40% of caloric intake) and a higher intake of dietary cholesterol. This anomaly was referred to as the 'Eskimo Paradox' (Bang et al., 1980). Epidemiological studies carried out in the 1970s suggested a strong correlation between the low incidence of coronary heart disease in Greenland Inuits and their high consumption of fish and marine mammals, both being rich in long chain n-3 fatty acids (Fischer & Weber, 1984). Other cross-cultural epidemiological studies among coastal Japanese and Alaskan populations lend further support to these findings, showing inverse relationships between long chain n-3 PUFA intake and cardiovascular disease (Hirai et al., 1989; Davidson et al., 1993).

Although the biochemical basis for cardioprotective effects of n-3 PUFAs remains unknown, it may be multifactorial and could collectively result in anti-arrhythmic, anti-atherogenic and decreased platelet reactivity/aggregation (anti-thrombotic). Investigations on the link between fish oils and cardiovascular disease in both animal and human models have concluded that this effect may be mediated by substrate competition between n-3 PUFA and arachidonic acid (AA, 20:4n-6) for cyclooxygenase (COX) enzymes that produce prostaglandins and thromboxanes. Competition between n-3 PUFA and AA could result in positive health benefits because a) n-3 fatty acids inhibit the production of AA through substrate competition for the DELTA-6 desaturase, b) long chain n-3 PUFA compete with AA for incorporation into the sn-2 position of membrane phospholipids thereby reducing membrane AA levels (Siess et al., 1980), and c) eicosanoids produced from EPA have anti-inflammatory and anti-aggregatory effects, for example increasing the membrane EPA/AA ratio shifts eicosanoid production from the pro-aggregatory eicosanoids PGI_2 and TXA_2 towards the anti-aggregatory TXA_3 in platelets (Coker & Parratt, 1985) and PGI_3 in endothelial cells (Fischer & Weber, 1984). These actions would result in vasodilation and decreased platelet aggregation, both having anti-thrombotic effects.

The potential anti-arrhythmic properties of n-3 PUFAs, such as ALA, EPA, and DHA, have been examined in animal models. In one such study, intravenous infusion of either fish oil or pure n-3 fatty acids in exercising dogs before an experimentally induced coronary artery obstruction reduced sudden cardiac death by preventing ventricular fibrillation (Billmann et al., 2000). The mechanisms of the antiarrhythmic effects of n-3 PUFAs have been further explored using spontaneously contracting cultured cardiac myocytes isolated from rats (Kang & Leaf, 2000). The cultured rat myocytes were induced to fibrillation using various toxic agents such as ouabain, beta-adrenergic agonists, and high Ca^{2+} concentrations added to the bathing medium. EPA and DHA added at low doses of 5–15 mu mol/L in the bathing medium inhibited the expected fibrillation when the toxic agents were added to the media (de Lorgeril et al., 1998). Interestingly, when the added EPA and DPA were removed from the culture media using delipidated bovine serum albumin, the cultured myocytes returned to fibrillation. Thus, n-3 PUFAs at low concentrations are able to modulate the activity of specific ion channels in myocardial sarcolemma.

One of the most effective ways to protect the myocardium from ischemic/reperfusion injury is by inhibition of the transmembrane Na^+/H^+ antiport exchanger. This transmembrane antiport system maintains the myocardial cell's pH, but during ischemia this system paradoxically participates in cell necrosis. The importance of transmembrane Na^+/H^+ antiport exchanger in ischemic heart disease was shown in a clinical trial using a specific Na^+/H^+ exchange inhibitor caporide (Theroux et al., 2000). The inhibitor showed a potential benefit in reducing the risk of cardiac death as long as it was taken prior to the ischemic event (Amusquivar et al., 2000). It has recently been shown that EPA and DHA at concentrations of 25–100 mu M inhibited the Na^+/H^+ antiport exchanger in isolated cardiomyocytes and, thus, could protect the myocardium from arrhythmias and cell death during ischemic events. This effect was limited to long chain n-3 fatty acids, as LA and ALA showed no significant effects on the Na^+/H^+ exchanger (Goel et al., 2002). Although the role of ALA in the prevention of arrhythmia caused by ischemia or during reperfusion remains unclear, the metabolic conversion of ALA to EPA is thought to mediate any of the cardioprotective effects of ALA, and at least one study using a canine model revealed cardioprotective effects with ALA (Chaudry et al., 1994).

The "diet and reinfarction trial (DART)" was the earliest controlled trial that examined the effects of dietary intervention in the secondary prevention of myocardial infarction (Burr et al., 1989). The study included 2033 Welsh men who had recovered from an earlier heart attack and were allocated to one of two groups that received or did not receive advice on each of three dietary factors, namely reduced fat intake and an increased ratio of polyunsaturated to saturated fat, increased intake of fatty fish, and increased consumption of cereal fiber. The results of this landmark trial revealed that men receiving at least 2 fatty fish meals per week (200–400 g of fish meat or 1.5 g fish oil per day through supplement) had a 29% reduction in fatal heart attack risk, which was apparent after 4 months of intervention, when compared to the non-fish consuming groups, but there was no significant reduction in the incidence of non-fatal heart attacks. The results of this study strongly suggest that marine n-3 fatty acids have a specific antiarrhythmic effect rather than anti-atherogenic or anti-thombotic effects (Burr et al., 1989).

The GISSI (1999), or Gruppo Italiano per lo Studio della Sopravvienenza nell 'Infarto Miocardio Prevenzione, study was initiated in 1993 and carried out over 3.5 years. This multicentered trial was conducted in Italy (172 centres) and included 11,324 patients who had suffered a heart attack less than 3 months prior to recruitment. Just as in the DART study (Thun et al., 1991), the GISSI Prevenzione study revealed that marine n-3 fatty acid intake conferred early and progressive risk reductions for cardiovascular disease.

The Lyon Diet Heart Study (de Lorgeril et al., 1998) conducted in France was one of the earliest intervention trials (with 204 control subjects and 219 experimental subjects), making the hypothesis that a Mediterranean diet high in ALA could reduce the relative risk of cardiovascular events and/or death in previous heart attack victims. After 1 year of study, total cardiovascular events were reduced from 34.7% in the control group to 24.5% in the fish oil group and 28% in the mustard oil group (Narayanan et al., 2001).

Omega-3 Fatty Acids and Inflammatory Diseases

Chronic inflammation associated with diseases such as inflammatory bowel disease, psoriasis, atherosclerosis, and rheumatoid arthritis may be caused or attenuated by alterations of normal cytokine pathways resulting in cytokine overproduction. Many anti-inflammatory pharmacotherapies have been developed to inhibit the production of pro-inflammatory cytokines. Omega-3 fatty acids may be used in the treatment and/or management of inflammatory diseases because they alter the cytokine biosynthesis (Wanasundara & Shahidi, 1997).

Arachidonic-acid-derived cytokines have proinflammatory actions in-vivo, whereas those derived from EPA are significantly less proinflammatory (Shoda et al.,

1995; Bagga et al., 2003). Studies investigating the effects of n-3 PUFAs on *ex-vivo* cytokine production by leukocytes is an active area of research, but these studies have produced inconsistent results. Mantzioris et al. (2000) showed a 20% decrease in *ex-vivo* IL-1 beta production in healthy men after 4 weeks of supplementation with 1.8 g of fish oil per day, demonstrating that n-3 PUFAs affect cytokine production by leukocytes, but this study was not a controlled trial. Results of a recent placebo-based, double-blind, parallel study involving 150 healthy men and women aged 25–72 years who were supplemented with ALA or fish oil revealed no significant differences in *ex-vivo* cytokine production (TNF-alpha, IL-6, IL-1beta and IL 10) between the placebo and intervention groups after 6 months of supplementation (Kew et al., 2003). However, they did show that monocytes had significantly increased levels of ALA in participants supplemented with α-linolenic acid, and increased monocyte EPA and DHA levels in those given fish oil. Both n-3 groups also had lowered monocyte AA levels compared to the control group, which might lead to decreased synthesis of the pro-inflammatory LTB_4 in-vivo (Tilley & Maurice, 2002). Although the mechanisms by which n-3 fatty acids suppress the production of inflammatory cytokines are unknown, the suppression of inflammatory eicosanoid production by EPA is likely to be involved.

Inflammatory bowel disease (IBD) is a general term for chronic inflammatory diseases of the gastrointestinal tract and mainly includes ulcerative colitis and Crohn's disease. The incidence of ulcerative colitis and Crohn's disease is higher and rising in Western countries compared to Asian countries, and epidemiological studies have attributed this trend to high intake of saturated and n-6 PUFA in typical western diets (Shoda et al., 1996).

The effects of n-3 fatty acid supplementation in IBD have been studied using many animal models. Thus, the effects of perilla oil (n-3, alinolenic acid rich), fish oil (n-3 long chain fatty acid rich), and safflower oil (n-6 fatty acid rich) supplementation on ulcer formation and pro-inflammatory cytokine production in rats have been examined. These results suggest that ALA may be superior to EPA and DHA for controlling intestinal inflammation in experimentally induced Crohn's disease, but the authors could not rule out the possibility of synergistic effects between n-3 fatty acids and other bioactives in perilla oil (Shoda et al., 1995). Nieto et al. (2002) used the trinitrobenzenesulfonic acid model to study the effects of n-3 fatty acid supplementation on ultrastructural and histological changes during experimentally induced ulcerative colitis in rats and revealed that rats given an n-3 fatty acid rich diet had significantly less macroscopic and microscopic colonic damage when compared to both the n-6 group and the n-6 plus n-3 group. In addition, the n-3 group had significantly lower inflammatory marker levels when compared to both other groups, which strongly suggest that n-3 fatty acids are therapeutic, whereas n-6 fatty acids exacerbate experimentally induced ulcerative colitis. Another model study uses 4% acetic acid feeding to induce IBD. Thus, using this model, rats given a fish oil (EPA) enriched diet for 6 weeks after treatment had improved intestinal function and considerably less histologic injury compared to rats given low n-3 fatty acid diets after treatment, demonstrating that n-3 PUFAs, especially EPA, have protective effects against acetic acid induced colitis (Empey et al., 1991).

Although animal models provide strong evidence for the protective effects of n-3 fatty acids against induced IBD, such models may not accurately portray the human etiology of this disease since it is induced using noxious chemicals. Several epidemiological studies have shown an inverse relationship between n-3 fatty acid intake and the risk of IBD. In addition, some intervention studies have shown that n-3 fatty acid supplementation is an effective therapeutic means to manage these diseases (Alsan & Triadafilopoulos, 1993). A 24-year study showed that the Greenlandic people who consumed a diet rich in marine-derived n-3 fatty acids exhibited a significantly lower incidence of inflammatory bowel disease when compared to Western populations. Furthermore, the incidence of Crohn's disease and dietary habits among Japanese men and women over a 19-year period was examined (Shoda et al., 1996). This study showed that individuals with lower dietary n-6/n-3 ratios were 21% less likely to suffer from Crohn's disease (RR 0.79). IBD sometimes exhibits alternating relapses and remissions, and some clinical studies have investigated the potential of n-3 fatty acids to prolong periods of remission. Belluzzi et al. (1996) carried out a double blind, placebo-based study to investigate the effects of 2.7 g per day of fish oil supplements in 78 patients with Crohn's disease who were at high risk for relapse as assessed by the Crohn's Disease Activity Index. After one year, 59% of patients in the fish oil group remained in remission (23 out of 39) compared to 26% in the placebo group (10 out of 39). Further analysis revealed the difference in relapse rate between the two groups to be due to fish oil supplementation only; cigarette smoking, gender, previous surgery, age, and duration of the disease did not affect the likelihood of relapse. In addition, examination of blood for indicators of inflammation (serum as–acid glycoprotein, serum a_2-globulin) revealed that the fish oil group had a significant decrease in all inflammatory markers assayed compared to the control group after one year.

Not all studies have supported the therapeutic effects of n-3 fatty acids in inflammatory bowel disease sufferers. For example, Lorenz-Meyer et al. (1996) performed a double-blind, placebo-based trial on 204 Crohn's disease patients in remission to study the effects of highly concentrated n-3 PUFAs on the maintenance of remission over a 12-month period. At the end of this trial there was no difference between the n-3 and the control groups; specifically, 30% of patients in both groups remained in remission. However, at the end of this study it was noticed that the n-3 group required less drug therapy (prednisolone) to manage the disease compared to the control group. This result implies that n-3 fatty acid supplementation may be somewhat helpful in the treatment of Crohn's disease. A recent clinical trial by Middleton et al. (2002) of 63 ulcerative colitis patients studied the effects of a combination of fish- and plant-derived n-3 fatty acids on disease remission. After 12 months, duration of remission was not significantly different (P:0.05) between groups (n-3 group: 55% remained in remission, control group: 38% remained in remission).

Based on these results, although there was a 17% increase in disease remission in the n-3 group, Middleton et al. (2002) were unable to support the therapeutic benefits of n-3 fatty acid supplementation. The insignificant effects of n-3 fatty acids may be due to the relatively low doses of n-3 fatty acids used in the study (1.9 g total).

There is a wealth of evidence both supporting and refuting the therapeutic potential of n-3 fatty acids for inflammatory bowel diseases. The conflicting results are most likely due to differences in study size, duration, source of n-3 fatty acids, and the amount of n-3 fatty acids provided. Therefore, more animal studies are needed to develop a comprehensive biochemical basis for the theorized effects of n-3 fatty acid supplementation in the treatment of inflammatory bowel diseases.

The effects of n-3 fatty acid supplementation in patients with arthritis, particularly rheumatoid arthritis, have been investigated and the effects of manipulating dietary fat intake on clinical outcomes in patients with rheumatoid arthritis have been examined (Kremer et al., 1985). The n-3 group reported noticeable reduction in morning stiffness and number of tender joints. The beneficial results were attributed to the intervention regimen conducted on the n-3 supplemented group by this group. Volker et al. (2000) carried out a randomized, placebo-based, double-blind clinical study to determine the effects of fish oil supplementation on clinical variables. After 15 weeks of supplementation, there was a significant improvement ($p<0.02$) in the clinical status of patients in the n-3 group compared to the placebo group. Although trials by Kremer et al. (1985) and Volker et al. (2000) do provide evidence about therapeutic benefits for n-3 fatty acid supplementation in rheumatoid arthritis, neither of these clinical trials were long-term studies, lasting 12 and 15 weeks, respectively. Geusens et al. (1994) studied the long-term effects of n-3 fatty acid supplementation in patients with active rheumatoid arthritis in a 12-month double-blind, randomized study. Ninety subjects who did not receive dietary interventions were supplemented daily with either 2.6 g of fish oil, or 1.3 g of fish oil and 3 g of olive oil, or 6 g of olive oil. After a 12-month supplementation period only the fish oil group (2.6 g per day) exhibited significant clinical improvements in both the patient's evaluation of pain and the physician's assessment of pain. In addition, a significant number of patients in this group had reduced antirheumatic medication use throughout the 12-month trial.

No significant improvements occurred in the combined fish and olive oil group or in the olive oil only group, implying that the observed therapeutic benefits of fish oil supplementation in patients with rheumatoid arthritis were dose dependent, with doses less than 2.6 g per day being ineffective.

Considerable evidence from in vitro and human studies suggest that n-3 fatty acids serve as effective therapeutic agents for the management of inflammatory arthritic diseases, but the biochemical basis for these observations are not well understood. However, it is likely that n-3 fatty acids exert their antiarthritic affects through modulation of inflammatory cytokine production. More in-depth knowledge of the roles of cytokines in inflammatory arthritic diseases is needed to understand how n-3 fatty acids influence this disease. Also, more long term and large-scale intervention studies investigating the effects of n-3 fatty acid supplementation on arthritis symptoms are needed in order to strengthen the proposed inverse relationship between n-3 fatty acids and inflammatory arthritic diseases. Several human studies have investigated the immunosuppressive effects of n-3 fatty acids in transplant patients. Thus, the effects of fish oil supplementation on kidney transplant acceptance and renal function was examined (Homan van der Heide et al., 1993). After 1 year there was an overall improvement of renal function in the fish oil supplemented group. The total number of rejection episodes was lower in the fish oil group compared to the control, as was mean arterial blood pressure. The observed hemodynamic and immunomodulatory effects of fish oil were speculated to be due to a shift away from the vasoactive and proinflammatory AA eicosanoids to EPA derived eicosanoids. However, a similar but more sophisticated study by Hernández et al. (2002) investigated the effects of fish oil supplementation (6 g per day) on kidney function and kidney rejection rate, as well as on proinflammatory cytokine production in 86 kidney transplant patients. After 12 months no differences were observed in the above parameters between the n-3 group (fish oil, experimental group) and the control group (6 g per day of soybean oil). However, this study may have been complicated by the choice of soybean oil as the placebo fatty acid source because soybean oil contains approximately 8% ALA, which may reduce the significance of differences between the experimental and the control groups.

Studies on animals and humans investigating the potential immunosuppressive effects of postoperative n-3 PUFA supplementation in organ transplant patients have been inconsistent, possibly due to the existing differences in study design such as amounts and sources of fatty acid supplements, the duration of study, and the type of organ transplant surgery studied. More clinical trials are needed to clearly support the beneficial effects of n-3 fatty acid immunonutrition in organ transplant patients.

The frequency of inflammatory lung diseases is increasing in western societies (Burney et al., 1990). Some have speculated that this trend may be due to high n-6 to n-3 fatty acid ratios in typical western diets. This may cause increased proinflammatory cytokine production and lead to bronchial inflammation in those prone to inflammatory lung diseases. Case control studies indicate that children who do not consume fish early in life are 3 times more likely to have asthma than those who do (Peat et al., 1992). EPA and DHA in fish tend to reduce the incorporation of arachidonic acid into membrane phospholipids and have been shown to decrease the production of proinflammatory arachidonic acid derived eicosanoids. Thus, fish oil may have therapeutic effects on inflammatory lung disease symptoms.

Koch et al. (1993) studied the effects of short-term infusions with polyunsaturated fatty acid emulsions on the pulmonary response to inflammatory stimulation (increased vascular resistance and permeability) in perfused rabbit lungs. Results for pulmonary artery pressure and lung weight gain (indicating edema formation) were significantly lower in the n-3 fatty acid group than in the control and n-6 fatty acid groups.

The therapeutic potential of n-3 fatty acids on bronchial inflammation in asthmatics has been examined in several clinical and prospective studies; most of which studied asthmatic children, or children at high risk of developing asthma. Hodge et al. (1996) investigated the association between consumption of oily fish and recurrence of pulmonary wheeze in 584 previously diagnosed asthmatic children (8–11 years old) living in Sydney, Australia. There was a small but significantly decreased (P<0.05) risk for current wheeze in children who consumed any amount of fresh fish or fatty fish (1 or more fresh/fatty fish meals per week). These results remained significant after adjustment for other possible risk factors such as parental asthma, parental smoking, ethnicity, early respiratory illness, and sex. Although Hodge et al. (1996) reported that oily fish consumption reduced the risk of asthma, this study only assessed asthma risk when questionnaires were distributed; it did not assess the prevalence or extent of asthma symptoms in subjects throughout the 5-month study period. Troisi et al. (1995) examined the association between several dietary factors and adult onset of asthma in the 10-year Nurses Health Study (77,866 women). Their results revealed no association between n-3 or n-6 fatty acids and asthma, but positive associations were observed between asthma and antioxidant vitamin intake (vitamins C and E, beta-carotene). These data suggest that n-3 fatty acid intake during adulthood is not an important determinant of asthma.

The symptoms of asthma are quite variable—questionnaires cannot adequately assess the prevalence and severity of asthma. Therefore, epidemiological studies are less reliable than clinical trials. Nagakura et al. (2000) performed a 10-month, placebo-based, randomized trial to evaluate the effects of fish oil capsules in 29 young patients (4–17 years old) with asthma who were receiving long-term treatment at the Department of Paediatrics-Higashi Saitama Hospital, Japan. The results indicate that EPA reduces the bronchoconstrictive effects of acetylcholine but not the frequency or severity of asthma attacks. The authors of this study speculate that their results may be due to the ability of n-3 fatty acids to reduce the production and release of proinflammatory eicosanoids but not of histamine, which may explain the decreased response to inhaled allergens but unchanged overall asthma symptoms. Emelyanov et al. (2002) performed an 8-week, placebo-based, randomized clinical trial to examine the effects of a lipid extract from New Zealand green-lipped mussel (rich in EPA and DHA) on the symptoms and biochemical markers of asthma in adults (18–56 years old). There were no significant differences in forced expiratory volume between the n-3 and placebo groups after 8 weeks of supplementation, but significant decreases in mean expired H_2O_2 (P=0.0001) and mean daytime wheeze (P=0.026) were observed after 8 weeks in the Lyprinol® group compared to placebo. Based on these results, it was concluded that Lyprinol® supplementation (100 mg of n-3 fatty acids daily) improved symptom management in adult asthma sufferers (Emelyanov et al., 2002). It was also noted that throughout this study no significant changes in blood pressure, serum creatinine, bilirubin, liver transaminase or alkaline phosphatase occurred in either group, which indicates no ill effects of Lyprinol® supplementation.

Based on the above summary, it is clear that a considerable body of evidence exists that both supports and refutes the potential therapeutic benefits of n-3 fatty acid supplementation in asthma. To date, only one large-scale intervention study investigating n-3 fatty acid supplementation in asthmatics has been initiated; the results of this 5-year study are highly anticipated because most previous trials investigating n-3 fatty acid supplementation in asthmatics have been on a small scale.

Omega-3 Fatty Acids and Cancer

Cancer is a general term for the more than 100 diseases that are characterized by uncontrolled and abnormal growth of cells (neoplasia) that are derived from normal tissues. The first description of these symptoms was in relation to breast carcinoma. Experimental and epidemiological studies have demonstrated that the composition of dietary fat affects the incidence and progression of some cancers; n-3 fatty acids have been shown to have anti-carcinogenic effects while saturated and n-6 fatty acids may promote cancer development (Prener et al., 1996). Several cancerous cell lines have been developed from animal and human tumors; these are established cell lines (e.g., Hep-G2 cells, caco-2 cells, LNCaP cells, PC-3 cells) derived from malignant tumors that proliferate indefinitely in culture under the appropriate conditions. Cell lines serve as excellent in vitro models for cancer studies because the biochemical processes that occur within these cells are remarkably similar to those within their parent tumors.

Epidemiological studies indicated that n-3 fatty acids might be protective against prostate cancer. Cross-cultural studies among the Inuit and non-Inuit peoples of Canada, Alaska and Greenland from 1969 to 1988 showed that the incidence rate of prostate cancer among the Inuit populations was 70%–80% less than the non-Inuit populations (Bhagavathi et al., 2003). This observation was attributed to dietary differences between the two populations, in particular the traditional seafood diet of Inuit peoples that are exceptionally rich in n-3 fatty acids were speculated as having anti-carcinogenic effects (Bhagavathi et al., 2003). The association between fatty fish consumption and prostate cancer in a long-term prospective cohort of 6,272 Swedish men was carried out by Terry et al. (2001). After adjustment for other dietary and lifestyle habits (multivariate analysis), it was observed that there were significant inverse associations between fatty fish consumption and prostate cancer incidence as well as prostate cancer death. Mamalakis et al. (2002) examined adipose tissue and prostate tissue fatty acid composition in 71 prostate cancer and benign hyperplasia patients from the island of Crete. Relative to benign hyperplasia patients, cancer patients had elevated adipose tissue levels of saturated fatty acids and reduced adipose tissue levels of MUFA. Compared to hyperplasia patients, cancer patients had reduced prostate tissue stearic acid to oleic acid ratios, and total stearic acid levels. Relative to benign hyperplasia patients, cancer patients had reduced prostate tissue levels of AA, DHA, EPA, total n-3 fatty acids, and n-3/n-6 fatty acid ratios. The pronounced elevation in adipose tissue saturated fatty acid levels in cancer patients highlights a possible role of dietary saturated fats in neoplastic processes, since adipose tissue fatty acid composition mimics the composition of dietary fats ingested (Amusquivar et al., 2000). The decreased prostate tissue level of C20 and C22 PUFAs in cancer patients possibly stems from enhanced metabolism of these fatty acids via lipoxygenase and cyclooxygenase pathways. Augustsson et al. (2003) analyzed data from the 12-year Health Professionals Follow-up Study to investigate whether high dietary intake of fish and long chain n-3 PUFAs reduced the risk of prostate cancer in 47,882 male American participants. Consumption of 3 or more fish meals per week was associated with a reduced risk of prostate cancer compared to infrequent fish consumption (less than 2 fish meals per month), and the strongest association was for metastatic cancer (RR=0.56, 95% CI: 0.37–0.86). Intake of n-3 PUFAs from foods other than fish showed a similar but weaker association. Each additional daily n-3 PUFA intake of 0.5 g from food was associated with a 24% decreased risk of metastatic prostate cancer. The results of this study show dietary n-3 PUFAs from fish and other sources reduce the risk of prostate cancer, especially advanced forms of prostatic carcinomas. These results imply that long term consumption of fish meat and n-3 PUFAs may slow the progression of prostate cancer towards metastasis, as evidenced by the significantly lowered relative risk for metastatic prostate cancer among the participants of this study (Rose & Connolly, 1999).

Experimental and epidemiological studies suggest that n-3 PUFAs have anti-tumor effects during the initiation and post-initiation stages of colon carcinoma (Reddy, 1994). Western populations exhibit significantly higher colon cancer incidence and mortality rates compared to Asian populations, which experts have long associated with high dietary fat and animal fat consumption by Western populations (Wynder et al., 1969). Caygill (1996) examined colon cancer mortality data from 24 European countries, showing a significant inverse correlation between colon cancer mortality and fish meat/fish oil consumption. This inverse correlation was significant for both men and women who consumed fish or fish oil for 1 year, 10 years, or 23 years before cancer mortality. This study strongly suggests that fish oil consumption can significantly reduce colorectal cancer mortality. Unfortunately, dietary amounts of fish or fish fatty acids were not adequately assessed in this study, making it impossible to critically assess these findings.

Anti et al. (1992) studied the effects of n-3 PUFAs on colonic cell proliferation in subjects at high risk for colon cancer. The results of this short-term study show that n-3 PUFAs reduce the proliferation of early stage colonic cancers, which may reduce the progression colorectal polyps to colorectal carcinoma and may protect high-risk individuals from colon cancer.

The vast majority of research on n-3 PUFA and colorectal cancer has been carried out using animal models and in-vitro studies. Takahashi et al. (1997) studied the effects of DHA supplementation on colon cancer using a rat model. The results of this study do not strongly support the premise that n-3 PUFAs, especially DHA, protect against colon carcinoma, but the aberrant results may be due to the low amount of DHA supplement. It is also possible that the protocol used to supplement DHA, specifically intragastric injection rather than dietary supplementation, may have caused additional stress in the animals and compromised their immune system, which could have enhanced the tumor-promoting effects of AOM regardless of DHA supplementation. Dwivedi et al. (2003) used a similar protocol to Takahashi et al. (1997) to study the effects of dietary n-3 and n-6 fatty acid supplementation on colon cancer development in male Wistar rats. This study also shows that oils rich in EPA and DHA are not as chemoprotective against colon carcinogenesis when compared to oils rich in ALA. The corn oil supplementation had the least chemoprotective effects against colon carcinogenesis. Since no control group or saturated fatty acid group were employed in this study, it is impossible to assess the promotional or inhibitory effects of corn oil and n-6 PUFAs on colon cancer development, except to say that n-6 PUFAs are less inhibitory than their n-3 counterparts.

Several studies have investigated the effects of n-3 and n-6 PUFAs on colon carcinoma cell line development and cyclooxygenase expression/activity. In one such study (Dommels et al., 2003) the effects of n-3 and n-6 fatty acids on the proliferation of two COX overexpressing colon cancer cell lines in culture (Caco-2 cell line and HT-29 cell line) were investigated; this group also assessed COX activity by measuring PGE_2 production in response to fatty acid treatments. Narayanan et al. (2001) studied the effects of DHA on Caco-2 cell growth, proliferation, and transcription rates of 3,800 genes belonging to 156 functional categories using DNA microarrays. Functional gene groups examined included several tumor suppressors, apoptosis factors, growth factors, chemokines, lipooxygenases, cyclooxygenases, transcription factors, cellular receptors and nuclear receptors. Compared to control cultures, the DHA treatment resulted in a 30% decrease in Caco-2 cell proliferation after 48 h. Hybridization occurred in only 13% of the genes present on the microarray, implying that the array used in this study was improperly designed. To confirm the microarray results of several representative genes, Narayanan et al. (2001) amplified RNA extracts using RT-PCR followed by separation using a denaturing polyacrilamide gel and northern blotting. Their results showed that DHA down regulated the expression rate of several genes encoding transcription factors, transcriptional enhancers, RNA polymerases, lipooxygenases, COX-2, and the inducible nitric oxide synthase. DHA treatments enhanced the expression of peroxisome proliferator activated receptors a and y by over twofold. These changes seem to indicate that DHA promotes Caco-2 cell apoptosis through modulation of several biological activities and suggests that DHA is an effective chemopreventive agent against colon carcinogenesis (Narayanan et al., 2001). Another study, using the same protocol and cells, showed that DHA enhances the expression of several cell cycle inhibitors, which further illustrates the antitumor effects of DHA in colon carcinogenesis (Bhagavathi et al., 2003).

Very few epidemiological studies have assessed the effect of dietary n-3 PUFAs and breast cancer. Holmes et al. (1999) analyzed data from 1982 breast cancer patients (mean age 54 years) registered in the 18-year Nurses Health Study. The results of this assessment showed that n-3 fatty acid intake significantly reduced breast cancer mortality by 48% (RR-0.52; 95%C1: 0.30–0.93). Recently, Holmes et al. (2003) re-examined the data of the 121,700 female nurses registered in the Nurses Health Study to find associations between breast cancer and dietary intake of meat, fish and eggs. After the 18-year follow up period 4,107 cases of breast cancer were diagnosed. Women in the highest quintile for meat, egg and fish intake showed no difference in breast cancer risk; secondary analyses did not affect these results. Similar results have been observed in the 8-cohort international pooling project involving 350,000 women who were followed up for 15 years (Bougnoux, 1999). Maillard et al. (2002) evaluated the fatty acid composition of adipose tissue from 241 women patients from central France with non-metastatic breast cancer and 88 patients with benign breast tumors to assess the protective effects of dietary n-3 fatty acid intake against breast cancer, using adipose fatty acid composition as a biomarker of past dietary fatty acid composition. This study showed that adipose n-3 fatty acid levels and breast cancer risk were inversely associated. Women with the highest adipose levels of ALA were 6% less likely to develop breast cancer compared to women with the lowest adipose ALA levels (RR-0.39; 95% C1: 0.19–0.78). Similarly, adipose DHA level and n-3 PUFA to n-6 fatty acid ratio were inversely associated with breast cancer risk (RR=0.31 and 0.33, respectively; 95% Cl: 0.13–0.75 and 0.17–0.66, respectively). These results suggest that n-3 PUFAs have protective effects against breast cancer risk, and also show that n-3 and n-6 PUFAs affect breast cancer risk.

The proposed anti-tumorigenic effects of n-3 PUFAs in breast cancer have been studied using in vitro and animal models of this disease; several studies show n-3 fatty acids are able to modulate second messenger systems and cell signalling cascades in cancerous breast cells. The ability of n-3 fatty acids to inhibit breast tumor development has been shown in several tumor transplant studies (Ip, 1997). Kort et al. (1987) showed that female rats with transplanted mammary carcinoma tumors (BN472 cells) who were fed 25% fish oil after tumor transplantation for 6 weeks exhibited significantly less tumor development compared to rats fed 25% cacao butter. Rose et al. (1995) showed a similar suppressive effect of n-3 PUFAs on transplanted human mammary tumors (MDAMB-435 cells). Nude mice given 20% fish oil diets after tumor transplantation showed less metastatic growth of the implanted tumors into the lungs and overall suppression of tumor development. Robinson et al. (2002) showed that dietary fish oil supplementation (50 mg/g chow) for 21 days after tumor transplantation did not significantly effect tumor development.

Several non-human studies support the premise that n-3 PUFAs inhibit breast carcinoma development by influencing the biochemical events that follow tumor initiation. Unfortunately, these findings do not correlate well with human breast cancer studies. This may imply that n-3 PUFAs at attainable human dietary levels (1%–3% of total calories) do not affect breast cancer development. More recently, Shahidi & Zhong (2010) reported that esters of epigallocatechin gallate (EGCG) with DHA were able to totally arrest tumorigensis in mice subjected to azoxymethane (AOM). However, other esters of EGCG, such as those with butyric acid or stearic acid did not display such overwhelming and exceptional results.

Omega-3 Fatty Acids in Neural Function and Mental Health

The human nervous system has the highest lipid content compared to all other non-adipose tissues; 50%–60% of the total dry weight of the adult human brain is lipid (Mahadik et al., 2001) and approximately one third of these lipids are n-3 fatty acids, mostly DHA (Bourre & Dumont, 1991). DHA is especially important during pre-natal human brain development; incorporation of DHA into growing neurons is a prerequisite for synaptogenesis (formation of synapses) (Martin & Bazan, 1992). The period of greatest brain development occurs from the third trimester of pregnancy until 18 months after birth; this period correlates well with the accumulation of DHA in this organ (Martinez, 1992). The importance of n-3 fatty acids during pre-natal development is best indicated by the observation that deficiency of these fatty acids during development greatly increases the likelihood of diminished visual acuity, cerebellar dysfunction and several cognitive impairments and neurological disorders (Chamberlain, 1996). The importance of n-3 fatty acids during human development is also evident by the fact that both the placenta and mammary tissues supply large amounts of DHA to the developing fetus and infants (Neuringer, 1993).

The effects of n-3 PUFAs on the clinical symptoms of depression and schizo-phrenia have received considerable attention. Depression is the most prevalent psy-chiatric disorder in North America; in the United States 1 in 20 people suffer from unipolar depression and 1 in 100 experience bipolar or manic depression (Keller, 2002). Omega-3 fatty acid supplementation is receiving much attention as a pos-sible adjunct therapy for depression; epidemiological and clinical studies suggest inverse association between n-3 fatty acid consumption and depression. Recent stud-ies among Inuit populations show an overall decline in mental health characterized by increased rates of depression as well as other mental illnesses, which may be linked to the rapid changes to a westernized culture including a shift from traditional sea-foods to processed foods (McGrath-Hanna et al., 2003). Tanskanen et al. (2001) per-formed a large survey to assess depression symptoms and frequency of fish intake among a cohort of 3,204 Finnish adults aged 25 to 64 years old. The results of this survey showed that mild to severe depression symptoms were 31% more prevalent among infrequent fish consumers (<3 fish meals per month); gender based assess-ments reached significance among female participants but not in men (P<0.01, chi-square tests). Thus, results from this large cohort of adults show strong correlations between infrequent fish consumption and depression, but unfortunately this study could not investigate the effects of n-3 fatty acids since information of the type of fish consumed was not obtained. Recently, Marangell et al. (2003) evaluated the effectiveness of DHA supplementation for the treatment of depression. The difference between groups were not statistically significant (two-way t-tests), however these results may be due to the relatively low level of n-3 fatty acids supplemented. A similar but smaller clinical trial showed that daily supplementation with 9.6g fish oil for 8-weeks significantly reduced depressive symptoms compared to a placebo group ($P<0.05$, Wilcoxon signed rank test) (Su et al., 2003).

The etiology of depression is not fully understood, however, several pathophysiological features of this disease have been identified, including overproduction of inflammatory cytokines (Smith, 1991). The beneficial effects of n-3 PUFAs in depression may be due to modulation of eicosanoid production. The fact that EPA-derived eicosanoids are the least proinflammatory ones provides a possible explanation for the beneficial effects of n-3 PUFAs supplementation in depression, since depression has been linked to proinflammatory cytokine production. Because the eicosanoid products of n-3 PUFAs do not activate macrophages to any extent compared to those derived from n-6 PUFAs, replacement of membrane n-6 fatty acids with those of the n-3 family would reduce proinflammatory cytokine production, especially if cyclooxygenase activity is enhanced in depressive patients. Several lines of evidence support beneficial effects of n-3 fatty acids on depressive disorders, but this evidence is far from being conclusive.

Schizophrenia is a mental disorder that affects 1% of all people regardless of race or nationality. Previous family history of schizophrenia is the major risk factor for this disease, however oxidative injury to neuronal cells and abnormal neuronal membrane phospholipid composition have been observed in schizophrenic patients post mortem. Reduced DHA levels have been observed in neurons of schizophrenic patients that may be the result of phospholipase A2 overexpression (Horrobin, 1992), hence n-3 PUFA supplementation has received attention. Many schizophrenic patients show signs of excessive in vivo lipid peroxidation; these include increased plasma thiobarbituric acid reactive substances (TBARS) (McCreadie et al., 1995) and breath pentane (Phillips et al., 1995), which suggests that reduced DHA in schizophrenics may be due to increased oxidative stress. The potential role of dietary n-3 PUFAs on antioxidant enzymes and parameters of oxidative stress was recently studied in rat neurons (Sarsilmaz et al., 2003). Results showed that n-3 fatty acid treated rats had significantly lower TBARS in corpus striatum neurons ($P<0.001$, ANOVA) and significantly less corpus striatum nitric oxide and striatum xanthine oxidase activity levels compared to controls. These results indicate that n-3 PUFA can improve oxidant parameters in normal neural tissue, and thus may reduce the pro-oxidative symptoms observed in schizophrenia. Hibbeln et al. (2003) quantified the erythrocyte fatty acid compositions of 76 medicated schizophrenic patients before and after 16 weeks of EPA (3g/day) or placebo supplementation. Several schizophrenic indices were performed on each patient before and after the supplementation period. Although plasma EPA levels were increased in the n-3 fatty acid group (P<0.05, Mann-Whitney tests), these differences did not correlate with reduced schizophrenia symptoms.

Omega-3 Fatty Acids and Gene Expression

The genomic effects of long chain PUFAs are mediated through specific interactions with hydrophobic binding sites on transcription factors; the earliest of such studies began with the discovery that peroxisome proliferator activated receptors (PPARs) are regulated by long chain PUFAs (Gottlicher et al., 1992). Several other fatty acid-regulated transcription factors have since been identified, including the hepatic nuclear factor 4 alpha (HNF4alpha) (Hertz et al., 1998), retinoid X receptor (RXR) (de Urquiza et al., 2000), and liver X receptors (LXR) (Ou et al., 2001). Experimental studies show that the level of sterol regulatory element binding protein-1 (SREBP1) synthesis and activation is modulated by fatty acids.

The peroxisome proliferator activated receptors (PPARs isoforms: alpha, (beta, gamma, gamma, delta) are the best understood fatty-acid-specific transcription factors; currently they are regarded as intracellular monitors of non-esterified fatty acid levels and may be involved in other cellular functions (Pawar & Jump, 2003; Desvergne & Wahli, 2000). The PPAR family is nuclear receptors whose DNA binding affinities are enhanced when in complex with n-3 long chain PUFAs. Activated PPARs bind with peroxisome proliferator response elements (PPREs), which are promoter proximal regulatory elements located near initiator sequences of many eukaryotic genes. Activated PPARs bind to DNA as heterodimers with retinoic X receptors (RXRs); n-3 fatty acids have been shown to promote the dimerization of PPARs and RXRs (Muerhoff et al., 1992). All PPAR isoforms possess a fatty acid binding activity; C_{18} and C_{20} fatty acids bind with greatest affinity (Xu et al., 1999). These authors also found that the binding of EPA changed the three-dimensional conformation of PPARs and enhanced its DNA binding affinity. The cellular alterations mediated by EPA would lead to decreased expression of genes involved in fatty acid and triacylglycerol biosynthesis while inducing genes involved in fatty acid oxidation. The role of PPARalpha on fatty acid regulation of hepatic lipid metabolism was examined in an animal model by Ren et al. (1997), who assessed hepatic mRNA transcript levels for fatty acid synthase, acyl-CoA oxidase, and CYP4A2 (CYP4A2 is one enzyme in the biosynthetic pathway of bile acids from cholesterol). Results indicated that PPARalpha was required for the n-3 fatty acid mediated upregulation of the hepatic CYP4A2 gene and the acyl-CoA oxidase gene, and that fish oil-mediated suppression of lipogenic genes did not involve PPARalpha. Primary hepatocytes treated with EPA showed a twofold increase in acyl-CoA oxidase mRNA levels compared to the control cultures, implying that only long chain PUFAs are able to induce acyl-CoA oxidase gene transcription. These results collectively imply that long-chain n-3 PUFAs in fish oils promote hepatic fatty acid oxidation and bile acid synthesis through activation of

PPARalpha transcription factor. Over time these genomic effects of n-3 PUFAs may lead to reduced blood lipid levels and could affect the symptoms of hyperlipidemia and hypercholesterolemia.

Although several studies show that EPA enhances the transcriptional activity of the PPAR family of transcription factors, most of these studies are performed using in vitro models. Very few studies examine the genomic effects of dietary n-3 fatty acids, thus more live animal studies are needed to lend further support to the existing studies that show n-3 fatty acids, particularly EPA, activate PPARs. Clarification is also needed as to whether the upregulation of PPARs by n-3 fatty acids operate via transcriptional or posttranscriptional mechanisms.

The sterol regulatory elements binding proteins (SREBPs) are transcription factors that regulate the transcription of several genes involved in lipid, cholesterol, bile acid, and lipoprotein biosynthesis. Evidence from animal and cell culture studies indicates that SREBP1 isoforms regulate fatty acid and triacylglycerol biosynthesis while SREBP2 regulates cholesterol biosynthesis. Animal studies show that transgenic mice overexpressing SREBP1a or SREBP1b have higher transcription rates of hepatic genes involved in lipogenesis, triacylglycerol biosynthesis and very low-density lipoprotein secretion, and develop fatty liver (Shimano et al., 1997). Unlike PPARs, unsaturated fatty acids have not been shown to directly bind with SREBPs, but unsaturated fatty acids do modulate the activity and abundance of SREBP1, which in turn affects lipogenic gene expression. Several animal studies show that dietary n-3 PUFAs suppress hepatic lipogenesis by inhibiting SREBP1 gene transcription and proteolytic activation, as well as increasing SREBP1$_{mRNA}$ decay (Pawar et al., 2003). Mice fed either normal diets or diets rich in fish oils showed 90% lower hepatic levels of nSREBP1 (active transcription factor) and 75% lower hepatic levels of pSREBP1 when fish oils were used. The changes in the fish oil group were accompanied by a decrease in hepatic fatty acid synthase mRNA (Xu et al., 1999). Unlike mice in the fish oil group, mice on diets rich in saturated and monounsaturated fatty acids had normal hepatic SREBP1 levels and activity. Furthermore, the n-3 fatty acids inhibited the transcription of several hepatic genes involved in glucose metabolism and lipogenesis, including glucokinase, acetylCoA carboxylase, and the DELTA-5 and DELTA-6 desaturases, which may be explained by n-3 PUFAs mediated reductions in hepatic SREPB1 levels (Sessler & Ntambi, 1998). In vitro studies show that n-3 PUFAs reduce the nuclear content of activated SREBP1 in two separate ways by inhibition of the proteolytic cleavage of pSREBP which occurs within the first hour of n-3 fatty acid treatment or via reduction in SREBP1 gene transcription, which is accompanied by decreased nuclear/endoplasmic reticulum membrane levels of pSREBP 1. The underlying molecular mechanisms by which n-3 PUFAs reduce cellular SREBP1 levels are not understood, but it is believed that n-3 PUFAs somehow enhance the rate of SREBP1c$_{mRNA}$ decay. Using cultured liver cells, n-3 PUFAs were shown to reduce the half-life of SREBP1c$_{mRNA}$ from 11 hours to less than 6 hours (Xu et al., 2001) and also inhibited lipogenic gene expression in cultured adipocytes which was SREBP1 independent (Mater et al., 2001).

Liver X receptors (LXRalpha and LXRbeta) are members of the nuclear hormone receptor superfamily; LXRalpha have recently been identified as fatty acid regulated transcription factor. Liver X receptors are important regulators of cholesterol, bile acid, fatty acid, and triacylglycerol biosynthesis in tissues such as the liver, brain, and gonads (Lu et al., 2001). Studies using human embryonic kidney (HEK) 293 cells have shown that unsaturated fatty acids bind to LXRalpha, antagonizing oxysterol activation of LXRalpha. However, another study by Pawar and Jump (2003) showed that EPA had no effect on LXRalpha activity in two different cell lines or in rat primary hepatocytes. Animal model studies have not shown any change in LXRalpha activity or abundance in response to changes in dietary fatty acid composition. Evidently, there is no consensus among the existing literature assessing unsaturated fatty acid regulation of LXRs; only a few research groups have investigated this observation and more studies are needed to assess the possible regulatory actions of n-3 and/or n-6 fatty acids on LXRs.

Hepatic nuclear factor 4alpha (HNF4alpha) is a transcription factor that enhances expression of several hepatic genes including CYP4A2, transferrin, apolipoprotein-CII, -CIII, -AII, -AIV and pyruvate kinase (Hertz et al., 2001). HNF4alpha is a member of the steroid receptor superfactorily. Hertz et al. (2001) were the first to report fatty acid regulation of HNF4alpha by using cultured hepatocytes. They showed that long chain acyl-CoAs (C_{14} and up) at in-vivo concentrations of approximately 2.6 m were ligands for HNF4alpha. Binding of saturated fatty acids enhanced the transcriptional activity of HNF4alpha, as evidenced by increased transferrin and apolipoprotein CIII expression. On the other hand, ALA, EPA, and DHA and their CoA thioesters inhibited HNF4alpha activity. Long chain n-3 PUFAs exert antilipogenic effects and promote lipid oxidation in hepatocytes.

The n-3 PUFAs clearly modulate the activity and/or abundance of at least three transcription factor families (PPARs, SREBP, and HNF4alpha), which play important roles in hepatic fatty acid, cholesterol, apolipoprotein, and carbohydrate metabolism. Omega-3 fatty acid regulation of PPARs and SREBP1 isoforms has been well established through a number of animal and cell culture studies. The significance of fatty acid regulation of HNF4alpha in-vivo is debatable, however the evidence that highly unsaturated acyl-CoAs may indeed influence HNF4alpha activity is a subject that deserves further investigation. Based on the current literature, the genomic effects of n-3 fatty acids on hepatic metabolism involve a shift from triacylglycerol synthesis, storage and apolipoprotein secretion towards hepatic oxidation of lipids. In turn, this response may reduce blood levels of triacylglycerols and low-density lipoproteins, which are important risk factors for several chronic diseases. The mechanisms by which n-3 fatty acids enhance hepatic lipid oxidation while concurrently decreasing hepatic lipid storage needs to be further examined, as these pathways may provide evidence for novel therapeutic strategies for better understanding of blood lipid and cholesterol disorders.

References

Albert, C.M.; Campos, H.; Stampfer, M.J.; Ridker, P.M.; Manson, J.E.; Willett, W.C.; Ma, J. blood levels of long-chain n-3 fatty acids and the risk of sudden death. Engl. J. Med. 2002, 346, 1113–1118.

Alsan, A.; Triadafilopoulos, G. Fish oil fatty acid supplementation in active ulcerative colitis: A double blind, placebo controlled, crossover study. Gut 1993, 35, 345–357.

Amusquivar, E.; Ruperez, F.J.; Barbas, C.; Herrera, E. Low arachidonic acid rather than alpha-tocopherol is responsible for the delayed postnatal development in offspring of rats fed fish oil instead of olive oil during pregnancy and lactation. J. Nutr. 2000, 130, 2855–2865.

Anti, M.; Marra, G.; Amelao, F.; Bartoli, G.M.; Ficarelli, R.; Percepce, A.; De Vetis, I.; Maria, G.; Sofo, L.; Rapaccini, G.L. Effect of omega-3 fatty acids on rectal mucosa cell proliferation in subjects at high risk for colon cancer. Gastroenterology 1992, 103, 883–891.

Augustsson, K.; Michaud, D.S.; Rimm, E.B.; Leitzmann, M.F.; Stampfer, M.J.; Willett, W.C.; Giovannucci, E.A. Prospective study of intake of fish fatty acids and prostate cancer. Cancer Epidemiol. Biomarkers Prev. 2003, 12, 64–67.

Bagga, D.; Wang, L.; Farias-Eisner, R.; Glaspy, J.A.; Reddy, T.A. Differential effects of prostaglandin derived from omega-6 and omega-3 polyunsaturated fatty acids on COX-2 expression and IL-6 secretion. Proceedings of the National Academy of Sciences of the United States of America 2003, 100, 1751–1756.

Bang, H.O.; Dyerberg, J.; Sinclair, H.M. The Composition of the Eskimo food in North Western Greenland. Am. J. Clin. Nutr. 1980, 33, 2657–2661.

Barrow, C.; Shahidi, F. Marine nutraceuticals and functional foods, CRC Press, Boca Raton, FL, 2008.

Belluzzi, A.; Brignola, C.; Campieri. M.; Pera, A.; Boschi, S.; Miglioli, M. Effect of an enteric-coated fish oil preparation on relapses in Crohn's disease. Engl. J. Med. 1996, 334, 1557–1560.

Bhagavathi, A.; Narayanan, B.; Narayanan, K.; Simi, B.; Reddy, B.S. Modulation of inducible nitric oxide synthase and related proinflammatory genes by the omega-3 fatty acid docosahexaenoic Acid. Cancer Res. 2003, 63, 972 979.

Bhagavathi, A.; Narayanan, B.; Narayanan, K.; Simi, B.; Reddy, B.S. Modulation of inducible nitric oxide synthase and related proinflammatory genes by the omega-3 fatty acid docosahexaenoic Acid. Cancer Res. 2003, 63, 972–979.

Billman, G.E.; Kang, J.X.; Leaf, A. Prevention of ischemia-induced cardiac sudden death by pure omega-3 polyunsaturated fatty acids in dogs. Circulation 2000, 99, 2452–2457.

Bougnoux, P. n-3 Fatty acids and cancer. Curr. Opin. Clin. Nutr. Metab. Care 1999, 2, 121–126.

Bourre, J.M.; Dumont, O. Essentiality of n-3 fatty acids for brain structure and function. World Rev. Nutr. Diet. 1991, 66, 103–117.

Burney, P.; Chinn, S.; Rona, R.J. Has the prevalence of asthma increased in children? Evidence from the National Study of Health and Growth 1973–86. Br. Med. J. 1990, 300, 1306–1310.

Burr, M.L.; Fehily, A.M.; Gilbert, J.M.; Rodgers, S.; Holiday, R.M.; Sweetman, P.M.; Elwood, P.C.; Deadman, N.M. Effects of changes in fat, fish and fibre intakes on death and myocardial reinfarction. Diet and Reinfarction Trial (DART). Lancet 1989, 2, 757–761.

Caygill, C.P.; Charlett, A.; Hill, M.J. Fat, fish oil and cancer. Br. J. Cancer 1996, 74, 159–164.

Chamberlain, J.G. Fatty acids in human brain phylogeny. Persp. Biol. Med. 1996, 39, 436–445.

Chaudry, A.A.; Wahle, K.W.; McClinton, S.; Moffat, S.E. 1994. Arachidonic acid metabolism in benign and malignant prostatic tissue in vitro: Effects of fatty acids and cyclooxygenase inhibitors. Intern. J. Cancer Res. 1994, 57, 176–180.

Coker, S.J.; Parratt, J.R. AH23848, A thromboxane antagonist, suppresses ischemia and reperfusion induced in anaesthetized greyhounds. Br. J. Pharmacol. 1985, 86, 259–264.

Davidson. M.; Bulkow, L.R.; Gellin, B.G. Cardiac mortality in Alaska's indigenous and nonnative residents. Intern. J. Epidemiol. 1993, 22, 62–71.

de Lorgeril, M.; Salen, P.; Laporte, F.; de Leiris, J. Alpha-linolenic acid in the prevention and treatment of coronary heart disease. Eur. Heart J., 2002, 3, D26–32.

de Lorgeril, M.; Salen, P.; Martin, J.L.; Monjaud, I.; Boucher, P.; Mamelle, N. Mediterranean dietary pattern in a randomized trial prolonged survival and possible reduced cancer rate. Arch. Intern. Med. 1998, 158, 1181–1187.

de Urquiza, A.M.; Liu, S.; Sjoberg, M.; Zetterstrom, R.H.; Griffiths, W.; Sjovall, J.; Perlmann, T. Docosahexaenoic acid, a ligand for the retinoid X receptor in mouse brain. Science 2000, 290, 2140–2144.

Desvergne, B.; Wahli, W. Peroxisome proliferator activated receptors: nuclear control of metabolism. Endocrine Rev. 2000, 20, 649–688.

Dolocek, T.A.; Granditis, G. Dietary polyunsaturated fatty acids and mortality in multiple risk factor intervention trial (MRFIT). World Rev. Nutr. Diet. 1991, 66, 205–216.

Dommels, Y.E.M.; Haring, M.G.M.; Keestra, N.G.M.; Alink, G.M.; van Balderen, P.J.; van Ommen, B. The role of cyclooxygenase in n-6 and n-3 polyunsaturated fatty acid mediated effects on cell proliferation, PGE_2 synthesis and cytotoxicity in human colorectal carcinoma cell lines. Carcinogenesis 2003, 24, 385–392.

Dwivedi, C.; Muller, L.A.; Goetz-Parten, D.E.; Kasperson, K.; Mistry, V.V. Chemopreventive effects of dietary mustard oil on colon tumour development. Cancer Lett. 2003, 196, 29–34.

Emelyanov, A.; Fedoseev, G.; Kranoschekova, O.; Abulimity, A.; Trendeleva, T.; Barnes, P.J. Treatment of asthma with lipid extract of New Zealand Green-lipped Mussel: A Randomized clinical trial. Euro. Respir J., 2002, 20, 596–600.

Empey, L.R.; Jewell, L.D.; Garg, M.L.; Thomson, A.B.; Clandinin, M.T.; Fedorac, R.N. Fish oil-enriched diet is mucosal protective against acetic-acid induced colitis in rats. Can. J. Physiol. Pharmacol. 1991, 69, 480–487.

Fischer, S.; Weber, P.C. Prostaglandin 1_3 is formed in-vivo in man after dietary eicosapentaenoic acid. Nature, 1984, 307, 165–168.

Francois, C.A.; Conner, S.L.; Bolewicz, L.C.; Conner, W.E. Supplementing lactating women with flaxseed oil does not increase docosahexaenoic acid in their milk. Am. J. Clin. Nutr. 2003, 77, 226–233.

Geusens, P.; Wouters, C.; Nijs, J.; Jiang, Y.; Dequeker, J. Long-term effect of omega-3 fatty acid supplementation in active rheumatoid arthritis. A 12-month, double-blind, controlled study. Arthrit. Rheumat. 1994, 37, 824–829.

GISSI-Prevenzione study investigators. Dietary supplementation with n-3 polyunsaturated fatty acids and vitamin E in 11,324 patients with myocardial infraction: Results of the GISSI-Prevenzione Trial. Lancet 1999, 354, 447–455.

Goel, D.P.; Maddaford, T.G.; Pierce, G.N. Effects of omega-3 polyunsaturated fatty acids on cardiac sarcolemmal na(+)/(h+) exchange. Am. J. Physiol. Heart Circul. Physiol. 2002, 283, H1688–H1694.

Gottlicher, M.; Widmark, E.; Li, Q.; Gustafsson, J.A. Fatty acids activate a chimera of the clofibric acid-activated receptor and the glucocorticoid receptor. Proc. Natl. Acad. Sci. 1992, 89, 4653–4657.

Hernandez, D.; Guerra, R.; Milena, A.; Torres, A.; Garcia, G.; Garcia, C.; Abreu, P.; Gonzalez, A.; Gomez, M.A.; Rufino, M. Dietary fish oil does not influence acute rejection rate and graft survival after renal transplantation: A randomized placebo-controlled study. Nephrol. Dial. Transplant. 2002, 17, 897–904.

Hertz, R.; Magenheim, J.; Berman, I.; Bar-Tana, J. Fatty Acid CoA Thioesters are ligands of hepatic nuclear factor-4alpha. Nature 1998, 392, 512–516.

Hertz, R.; Sheena, V.; Kalderon, B. Suppression of hepatic nuclear factor-4alpha by acyl-CoA thioesters of hypolipidemic peroxisome proliferators. Biochem. Pharmacol. 2001, 61, 1057–1062.

Hibbeln, J.R.; Makino, K.K.; Martin, C.E.; Dickerson, F.; Boronow, J.; Fenton, W.S. Smoking, gender, and dietary influences on erythrocyte essential fatty acid composition among patients with schizophrenia or schizoaffective disorder. Biol. Psychiatry 2003, 53, 431–441.

Hirai, A.; Terano, T.; Tamura, Y.; Yoshida, S. Eicosapentaenoic acid and adult diseases in Japan: Epidemiological and clinical aspects. Intern. Med. Suppl. 1989, 225, 69–75.

Hodge, L.; Salome, C.M.; Peat, J.K.; Haby, M.M.; Xuan, W.; Woolcock, A.J. Consumption of oily fish and childhood asthma risk. Med. J. Aust. 1996, 164, 137–140.

Holmes, M.D.; Colditz, G.A.; Hunter, D.J.; Hakinson, S.E.; Rosner, B.; Speizer, F.E.; Willet, W.C. Meat, fish and egg intake and risk of breast cancer. Intern. J. Cancer Res. 2003, 104, 221–227.

Holmes, M.D.; Stampfer, M.J.; Colditz, G.A.; Rosner, B.; Hunter, D.J.; Willet, W.C. Dietary factors and the survival of women with breast carcinoma. Cancer 1999, 86, 826–835.

Homan van der Heide, J.J.; Bilo, H.; Donker, J.M.; Wilmink, J.M.; Tegzess, A.M. Effect of dietary fish oil on renal function and rejection in cyclosporine-treated recipients of renal transplants. Engl. J. Med. 1993, 329, 769–773.

Horrobin, D.F. The Relationship between schizophrenia and essential fatty acids and eicosanoid metabolism. Prostagl. Leukotr. Essent. Fatty Acids 1992, 46, 71–77.

Hu, F.B.; Bronner, L.; Willet, W.C.; Stampfer, M.J.; Rexrode, K.M.; Albert, C.M.; Hunter, D.; Manson, J.E. Fish and omega-3 fatty acid intake and risk of coronary heart disease in women. J. Am. Med. Assoc. 2002, 287, 1815–1821.

Ip, C. Review of the effects of trans fatty acids, oleic acid, n-3 polyunsaturated fatty acids and conjugated linoleic acid on mammary carcinogenesis in animals. Am. J. Clin. Nutr. 1997, 66, 1523S–1529S.

Kanayasu-Toyoda, T.; Morita, I.; Murota, S.-I. Docosapentaenoic acid (22:5, n-3) an elongation metabolite of eicosapentaenoic acid (20:5, n-3), is a potent stimulator of endothelial cell migration on pretreatment in vitro. Prostagland., Leukotri. Essen. Fatty Acids 1996, 54, 319–325.

Kang, J.X.; Leaf, A. Prevention of fatal cardiac arrhythmias by polyunsaturated fatty acids. Am. J. Clin. Nutr. 2000, 71, 202–207.

Keller, J.R. Omega-3 fatty acids may be effective in the treatment of depression. Topics Clin. Nutr. 2002, 17, 21–27.

Kew, S.; Banerjee, T.; Minihane, A.M.; Finnegan, Y.E.; Muggli, R.; Albers, R.; Williams, C.M.; Calder, P.C. Lack of effect of foods enriched with plant- or marine-derived n-3 fatty acids on human immune function. Am. J. Clin. Nutr. 2003, 77, 1287–1295.

Koch, T.; Duncker, H.P.; Klein, A.; Schlotzer, E.; Peskar, B.M.; Van-Ackern, K.; Neuhof, H. Modulation of pulmonary vascular resistance and edema formation by short-term infusion of a 10% fish oil emulsion. Infusionsther Transfusions med. 1993, 20, 291–300.

Kort, W.J.; Weijma, I.M.; Vergroesen, A.J.; Westbroek, D.L. Conversion of diets at tumour induction shows the pattern of tumour growth and metastasis of the first given diet. Carcinogenesis 1987, 8, 611–614.

Kremer, J.M.; Bigauoette, J.; Michalek, A.V.; Timchalk, M.A.; Lininger, L.; Rynes, R.I.; Huyck, C.; Zieminski, J. Effects of manipulation of dietary fatty acids on clinical manifestations of rheumatoid arthritis. Lancet, 1985, 1, 184–187.

Lorenz-Meyer, H.; Bauer, P.; Nicolay, C.; Schultz, B.; Purrmann, J.; Fleig, W.E.; Scheurlen, C.; Koop, I.; Pudel, V.; Carr, L. Omega-3 fatty acids and low carbohydrate diet for the maintainance of remission in Crohn's disease. A Randomized Controlled Multicenter Trial (German Crohn's Disease Study Group). Scand. J. Gastroenterol. 1996, 31, 778–785.

Lu, T.T.; Repa, J.J.; Mangelsdorf, D.J. Orphan nuclear receptors as elixirs and fixers of sterol metabolism. J. Biol. Chem. 2001, 276, 37735–37738.

Mahadik, S.P.; Evans, D.; Lal, H. Oxidative stress and role of antioxidant and n-3 essential fatty acid supplementation in Schizophrenia. Prog. Neuropsychopharmacol. Biol. Psychiatry 2001, 25, 463–493.

Maillard, V.; Bougnoux, P.; Ferrari, P.; Jourdain, J.L.; Pinault, M.; Lavillonniere, F.; Body, G.; Le Floch, O.; Chajes, V. N-3 and N-6 Fatty acids in breast adipose tissue and relative risk of breast cancer in a case control study in tours, France. Intern. J. Cancer Res. 2002, 98, 78–83.

Mamalakis, G.; Kafatos, A.; Kalogeropoulos, N.; Andrikopoulos, N.; Daskalopulos, G.; Kranidis, A. Prostate cancer vs hyperplasia: Relationships with prostatic and adipose tissue fatty acid composition. Prostaglandins, Leukot. Essent. Fatty Acids 2002, 66, 467–477.

Mantzioris, E.; Cleland, L.G.; Gibson, R.A.; Neumann, M.A.; Demasi, M.; James, M.J. Biochemical effects of a diet containing foods enriched with n-3 fatty acids. Am. J. Clin. Nutr. 2000, 72, 42–48.

Marangell, L.B.; Martinez, J.M.; Zboyan, H.A.; Kertz, B.; Kim, H.F.S.; Puryear, L.J. A double-blind, placebo-controlled study of the omega-3 fatty acid docosahexaenoic acid in the treatment of major depression. Am. J. Psychiatry 2003, 160, 996–998.

Martin, R.E.; Bazan, N.G. Changing fatty acid content of growth cone lipids prior to synaptogenesis. J. Neurochem. 1992, 59, 318–325.

Martinez, M. Abnormal fatty acid profiles of polyunsaturated fatty acids in the brain, liver, kidney and retina of patients with peroxisomal disorders. Brain Res. 1992, 583, 171–182.

Mater, M.K.; Thelen, A.P.; Pan, D.A.; Jump, D.B. Sterol response element binding protein-lc (SREBP1c) is involved in the polyunsaturated fatty acid suppression of hepatic s14 gene transcription. J. Biol. Chem. 2001, 274, 32725–32744.

McCreadie, R.G.; Macdonald, E.; Wiles, D.; Campell, G.; Patterson, J.R. Plasma lipid peroxide and serum vitamin e levels in patients with and without tardive dyskinesia and normal subjects. Br. J. Psychiatry 1995, 167, 1–8.

McGrath-Hanna, N.K.; Greene, D.M.; Tavernier, R.J.; Bult-Ito, A. Diet and mental health in the Arctic: Is diet an important risk factor for mental health in circumpolar peoples? A review. Intern. J. Circumpolar Health 2003, 62, 228–241.

Middleton, S.J.; Naylor, S.; Woolner, J.; Hunter, J.O. A double-blind, randomized, placebo-controlled trial of essential fatty acid supplementation in the maintenance of remission of ulcerative colitis. Aliment. Pharmacol. Therapeut. 2002, 16, 1131–1135.

Muerhoff, A.S.; Griffin, K.J.; Johnson, E.F. The peroxisome proliferator activated receptor mediates the induction of cyp4a6, a cytochrome p450 fatty acid omega hydroxylase, by clofibric acid. J. Biol. Chem. 1992, 267, 19051–19053.

Nagakura, T.; Matsuda, S.; Shichijo, K.; Sugimoto, H.; Hata, K. Dietary supplementation with fish oil rich in n-3 polyunsaturated fatty acids in children with bronchial asthma. Euro. Respir. J. 2000, 16, 861–865.

Narayanan, B.A.; Narayanan, N.K.; Reddy, B.S. Docosahexaenoic acid regulated genes and transcription factors inducing apoptosis in human colon cancer cells. Intern. J. Oncol. 2001, 19, 1255–1262.

Neuringer, M. Cerebral cortex docosahexaenoic acid is lower in formula fed than in breast fed infants. Nutr. Rev. 1993, 51, 238–241.

Nieto, N.; Torres, M.I.; Rios, A.; Gil, A. Dietary polyunsaturated fatty acids improve histological and biochemical alterations in rats with experimental ulcerative colitis. J. Nutr. 2002, 132, 11–19.

Ou, J.; Tu, H.; Shan, B.; Luk, A.; DeBose-Boyd, R.A.; Bashmakov, Y.; Goldstein, J.L.; Brown, M.S. Unsaturated fatty acids inhibit transcription of the sterol regulatory element binding protein-1c (SREBP-1c) gene by antagonizing ligand-dependent activation of the LXR. Proc. Natl. Acad. Sci. 2001, 98, 6027–6032.

Pawar, A.; Botolin, D.; Mangelsdorf, D.J.; Jump, D.B. The role of liver X receptor alpha in the fatty acid regulation of hepatic gene expression. J. Biol. Chem. 2003, 278, 40736–40743.

Pawar, A.; Jump, D.B. Unsaturated fatty acid regulation of peroxisome proliferator activated receptor alpha activity in rat primary hepatocytes. J. Biol. Chem. 2003, 278, 35931–35939.

Peat, J.K.; Salome, C.M.; Woolcock, A.J. Factors associated with bronchial hyper responsiveness in australian adults and children. Eur. Respir. J. 1992, 5, 921–929.

Phillips, M.; Erickson, G.A.; Sabas, N.; Smith, J.P.; Greenberg, J. Volitile organic compounds in the breath of schizophrenic patients. J. Clin. Pathol. 1995, 48, 466–469.

Prener, A.; Storm, H.H.; Nielsen, N.H. Cancer of the male genital tract in circumpolar inuit. Acta Oncologica Stockholm, Sweden 1996, 35, 589–593.

Reddy, B.S. Chemoprevention of colon cancer by dietary fatty acids. Cancer Metastasis Rev. 1994, 13, 285–302.

Ren, B.; Thelen, A.P.; Peters, J.M.; Gonzales, F.J.; Jump, D.B. Polyunsaturated fatty acid suppression of hepatic fatty acid synthase and S14 gene expression does not require peroxisome proliferator-activated receptor alpha. J. Biol. Chem. 1997, 272, 26827–26832.

Robinson, L.E.; Clandin, T.; Field, C.J. The role of dietary long-chain n-3 fatty acids in anticancer immune defence and R3230AC mammary tumour growth in rats: Influence of diet fat composition. Breast Cancer Res. Treatment 2002, 73, 145–160.

Rose, D.P.; Connolly, J.M. Omega-3 fatty acids as cancer chemopreventative agents. J. National Cancer Inst. 1999, 83, 217–244.

Rose, D.P.; Connolly, J.M.; Rayburn, J.; Coleman, M. Influence of diets containing eicosapentaenoic acid or docosahexaenoic acid on growth and metastasis in breast cancer cells in nude mice. J. National Cancer Inst. 1995, 87, 587–592.

Sarsilmaz, M.; Songur, A.; Ozyurt, H.; Kus, I.; Ozen, O.A.; Ozyurt, B.; Sogut, S.; Akyol, O. Potential role of dietary omega-3 fatty acids on some oxidant/antioxidant parameters in rats' corpus striatum. Prostagl. Leukotr. Essen. Fatty Acids 2003, 69, 253–259.

Sessler, A.M.; Ntambi, J.M. Polyunsaturated fatty acid regulation of gene expression. J. Nutr. 1998, 128, 923–926.

Shahidi, F. Maximising the value of marine by-products, Woodhead Publishing Ltd. Cambridge, UK, 2007.

Shahidi, F. Nutraceutical and specialty lipids and their co-products, CRC Press. Boca Raton, FL, 2006.

Shahidi, F.; Finley, J.W. Omega-3 fatty acids: Chemistry, nutrition and health effects. ACS Symposium Series 788. American Chemical Society, Washington, DC, 2001.

Shahidi, F.; Miraliakbari, H. Omega-3 (n-3) fatty acids in health and disease. Part 1–Cardiovascular disease and cancer. J. Med. Food 2004, 7, 387–401.

Shahidi, F.; Miraliakbari, H. Omega-3 Fatty Acids in Health and Disease in Part 2–Health effects of omega-3 fatty acids in autoimmune diseases, mental health and gene expression. J. Med. Food. 2005, 8, 133–150.

Shahidi, F.; Wanasundara, U.N. Omega-3 fatty acid concentrates: Nutritional aspects and production technologies. Trends Food Sci. Technol. 1998, 9, 230–240.

Shahidi, F.; Zhong, Y. Fatty acid derivatives of catechins and methods of their use. US Provisional Patent Application No. 61/322,004, 2010.

Shimano, H.; Horton, J.D.; Shimomura, I.; Hammer, R.E.; Brown, M.S.; Goldstein, J.L. Isoform lc of sterol regulatory element binding protein is less active than isoform la in livers of transgenic mice and in cultured cells. J. Clin. Invest. 1997, 99, 846–854.

Shoda, R.; Matsueda, K.; Yamato, S.; Umeda, N. Epidemiologic analysis of Crohn's disease in Japan: Increased dietary intake of n-6 polyunsaturated fatty acids and animal protein relates to the increased incidence of Crohn's disease in Japan. Am. J. Clin. Nutr. 1996, 63, 741–745.

Shoda, R.; Matsueda, K.; Yamato, S.; Umeda, N. Therapeutic Efficacy of N3 Polyunsaturated fatty acid in experimental Crohn's disease, J. Gastroenterol. 1995, 30, 98–101.

Siess, W.; Roth, P.; Scherer, B.; Kurzmann, I.; Bohlig, B.; Weber, P.C. Platelet membrane fatty acids, platelet aggregation, and thromboxane formation during a mackerel diet. Lancet 1980, 1, 441–444.

Sinclair, H.M. Deficiency of essential fatty acids and atherosclerosis. Etcetera. Lancet 1956, 267, 381–383.

Smith, R.S. The macrophage theory of depression. Med. Hypotheses 1991, 35, 298–306.

Su, K.P.; Huang, S.Y.; Chiu, C.C.; Shen, W.W. Omega-3 fatty acids in major depressive disorder. A preliminary double-blind, placebo-controlled trial. Euro. Neuropsychopharmacol. 2003, 13, 267–271.

Takahashi, M.; Fukutake, M.; Isoi, T.; Fukuda, K.; Sato, H.; Yazawa, K.; Sugimura, T.; Wakabayashi, K. Suppression of azomethane-induced rat colon carcinoma development by a fish oil component, docosahexaenoic acid (DHA). Carcinogenesis 1997, 18, 1337–1342.

Tanskanen, A.; Hibbelin, J.R.; Tuomilehto, J.; Uutela, A.; Haukkala, A.; Viinamaki, H.; Lehtonen, J.; Vartianen, E. Fish consumption and depressive symptoms in the general population in Finland. Psychiatric Services 2001, 52, 529–531.

Terry, P.; Lichtenstein, P.; Feychting, M.; Ahlbom, A.; Wolk, A. Fatty fish consumption and risk of prostate cancer. Lancet 2001, 357, 1764–1766.

Theroux, P.; Chaitman, B.R.; Danchin, N.; Erhardt, L.; Meinertz, T.; Schroeder, J.S.; Togoni, G.; White, H.D.; Willerson, J.T.; Jessel, A. Inhibition of the sodium/hydrogen exchanger with cariporide to prevent myocardial infraction in high risk ischemic situations. Main Results of the GUARDIAN Trial. Circulation 2000, 102, 3032–3038.

Thun, M.J.; Namboodiri, M.M.; Heath, C.W. Jr. Aspirin use and reduced risk of fatal colon cancer. Engl. J. Med. 1991, 325, 1593–1596.

Tilley, D.G.; Maurice, D.H. Vascular smooth muscle cell phosphodiesterase (PDE) 3 and PDE4 activities and levels are regulated by cyclic AMP In Vivo. Molecular Pharmacol. 2002, 62, 497–506.

Troisi, R.J.; Willet, W.C.; Weiss, S.T.; Trichopoulos, D.; Rosner, B.; Speizer, F.E.A. Prospective study of diet and adult-onset asthma. Am. J. Respir. Crit. Care Med. 1995, 151, 1401–1408.

Volker, D.; Fitzgerald, P.; Major, G.; Garg, M. Efficacy of fish oil concentrate in the treatment of rheumatoid arthritis. J. Rheumatol. 2000, 27, 2305–2307.

Wanasundara, U.N.; Shahidi, F. Structural characteristics of marine lipids and preparation of omega-3 concentrates, in: Flavor and lipid chemistry of seafoods, Shahidi, F. and Cadwallader, K.K. Eds., Series 674, American Chemical Society Symposium Washington DC, 1997, 240–254.

Wynder, E.L.; Kajitani, T.; Ishikawa, S.; Dodo, H.; Takano, A. Environmental factors of cancer of the colon and rectum. II. Japanese Epidemiological Data. Cancer 1969, 12, 1210–1220.

Xu, H.E.; Lambert, M.H.; Montata, V.G.; Parks, D.J.; Blanchard, S.G.; Brown, P.J.; Sternbach, D.D.; Lehmann, J.M. Molecular recognition of fatty acids by peroxisome proliferator activated receptors. Molecular Cell 1999, 3, 397–403.

Xu, J.; Teran-Garcia, M.; Park, J.H. Polyunsaturated fatty acids suppress hepatic sterol regulatory element binding protein-1 expression by accelerating transcript decay. J. Biol. Chem. 2001, 276, 9800–9807.

Xu, M.T.; Nakamura, M.T.; Cho, H.P.; Clarke, S.D. Sterol regulatory element binding protein-1 expression is suppressed by dietary polyunsaturated fatty acids: A mechanism for the coordinate suppression of lipogenic genes by polyunsaturated fats. J. Biol. Chem. 1999, 274, 23577–23583.

Zhong, Y.; Madhujith, T.; Mahfouz, N.; Shahidi, F. compositional characteristics of muscle and visceral oil from steelhead trout and their oxidative stability. Food Chem. 2007, 104, 602–608.

Clinical Effects of n-3 PUFA Supplementation in Human Health and Inflammatory Diseases

Jennifer M. Monk[1,4], David N. McMurray[1,2], and Robert S. Chapkin[1,3]

[1]*Program in Integrative Nutrition & Complex Diseases;* [2]*Department of Microbial & Molecular Pathogenesis, Texas A&M University Health Science Center, College Station, TX;* [3]*Vegetable Fruit Improvement Center, Texas A&M University, College Station, Texas;* [4]*Dr. Jennifer Monk is a recipient of a Postdoctoral Fellowship from the Natural Sciences and Engineering Research Council of Canada (NSERC, PDF-388466-2010). This work was supported in part by NIH grants DK071707, CA59034, and USDA 2010-34402-20875, "Designing Foods for Health" through the Vegetable & Fruit Improvement Center*

Introduction

Long chain n-3 polyunsaturated fatty acids (PUFA) derived from marine sources, namely eicosapentaenoic acid (EPA; $20:5^{\Delta5,8,11,14,17}$) and docosahexaenoic acid (DHA; $22:6^{\Delta4,7,10,13,16,19}$), have been shown to exhibit beneficial anti-inflammatory effects in multiple inflammatory disease states (Chapkin et al., 2007; Sijben & Calder, 2007). Within the typical American diet, the daily consumption of n-3 PUFA is between 0.7 and 1.6 g, which is equivalent to approximately 0.2–0.7% of total calories (Conquer & Holub, 1998; Kris-Etherton et al., 2000). Of this, the amount of fish-derived long chain n-3 PUFA (i.e., EPA and DHA) is reported to be less than 0.1–0.2 g per day with the majority being comprised of alpha-linolenic acid (ALA), the plant based form of n-3 PUFA (Kim et al., 2010b). In human clinical trials, n-3 PUFA intake levels, mainly in the form of EPA and DHA, are consumed at 1–9 g/day, which corresponds to 0.45–4.0% of calories (Kelley et al., 1998, 1999; Rees et al., 2006; Thies et al., 2001). This physiological range is comparable to levels consumed in traditional Japanese diets (containing 1–2% of daily energy as long chain n-3 PUFA) or by the Greenland Inuit, consuming 2.7–6.3% of daily energy (6–14 g/day) (Damsgaard et al., 2008; Feskens & Kromhout, 1993; Nagata et al., 2002; Okuyama et al., 1996).

Research conducted to elucidate the beneficial effects of n-3 PUFA on both physiological and pathophysiological processes highlights the attempt to determine if this bioactive food component is beneficial with respect to both preventing disease onset and/or improving the clinical outcomes in already established pathologies.

Our objective in this chapter is to highlight some of the recent findings pertaining to the impact of n-3 PUFA consumption on prevalent human diseases, which include an inflammatory dimension, and thus, are most likely to be impacted by an anti-inflammatory food component. Since this is an expansive topic that is both an active and prolific area of research, where appropriate, the reader is referred to more specific comprehensive review papers for further details. Additionally, in making comments on the outcomes from clinical trials, we preferentially selected meta-analyses, thereby systematically providing the findings from multiple studies to allow for the opportunity to make more conclusive comments on what is currently known.

Putative Mechanisms of n-3 PUFA Action

Following n-3 PUFA consumption, tissue enrichment readily occurs, and this is particularly apparent in diverse immunological cell populations wherein n-3 PUFA are incorporated into both plasma and intracellular membranes as summarized in detail elsewhere (Calder, 2007). Generally, increasing the dietary intake of n-3 PUFA from fish oil results in increased proportions of the respective n-3 PUFA (DHA and EPA) into cells and, typically, this occurs at the expense of n-6 PUFAs, especially linoleic and arachidonic acid, therefore the cellular level is readily influenced by diet (Calder, 2007; Chapkin et al., 2007;). Further, time-course studies indicate that the incorporation of EPA and DHA into human immune cells reaches its peak within 4 weeks post initiation of dietary intake in a dose-response manner, as reviewed elsewhere (Calder, 2007). The sub-cellular incorporation of n-3 PUFA is localized primarily within the membrane phospholipids at the *sn-2* position (Anderson & Sperling, 1971; Stillwell & Wassall, 2003). In this connection, dietary n-3 PUFA have been shown to specifically alter plasma membrane micro-organization (lipid rafts) at the immunological synapse between T cells and antigen-presenting cells ultimately suppressing signal transduction and nuclear translocation/activation of transcription factors (Fan et al., 2004; Kim et al., 2008, 2010; Yog et al., 2010). Interestingly, DHA and EPA enrichment has been shown to occur in both membrane lipid raft and non-raft membrane fractions isolated from CD4[+] T cells (Switzer et al., 2004). Therefore, the presence of long chain n-3 PUFA in the membrane imparts unique physiochemical properties to cellular membranes, and DHA-induced alterations in membrane structure and function have been proposed to underlie the beneficial effects of n-3 PUFA (Chapkin et al., 2007; Ma et al., 2004, 2004b; Seo et al., 2006; Stillwell & Wassall, 2003).

Generally, n-3 PUFA (DHA and EPA) exert anti-inflammatory and immune suppressive functions, which may explain the beneficial role of this bioactive food component in human clinical trials and supplementation studies. The mechanisms through which n-3 PUFA exert these effects are broadly based and represent an active and ongoing research front. Appropriately crafted studies in animal and cell culture models relevant to human disease states have provided insight into the multiple mechanisms through which n-3 PUFA appear to play a beneficial role in either preventing the onset or slowing the progression of many human pathologies. Interestingly, n-3 PUFA appear to mediate effects at multiple stages of cellular complexity and organization affecting signaling pathways generated at the level of tissue, the cell membrane, or the intracellular second messengers and transcription factors, thereby ultimately impacting gene expression (Fig. 2.1). Thus, the pleiotropic effects of n-3 PUFA on physiological processes demonstrate that this biologically relevant bioactive food component exerts a more diverse and potent effect compared to other nutraceuticals.

Putative mechanisms of action elicited by n-3 PUFA impact diverse physiological processes including cell membrane structure/function, eicosanoid signaling, nuclear receptor activation, and whole-body glucose and lipid metabolism, thereby providing significant protection against a variety of apparently unrelated human diseases (Chapkin et al., 2007; Hu et al., 2003; Jump et al., 2005; Lupton & Chapkin, 2003; Stulnig, 2003).

Immunmodulatory effects are elicited by n-3 PUFA through multiple mechanisms, including diminishing T-cell proliferative capacity in response to mitogenic

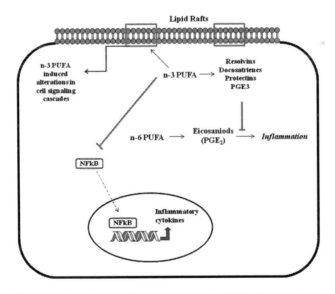

Fig. 2.1. n-3 PUFA mechanisms of action. The presence of n-3 PUFA in the membrane imparts unique physiochemical properties to cellular phospholipid domains (lipid rafts), ultimately affecting membrane structure and function, leading to alterations in signaling cascades. n-3 PUFA can modulate inflammation by suppressing nuclear receptor activation (e.g., NF-κB) and by opposing n-6 PUFA-derived eicosanoids. Solid arrows indicate direct effects, hammerhead arrows indicate inhibition, and dotted arrows indicate migration.

and antigenic stimulation (Anderson & Fritsche, 2004; Arrington et al., 2001, 2001b; Zhang et al., 2005, 2006). These suppressive effects have also been observed in the dendritic cell, endothelial cell, macrophage, and neutrophil components of the inflammatory response (Bagga et al., 2003; Hughes & Pinder, 2000; Massaro et al., 2008; Novak et al., 2003; Prescott, 1984; Zeyda et al., 2005). Additionally, dietary n-3 PUFA modulate components of intracellular signaling pathways regulating T-cell activation are summarized in detail elsewhere (Chapkin et al., 2009). Alterations in T-cell function by n-3 PUFA could reflect direct effects on the ability of target T-cell populations to respond to activating stimuli and/or indirect effects on the activity of accessory cells (non-T-cell populations), which promote T-cell activation or a combination of these two distinct mechanisms (Chapkin et al., 2009).

It is well documented that EPA and DHA (n-3 PUFA) supplant n-6 PUFA, principally linoleic acid and arachidonic acid (the major eicosanoid precursor), and can therefore dramatically alter both the spectrum and biological properties impacted by cyclooxygenase (COX) and lipoxygenase (LOX) derived metabolites (Chapkin et al., 2007; Smith, 2005). Arachidonic acid (n-6 PUFA) derived metabolites such as COX-derived prostaglandin (PG)-E_2 and LOX-derived leukotriene (LT)-B_4 are largely pro-inflammatory (Calder, 2006). n-3 PUFA can also act as substrates for both COX and LOX, giving rise to eicosanoids with modified structures from those derived from n-6 PUFA, primarily PGE_3, LTB_5, and LTE_5 (Calder, 2006). Some of the well-defined anti-inflammatory effects of n-3 PUFA and the underlying mechanisms supporting these effects are reviewed in detail elsewhere (Calder, 2006; Schmitz & Ecker, 2008). In brief, increasing the content of n-3 PUFA in cellular membranes results in reduced generation of n-6 PUFA-derived eicosanoids while concomitantly increasing n-3 PUFA-derived eicosanoid and resolving levels (Calder, 2006; Seki et al., 2009). Inhibiting the activation of NF-kB (via decreased phosphorylation of IkB) and altering the activity of other transcription factors decreases the generation of inflammatory cytokines, that is, tumor necrosis factor (TNF)-alpha, interleukin (IL)-1beta, IL-6, and IL-8 (Calder, 2006; Seki et al., 2009).

It is important to retain perspective with respect to the beneficial effects of n-3 PUFA. Although n-3 PUFA appear to exert primarily positive effects in many disease states, current research has demonstrated that n-3 PUFA can exert a detrimental effect on immunity to some infections, such as *M. tuberculosis* (Bonilla et al., 2010, 2010b; McFarland et al., 2008). *Fat-1* mice, which express the *fat-1* gene from *Caenorhabditis elegans* that encodes an n-3 desaturase that catalyzes n-3 PUFA synthesis from n-6 PUFA substrates, exhibited increased susceptibility to mycobacterial infection, which was correlated with reduced inflammatory cytokine secretion and impaired activation of antimicrobial macrophage functions (Bonilla et al., 2010). Thus, n-3 PUFA may suppress anti-microbial resistance, raising concerns regarding the safety of n-3 PUFA supplementation in some human populations. This is an issue that requires further investigation.

Human Inflammatory Pathologies and n-3 PUFA

Inflammatory processes and responses are regulated and shaped, at least in part, by PUFA, and the balance between n-6 and n-3 PUFA appears to be important in determining the development and severity of inflammatory responses (Calder, 2006; Deckelbaum & Calder, 2010; Tjonneland et al., 2009). Higher dietary intakes of n-6 PUFA, particularly arachidonic acid, could potentate inflammatory processes and therefore either predispose individuals to or exacerbate inflammatory diseases (Sijben & Calder, 2007). Therefore, increasing the dietary intake of anti-inflammatory n-3 PUFA (DHA and EPA) may alter the ratio between n-6 and n-3 PUFA systemically and decrease the risk of inflammatory conditions and, thus, provide a benefit to patients with inflammatory diseases (Sijben & Calder, 2007).

Many randomized controlled trials have been conducted with n-3 PUFA in various disease states, including cardiovascular, inflammatory, immunological, psychological, neurological, and aging disorders. The presumed mechanisms of beneficial effects relate to the anti-inflammatory, anti-thrombic, anti-arrhythmic, hypolipidemic, and vasodilatory properties of n-3 PUFA (Turner et al., 2010). Interestingly, chronic inflammation represents a common thread that links many human diseases. In this regard, some of the most common causes of death and disability in humans are characterized by exaggerated inflammation and excessive formation of inflammatory eicosanoids and cytokines, whereas other pathologies include an inflammatory component that may be less pronounced (Sijben & Calder, 2007). Thus, the anti-inflammatory and immunomodulatory effects of n-3 PUFA may provide the common mechanism that is likely to affect multiple complex mediators that ultimately impact, and potentially improve, diverse disease outcomes. Research to elucidate the specific mechanisms of n-3 PUFA action has yielded enough supportive data to warrant the use of n-3 PUFA supplementation in clinical and epidemiological studies.

In some diseases, the onset is associated with low dietary consumption of fish (i.e., long chain n-3 PUFA), such as cardiovascular disease (CVD) (Daviglus et al., 1997; Hallgren et al., 2001; Oomen et al., 2000), depression (Hibbeln, 1998, 2002) and Alzheimer's disease (Morris et al., 2003; Palacios-Pelaez et al., 2010). The role of excess inflammation in neural disorders remains poorly understood; however, the essentiality of n-3 PUFA (particularly DHA) as a structural component in normal neural function may provide insight into these findings (Sijben & Calder, 2007). For other diseases, a clear association between risk of onset and a shifted balance of n-6 PUFA and n-3 PUFA intake is lacking (Sijben & Calder, 2007). Despite this, many investigators have recognized the potential of n-3 PUFA to dampen excessive inflammation in most inflammatory diseases and conditions (Calder, 2006; Sijben & Calder, 2007). Inflammatory mediators, predominantly inflammatory cytokine levels, have been reduced by n-3 PUFA supplementation in many randomized, double-blinded, placebo-controlled clinical trials. Beneficial decreases in inflammatory biomarker levels resulting from n-3 PUFA supplementation have been documented in a wide range of inflammatory pathologies, namely esophageal (Tashiro et al., 1998) and pancreatic cancer (Wigmore et al., 1997), Crohn's disease (Trebble et al., 2004), procto-colitis (Almallah et al., 2000), rheumatoid arthritis (Kremer et al., 1990, 1995), chronic obstructive pulmonary disease (COPD) (Matsuyama et al., 2005), exercise-induced broncho-constriction in asthma (Mickleborough et al., 2006), human immunodeficiency virus (HIV) (Virgili et al., 1997), multiple sclerosis (Gallai et al., 1995) and chronic renal disease (Cappelli et al., 1997). Although not all clinical trials of n-3 PUFA supplementation in the aforementioned pathologies have reported improvement in inflammatory biomarker levels, as reviewed elsewhere (Sijben & Calder, 2007), it is important to note that inflammation was not exacerbated in any of the trials and, therefore, was either beneficial or showed no effect.

Ultimately, due to the vast array of putative mechanisms through which n-3 PUFA can influence biological processes, the potential for n-3 PUFA to impact multiple pathologies exists. Therefore, in this review we arbitrarily selected some major chronic diseases with an inflammatory component in which the anti-inflammatory effects of n-3 PUFA are likely to have an impact: CVD, inflammatory bowel disease (IBD), obesity, colorectal cancer, and rheumatoid arthritis (RA).

Cardiovascular Disease (CVD) and n-3 PUFA

Research interest into the therapeutic value of marine oils was initiated following the observation that the Greenland Inuit populations exhibit a low incidence of CVD (Bang et al., 1976). The exact cardio-protective mechanisms of n-3 PUFA remain unknown but are hypothesized to involve a reduction of arrhythmias, heart rate, ischemia/reperfusion-induced injury, serum triglyceride levels, inflammation, or improved endothelial function (Filion et al., 2010). Previously, meta-analyses conducted to elucidate the effect of n-3 PUFA on cardiovascular outcomes have reached inconclusive outcomes reaching either positive (Bucher et al., 2002) or negative (Yzebe & Lievre, 2004) conclusions. Some clinical trials have demonstrated that n-3 PUFA consumption reduces serum triglyceride levels (Eslick et al., 2009), lowers systolic and diastolic blood pressure (Geleijnse et al., 2002) and resting heart rate (Mozaffarian et al., 2005). In this context, the US Food and Drug Administration has approved pharmaceutical n-3 PUFA products for the treatment of hypertriglyceridemia (Park & Mozaffarian, 2010). Further evidence from observational studies and some limited clinical trials suggest that the consumption of fish or n-3 PUFA may also exert beneficial effects on other well-established cardiovascular disease risk factors, including inflammation, endothelial function, heart rate variability and, at high doses, platelet aggregation (Chin et al., 1993; Kristensen et al., 1989; Lopez-Garcia et al., 2004; Mozaffarian et al., 2008; Park & Mozaffarian, 2010).

The GISSI Prevenzione study, which consisted of a large intervention trial of secondary prevention after myocardial infarction, identified a substantial reduction in all-cause and cardiovascular mortality with 1 g/day of n-3 PUFA supplementation (GISSI, 1999). Further, sudden cardiac death was greatly reduced in this population within four months of initiating n-3 PUFA supplementation; however, no benefit was apparent in the occurrence of non-fatal myocardial infarctions or stokes (GISSI-Prevenzione Investigators, 1999). A beneficial role of n-3 PUFA in secondary prevention of cardiovascular diseases was reported in patients treated with chronic hemodialysis, supplementation with 1.7 g/day of n-3 PUFA over a 2-year period. Supplementation reduced the incidence of myocardial infarction but had no effect on cardiovascular mortality (Svensson et al., 2006). The majority of evidence from observational studies, clinical trials, meta-analyses, and systematic reviews suggests that supplementation with n-3 PUFA reduces overall mortality due to myocardial infarction and sudden cardiac death (Albert et al., 2002; Bucher et al., 2002; Gapinski et al., 1993; GISSI-Prevenzione Investigators, 1999; Leon et al., 2008; Saravanan et al., 2010; Streppel et al., 2008; Yokoyama et al., 2007; Zhao et al., 2009). Readers are referred to a recent systematic review highlighting the major effects of n-3 PUFA supplementation on multiple aspects of the etiology of cardiovascular disease for further information (Saravanan et al., 2010).

Based on a substantial number of studies and an array of evidence, the joint American College of Cardiology and American Heart Association Statement on n-3 PUFA Use recommends an intake of at least two fish meals per week in patients with coronary artery disease and supplemental therapy for 1 year with 1 g/day of n-3 PUFA ethyl esters for those who have had a myocardial infarction (Kris-Etherton et al., 2000; Saravanan et al., 2010).

At this time, the anti-inflammatory, anti-atherosclerotic, and anti-immunomodulatory effects of n-3 PUFA in relation to cardiovascular disease have not been proven to translate into clinical benefits (Massaro et al., 2008); therefore, further focused studies are still required to explore these properties.

Inflammatory Bowel Disease (IBD) and n-3 PUFA

Inflammatory bowel disease (IBD) is a generic term given to a heterogeneous group of gastrointestinal disorders that are characterized by chronic inflammation. The two major forms of this chronic relapsing and remitting disorder are ulcerative colitis (UC) and Crohn's disease (CD). Inflammation is primarily restricted to the colonic mucosa in UC, whereas inflammation is widespread in CD, extending into the small intestine and other organs (Chapkin et al., 2007; Podolsky, 2002). From a mechanistic stand point, n-3 PUFA can influence the mucosal inflammatory response by antagonizing the production of inflammatory eicosanoid mediators derived from n-6 PUFA (arachidonic acid), suppress the production of some inflammatory cytokines, and down-regulate the expression of a number of genes involved in inflammation (Chapman-Kiddell et al., 2010; Gil, 2002). Interestingly, the incidence of CD has been demonstrated to be strongly correlated with the ratio of n-6 to n-3 PUFA intake, but significantly less correlated with the intake of fish oil (Shoda et al., 1996). This suggests that the higher the intake of n-6 PUFA in the diet, the higher the incidence of CD. Short-term supplementation (two weeks) with cod liver oil (2.3 g EPA, 0.3 g DPA, and 3.7 g DHA) to patients with both CD and UC resulted in improved IBD activity scores over baseline values (Brunborg et al., 2008); however, longer studies would provide more insight into the sustainable effects of n-3 PUFA supplementation in IBD.

The traditional principles of management of IBD are the induction and maintenance of remission; therefore, the aim of treatment is preventing relapses and prolonging the duration of remission intervals while minimizing adverse reactions (Belluzzi, 2002; Belluzzi et al., 1996; Turner et al., 2010). A recent 2010 meta-analysis conducted by Turner et al. (2010) summarized the findings of six double-blinded, placebo-controlled trials wherein n-3 PUFA was supplemented daily in the range of 1.8–3.3 g of EPA and 0.6–1.8 g of DHA to patients with quiescent CD. Supplementation with n-3 PUFA reduced the risk of relapse within one year in CD patients (pooled risk ratio, RR=0.77; 95% confidence interval, CI=0.67–0.98) (Turner et al., 2010). Reassessment of five studies only administering enteric-coated n-3 PUFA formulations retained the significant reduction in the risk of disease relapse compared with placebo (RR=0.71; 95% CI=0.54–0.93) (Turner et al., 2010). The studies included in this analysis were heterogeneous and prompted the authors to go on to compare outcomes from only two high quality, adequately powered, large multicenter randomized double-blind placebo-controlled studies—the Epanova Program in Crohn's studies(EPIC)-1 and EPIC-2 (Feagan et al., 2008). These studies were rigorously performed but failed to show any clinical benefit of supplementation with n-3 PUFA in either CD patients in remission (EPIC-1) or patients with active disease (EPIC-2) (Feagan et al., 2008). Interestingly, when individual countries (e.g., Israel, Italy) were separately examined, there was a protective effect. Therefore, despite the beneficial pooled effect of n-3 PUFA supplementation in maintaining CD remission that was apparent in the larger meta-analysis (Turner et al., 2010), conflicting clinical findings exist. Further, assessment of three studies independently assessing the role of n-3 PUFA supplementation on prevention of ulcerative colitis remission found no effect, and thus, overall, there is inconclusive evidence to support a role for n-3 PUFA supplementation to maintain IBD remission (Stenson et al., 1992; Turner et al., 2010).

Obesity and n-3 PUFA

Obesity, defined as a body mass index ≥ 30, is a low-grade chronic inflammatory state that leads to several chronic morbidities including insulin resistance, hyperglycemia, dyslipidemia, atherosclerosis, and hypertension, all of which are components of metabolic syndrome and constitute important risk factors for cardiovascular disease and type 2 diabetes (Olufadi & Byrne, 2008). The causative connection between obesity and the metabolic syndrome remains incompletely understood; however, the emerging consensus is that inflammation links these pathologies together (Zuniga et al., 2010), thereby providing an intriguing mechanism through which n-3 PUFA supplementation could be beneficial. In this connection, a recent study conducted in Puerto Rican adults living in the Greater Boston area found that individuals consuming a diet high in marine-derived n-3 PUFA had a lower likelihood of developing metabolic syndrome (Noel et al., 2010). This finding was corroborated in a separate study conducted in adult men, wherein higher n-3 PUFA intakes were negatively associated with metabolic syndrome development (Warensjo et al., 2006).

In obesity, white adipose tissue (WAT) mass increases and its function apart from energy storage includes a role in the integration of endocrine, metabolic, and inflammatory signals (Ahmed & Gaffen, 2010). Further, adipose tissue possesses a secretory role, wherein adipokines (i.e., WAT-derived bioactive molecules) can exert their physiological functions locally (autocrine/paracrine) or systemically (endocrine), thereby allowing WAT to participate in a wide range of biological processes (Moreno-Aliaga et al., 2010; Trayhurn & Wood, 2005). Of particular interest, the systemic inflammatory nature of obesity is underscored by increased blood levels of primarily adipocyte-derived inflammatory mediators, such as leptin, serum amyloid A (SAA), C-reactive protein (CRP), plasminogen activator inhibitor-1 (PAI-1) and inflammatory cytokines (IL-6, TNF-alpha, and IL-1), while concentrations of the anti-inflammatory hormone, adiponectin, are decreased (Moreno-Aliaga et al., 2010; Trayhurn & Wood, 2004, 2005). Overall, this scenario supports a systemic pro-inflammatory phenotype, which is observed characteristically in obese subjects (Fain, 2006; Fantuzzi, 2005). Further, the circulating concentrations of these inflammatory mediators have been shown to decrease following weight loss in obese subjects (Dietrich & Jialal, 2005; Esposito et al., 2003). In obesity, the adipose tissue is infiltrated by immune cells (i.e., macrophages and T cells), which accumulate and secrete inflammatory factors, which complicates the identification of the cellular source of inflammatory cytokines and regulatory factors (i.e., TNF-alpha, IL-6, IL-1, MCP-1 and MIP-1alpha) (Weisberg et al., 2003, 2006; Zeyda et al., 2007). Further, infiltrating T cells accumulate in the adipose tissues as obesity progresses, thereby increasing the number of Th1 CD4[+] T cells (interferon (IFN)-gamma-producing) while concomitantly decreasing the frequency of Tregs, which can exert positive effects on metabolic parameters (Feuerer et al., 2009; Nishimura et al., 2009).

The circulating concentrations of IL-6 are high in obesity (Bastard et al., 2000; Naugler & Karin, 2008), and this cytokine is required for the differentiation of naïve CD4[+] T cells into the Th17 cell lineage (Harrington et al., 2005; Park et al., 2005). Th17 cells have been shown to play important roles in the pathogenesis of numerous inflammatory and autoimmune diseases and secrete the cytokine for which they are named (i.e., IL-17). Furthermore, the expansion and maintenance of the Th17 cell population and its functions are supported by IL-23, thereby stabilizing the Th17 phenotype (Ouyang et al., 2008; Sheibanie et al., 2007b). In this connection, obese patients exhibit up-regulated circulating levels of both IL-17 and IL-23 (Sumarac-Dumanovic et al., 2009) and obesity is positively correlated with enhanced IL-17 expression in animal models (Winer et al., 2009). Additionally, diet-induced obesity in mice results in elevated Th17 cell frequency, indicating that obesity specifically predisposes naïve T cells toward differentiation into Th17 cells (Ahmed & Gaffen, 2010). Thus, the chronic systemic inflammation associated with obesity predisposes the cytokine microenvironment to the generation of Th17 responses, at least in part via IL-6, and these interesting findings provide the basis for an emerging area of research centered on evaluating the connection between chronic inflammation and Th17 cells. The association is not limited to the obese state but, in fact, is highly relevant with respect to other chronic inflammatory pathologies as described in connection with inflammatory bowel disease elsewhere (Mizoguchi & Mizoguchi, 2010; Sarra et al., 2010).

PGE_2, one of the most abundant metabolites of arachidonic acid, may link systemic inflammation and Th17 cells to a potential beneficial mechanism mediated by n-3 PUFA. PGE_2 has been shown simultaneously to reduce IL-12 and enhance IL-23 production by dendritic cells and, presumably, in other inflammatory cell types, as well, thereby predisposing adaptive immunity toward eliciting Th17 responses, as seen in mouse models of arthritis and inflammatory bowel disease (Sheibanie et al., 2007, 2007b). Furthermore, recent publications provide evidence that PGE_2 can directly influence the capacity of human T cells to produce cytokines that favor the priming and expansion of Th17 cells (Boniface et al., 2009; Chizzolini et al., 2008; Napolitani et al., 2009). Marine oil derived n-3 PUFA supplementation ultimately leads to the decreased PGE_2, providing a mechanism by which n-3 PUFA may suppress Th17 cell development.

Obesity is also characterized by a decreased circulating level of adiponectin, which has been shown to elicit anti-inflammatory effects by induction of IL-10 and IL-1 receptor antagonist (IL-1RA) (Wolf et al., 2004). Further, adiponectin inhibits endothelial cell NFkB signaling, thereby providing a mechanism for reducing inflammatory cytokine production that could be extended to other cell types, although further research is required (Ouchi et al., 2000; Yokota et al., 2000). Inflammatory cytokines, TNF-alpha and IL-6, inhibit adiponectin gene expression in adipocytes (Fasshauer et al., 2003), thereby potentiating an obese inflammatory phenotype. Interestingly, circulating levels of adiponectin are increased by consumption of n-3 PUFA in mice (Flachs et al., 2006) and insulin-resistant rats (Rossi et al., 2005). Further, leptin deficient mice (ob/ob) exhibit increased adiponectin mRNA expression in adipose tissue (Gonzalez-Periz et al., 2009). The effects of n-3 PUFA on adiponectin levels could be mediated through various signaling pathways as reviewed elsewhere (Moreno-Aliaga et al., 2010). Very recently, it was demonstrated that n-3 PUFA activation of GPR120 receptor on macrophages and fat cells reverses insulin resistance in obese mice (Oh da et al., 2010). These data suggest a link/cross talk between adipose tissue, immune cells in obesity and n-3 PUFA.

Despite these interconnected inflammatory pathways that are elevated in obesity, two recent clinical studies supplementing n-3 PUFA to obese subjects (BMI > 29 kg/m²) at 1.1–3.4 g/day found no effect on circulating levels of IL-6, TNF-alpha, or the soluble TNF-alpha receptor (Chan et al., 2002; Jellema et al., 2004). In a separate clinical trial, overweight and obese women (BMI > 25 kg/m²), supplemented with 1.3 g/day of EPA plus 2.9 g/day of DHA, exhibited decreased circulating levels of IL-6 and CRP compared to baseline levels (Browning et al., 2007). Adding to the conflicting reports, in a recent randomized, placebo controlled, dietary intervention study, four isoenergetic diets were administered to obese individuals, and 1.2 g/day of n-3 PUFA was added to all dietary groups. There was no change in the circulating levels of any inflammatory mediators assessed (CRP, IL-6, TNF-alpha, adiponectin or leptin) in any dietary group (Tierney et al., 2010). However, in a separate clinical setting, n-3 PUFA administration to obese subjects was shown to reduce the circulating levels of inflammatory cytokines and acute phase proteins (White & Marette, 2006). Overall, these results indicate that n-3 PUFA supplementation exerts inconsistent effects on the systemic inflammatory status of obese individuals and further investigation is required.

Thus, despite compelling evidence supporting the underlying inflammatory nature of obesity and therefore the existence of multiple pathways in which n-3 PUFA could be beneficial and potentially decrease inflammation, there is a lack of clinical evidence to support adoption at this time. This lack of effect may be explained, in part, by the absence of measurements taken within the target tissue (i.e., the adipose tissue). Instead, these studies typically use blood levels alone to assess systemic inflammatory status following n-3 PUFA supplementation, which may not be sufficiently sensitive to identify subtle changes that may be occurring specifically within the adipose tissue. Therefore, further mechanistic studies are necessary to determine if n-3 PUFA exerts beneficial anti-inflammatory effects within the adipose tissue of the obese and to discern the appropriate dosage level. If these types of studies do indeed demonstrate a beneficial effect, then clinical intervention trials assessing more refined and biologically relevant endpoints both systemically and in the target tissue (adipose tissue biopsies) would be merited.

Cancer, Inflammation, and n-3 PUFA

The link between abdominal obesity and malignant disease is a growing area of research supported by many epidemiologic studies that have shown a positive association between obesity and multiple types of cancers including endometrial (Schapira et al., 1991), breast (Folsom et al., 1990), and prostate (Kolonel, 1996). Further, within the gastrointestinal tract, the positive association between obesity and cancer has been observed in the colon (MacInnis et al., 2006; Potter, 1996), gallbladder (Zatonski et al., 1997), gastric cardia (Chow et al., 1998; Hansson et al., 1994), and the esophagus (Lagergren et al., 1999). Combining data from 39 prospective cohort studies conducted in the Asia-Pacific region that encompassed more than 400,000 individuals revealed that the relative risk of dying from cancer (irrespective of location) was increased in both individuals who were overweight and obese (i.e., BMI of 25.0–29.9 and ≥ 30 kg/m^2, respectively), and that all cancer mortality was driven mainly by cancers of large intestine, breast, ovary, cervix, prostate, and leukemia (Parr et al., 2010). Interestingly, there was little evidence of regional differences in the relative risks between cohorts from Asia and Australia/New Zealand (Parr et al., 2010). More recently, a meta-analysis including thirty prospective studies discerned that a 5-unit increase in BMI was related to increased risk of colon cancer in both men and women (RR: 1.30; 95% CI: 1.25–1.35 and RR 1.12; 95% CI: 1.07–1.18, respectively) (Larsson & Wolk, 2007). Further, BMI was found to be positively associated with rectal cancer only in men (RR: 1.12; 95% CI: 1.09–1.16), with no significant association in women (Larsson & Wolk, 2007). Utilizing a different measurement to assess changes in body size indicative of weight gain and obesity, increasing waist circumference and waist-hip ratio were associated with increased colon cancer risk in both sexes (Larsson & Wolk, 2007). This finding has been confirmed in a separate meta-analysis which compared obese (BMI ≥ 30 kg/m^2) and non-obese individuals (RR: 1.35, 95% CI: 1.24–1.46) (Moghaddam et al., 2007). Readers are referred to a comprehensive review examining the underlying mechanisms between obesity and cancer for more information (Calle & Kaaks, 2004).

Overall, these observations strongly suggest the existence of a mechanistic link between adiposity and carcinogenesis, which requires more in-depth consideration and one that is likely to include aspects of inflammation, thereby providing a means through which n-3 PUFA could modulate these physiological processes. Kitayama et al. (2009) reported that patients with advanced gastric cancer exhibited reduced serum levels of an anti-inflammatory mediator, namely adiponectin, thus supporting a proinflammatory phenotype (Kitayama et al., 2009). This finding was supported by a case-control study, wherein adiponectin levels measured prospectively were inversely associated with risk for colorectal cancer in men, indicating that low adiponectin levels were not simply a consequence of cancer (Wei et al., 2005). In addition to the effects of adiponectin on insulin and glucose metabolism (low circulating hormone levels in conditions characterized by insulin-resistance), this hormone also promotes apoptosis by activation of apoptotic enzymes in the caspase cascade and exerts anti-inflammatory properties (Wei et al., 2005). Therefore, decreased circulating levels of adiponectin would promote both inflammation and carcinogenesis and represents one of many bioactive proteins that may link inflammation, cancer, and obesity. Further studies are required to investigate this and other mechanistic links between chronic diseases and the potential for modulation of these factors by n-3 PUFA.

A recent study determined that systemic inflammation may be involved in the early development of colorectal neoplasia, as indicated by the positive association of colorectal adenomas detected by colonoscopy and circulating levels of CRP, IL-6, and TNF-alpha (Kim et al., 2008b). Interestingly, in older adults (aged 70–80 years) these three inflammatory mediators were demonstrated to be positively associated with incident cancers and cancer deaths (Il'yasova et al., 2005). Other clinical studies assessing the relationship between inflammatory mediators and colorectal cancer have reached a range of conclusions. Specifically, a positive association was detected between colorectal cancer and CRP levels in prospective studies assessing this relationship in differing ethnic populations (Erlinger et al., 2004; Gunter et al., 2006; Otani et al., 2006). In contrast, other studies have shown no association (Ito et al., 2005; Zhang et al., 2005).

Accumulating evidence supports the notion that systemic inflammation, mediated at least in part by classic inflammatory cytokines (i.e., IL-6 and TNF-alpha) may provide a plausible mechanism for colon carcinogenesis. In this connection, IL-6 can enhance tumorigenesis by stimulating cell growth and inhibiting apoptosis (Pais et al., 2009), and the concentration of this cytokine is commonly associated with metastatic disease (Chung & Chang, 2003). TNF-alpha activates NF-κB through phosphorylation of its inhibitor (IκB), thereby increasing production of nitric oxide (NO), which is a substrate for reactive oxygen species (ROS) and is capable of stimulating other inflammatory cytokine gene expression (Pais et al., 2009; Sonnenberg et al., 2004). Additionally, ROS can damage DNA through multiple mechanisms, including DNA base modification, deletions, frame shifts, strand breaks, DNA-protein cross-links, and chromosomal rearrangements (Pais et al., 2009). Reduction in ROS and inflammatory cytokines mediated by n-3 PUFA provides a mechanistic basis supporting a beneficial role of n-3 PUFA in impairing or improving cancer risk; however, further studies conducted in humans are warranted.

The risk of developing colorectal cancers increases by approximately 0.5–1% each year after 7 years in patients with chronic intestinal inflammation (Itzkowitz & Yio, 2004; Rubin & Kavitt, 2006). Despite compelling data indicating a functional link between inflammation and colon cancer, the pathways regulating the initiation and maintenance of inflammation during cancer development remain poorly understood and, therefore, highlight the importance of identifying the overlapping regulatory relationships among genes believed to drive inflammation-associated colonic tumor development (Jia et al., 2008). Bioactive food components containing n-3 PUFA have been shown to modulate important determinants that link inflammation and either cancer development or progression (Belluzzi et al., 2000; Davidson et al., 2004; Hudert et al., 2006; Prescott & Stenson, 2005; Stenson et al., 1992). Further, clinical and experimental data indicate a protective effect of n-3 PUFA on colon cancer (Anti et al., 1992, 1994; Chang et al., 1997, 2003; Jia et al., 2008; Reddy, 1994), whereas dietary lipids rich in n-6 PUFA found in vegetable oils (e.g., linolenic acid and arachidonic acid) have been shown to enhance the development of colon tumors (Chang et al., 1997; Whelan & McEntee, 2004). COX-2 derived PGE_2 can promote tumor initiation and progression by enhancing cell proliferation, angiogenesis, and cell migration and invasion while inhibiting apoptosis (Buchanan & DuBois, 2006; Holla et al., 2006; Shao et al., 2005; Wang et al., 2006). In contrast to arachidonic acid (n-6 PUFA) derived PGE_2, EPA-derived PGE_3 exhibits antiproliferative and, therefore, chemoprotective properties (Bagga et al., 2003; Yang et al., 2004). In a major recent finding it was demonstrated that EPA reduced rectal polyp number and size in patients with familial adenomatous polyposis (FAP) (West et al., 2010). Most impressive was the fact that fish oil derived n-3 PUFA suppressed FAP to a degree similar to the selective COX-2 inhibitor celecoxib. Collectively, these data indicate that n-3 PUFA holds promise as a chemoprevention agent for FAP and sporadic colon cancer.

Another putative mechanism of n-3 PUFA action is via the effect of EPA and DHA on nuclear receptors. For example, NF-κB and STAT3, which has been linked to inflammatory colonic pathologies (Naugler & Karin, 2008; Yu et al., 2009), may be modulated by n-3 PUFA (Chapkin et al., 2007). Moreover, the effect of n-3 PUFA on NF-κB could be extended to other inflammatory pathologies beyond the intestine. Generally, NF-κB is capable of influencing both mucosal inflammation and repair, which highlights the important role of the stimulatory environment, which will largely determine if its activation is ultimately protective or deleterious for the host (Chapkin et al., 2007; Schottelius & Dinter, 2006; Yan et al., 2005). Interestingly, n-3 PUFA and perhaps their bioactive metabolites are capable of suppressing NF-κB, whereas n-6 PUFA have been shown to promote activation. These data suggest that the two different classes of PUFA (n-3 versus n-6) elicit opposing effects on the activation of potent pro-inflammatory transcription factors (Becuwe et al., 2003; Camandola et al., 1996; Chapkin et al., 2007; Fan et al., 2003, 2004; Marcheselli et al., 2003; Novak et al., 2003). Cancer-related inflammation-based mechanisms involving NF-κB, Toll-like receptors (TLR), and n-3 PUFA are elaborated upon elsewhere (Lee et al., 2010; Rakoff-Nahoum & Medzhitov, 2009; Rodriguez-Vita & Lawrence, 2010).

Human clinical trials have demonstrated that dietary n-3 PUFA can confer resistance to toxic carcinogenic agents by favorably modulating the balance between colonic epithelial cell proliferation and apoptosis (Anti et al., 1992, 1994; Cheng et al., 2003; Courtney et al., 2007). Animal models that reproduce colonic colitis-associated carcinogenesis have demonstrated that the anti-tumorigenic effect of n-3 PUFA may be mediated, in part, via its anti-inflammatory properties (Jia et al., 2008), thereby providing evidence to support further investigation of this finding in the context of human clinical trials.

Rheumatoid Arthritis (RA) and n-3 PUFA

In the United States, 30% of the population exhibit symptoms of arthritis, 5–10% arthritis-associated disabilities and 0.5% are disabled as a result of the disease (Hurst et al., 2010). Joint stiffness and pain are the usual symptoms, and the morbidity associated with arthritis exacts a massive economic cost (Hurst et al., 2010). Typically, joint lesions are characterized by infiltration of activated macrophages, T lymphocytes, and plasma cells into the synovium and by proliferation of the synovial cells themselves (Calder, 2006). Generally, in arthritis, inflammation induces the expression COX-2, supporting ongoing inflammatory cytokine release and exhibits high levels of matrix metalloproteinases (MMPs), which exacerbates protein degradation (Hurst et al., 2010). Further, synovial biopsies and fluid from rheumatoid arthritis (RA) patients and ex vivo cultured synovial cells either contain or produce high concentrations of inflammatory mediators [TNF-alpha, IL-1beta, IL-6, IL-8, granulocyte-macrophage colony-stimulating factor (GM-CSF), PGE_2, leukotriene B_4 (LTB_4), and platelet-activating factor] and increased expression of COX-2 (Feldmann & Maini, 1999; Sano et al., 1992; Sperling, 1995).

Mechanistic evidence supporting the beneficial effect of n-3 PUFA in RA is largely provided by in vitro culture systems that utilize chondrocytes or cartilage explants derived from bovine metacarpo- or metatarso-phalangeal joints or human patients undergoing knee replacement surgery. Collectively, these culture systems have been demonstrated to mimic many of the degenerative and inflammatory features of arthritis (Hurst et al., 2010; Caterson et al., 2000; Little et al., 2002). In these culture systems, addition of either n-3 PUFA or EPA alone reduced mRNA and protein expression of COX-2, as well as mRNA expression of inflammatory cytokines (IL-1alpha, IL-1beta, and TNF-alpha) and key cartilage-degrading proteinases, specifically ADAMTS-4, ADAMTS-5, MMP-3, and MMP-13 (Hurst et al., 2009; Zainal et al., 2009). Similar to other inflammatory pathologies, in arthritis, n-3 PUFA compete with n-6 PUFA in metabolism to eicosanoids, thereby resulting in the decreased production of inflammatory eicosanoids (LTB_4), and support the production of anti-inflammatory resolvins and docosatrienes (Calder, 2006; Hurst et al., 2010; Seki et al., 2009). Additionally, n-3 PUFA reduce the expression of COX-2, inflammatory mediators (i.e., C-reactive protein and cytokines, namely, IL-1alpha, IL-1beta and TNF-alpha) and cartilage-degrading proteinases, and ultimately modulate signaling pathways, transcription factors (e.g., NFκB, STAT-3), and reduce lymphocyte proliferation and macrophage activation (Hurst et al., 2010; Groeger et al., 2010). Recently, it has been demonstrated that electrophilic oxo-derivatives (EFOX) generated from n-3 PUFA (DHA and DPA) can be produced by aspirin-enhanced COX-2 oxidation reactions (Groeger et al., 2010). These compounds appear to have anti-inflammatory properties.

Clinical studies supplementing RA patients with 1–7 g of n-3 PUFA/day (average of approximately 3.5 g/day) have demonstrated improvement in subjective outcome measurements such as morning stiffness, onset of fatigue, tender or swollen joints, and pain reported by patients and/or physicians (Geusens et al., 1994; Kremer et al., 1990, 1995; Nielsen et al., 1992; Sperling et al., 1987; van der Tempel et al., 1990). Additionally, biochemical measurements demonstrate that n-3 PUFA decrease neutrophil derived LTB_4 and IL-1 production by macrophages isolated from RA patients following supplementation (Kremer et al., 1990, 1995). Short-term supplementation (two weeks) with cod liver oil (2.3 g EPA, 0.3 g DPA, and 3.7 g DHA) has been shown to improve joint pain intensity in patients (Brunborg et al., 2008). Although these reports are subjective in nature with respect to the clinical outcomes, in support of these findings, the serum ratio of n-6 to n-3 fatty acid profile and plasma levels of LTB_4 were reduced, following short-term supplementation (Brunborg et al., 2008). Interestingly, some RA patients consuming n-3 PUFA supplements were able to lower or discontinue their use of anti-inflammatory drugs, which is indicative of a beneficial therapeutic effect (Kremer et al., 1987; Kremer, 2000). Overall, the results from at least 14 randomized placebo-controlled double-blinded studies of n-3 PUFA-rich fish oil supplementation in patients with RA support a beneficial role of n-3 PUFA, although definitive studies are needed in order to make recommendations for clinical practice (Calder, 2006).

Conclusion and Future Directions

The inconclusive effects of n-3 PUFA supplementation in human clinical trials in connection with various pathologies, overall, suggest that further research is necessary to elucidate the mechanism(s) through which n-3 PUFA may reduce human disease incidence/severity either through prevention, delayed onset, or in extending disease remission periods. The basic scientific literature provides strong and conclusive evidence that supports the beneficial roles for the anti-inflammatory and immunomodulatory effects of n-3 PUFA on both physiological and pathophysiological processes. However, to date, very few studies have been able to recapitulate these findings in a clinical setting. This may be a byproduct of the limitations of human clinical studies or simply reflect the need for more in-depth mechanistic studies to help redefine precise biomarkers as outcome measurements for future clinical trials. The majority of human clinical data are limited to biochemical measurements based in blood. Although these levels may reflect spillover from extravascular tissue sites and provide insight into the systemic disease phenotype, they are not always interpretable or reflective of the local tissue environment in which disease processes arise. In some clinical settings the systemic impact of n-3 PUFA supplementation was not observed; however, these findings do not discount the possibility of an impact in relevant local tissue sites that was not assessed. Therefore, future clinical studies should make use of tissue biopsies and exfoliated cells, thereby allowing the detection of n-3 PUFA-mediated effects within the diseased tissue or organ site. Further, determination of the timing of supplementation with respect to the progression of the disease, optimal dosage level, n-6/n-3 ratio, method of administration, bioavailability, duration of intervention, and other patient related attributes (age, sex, gene polymorphisms, behavioral considerations, compliance, health status, and existence of co-morbidities) are all likely to influence the efficacy of n-3 PUFA in clinical settings. Therefore, for the time being, the future research emphasis should focus on mechanistic studies conducted in biologically relevant animal and tissue culture models that reproduce the critical features of various pathologies. This will allow scientists to determine the optimal treatment conditions, patient inclusion criteria, and sensitive biomarkers required in clinical settings to adequately assess the impact of n-3 PUFA supplementation on disease status.

References

Ahmed, M.; Gaffen, S.L. IL-17 in obesity and adipogenesis. Cytokine Growth Factor Rev. 2010, 21, 449–453.

Albert, C.M.; Campos, H.; Stampfer, M.J.; Ridker, P.M.; Manson, J.E.; Willett, W.C.; Ma, J. Blood levels of long-chain n-3 fatty acids and the risk of sudden death. N. Engl. J. Med. 2002, 346, 1113–1118.

Almallah, Y.Z.; El-Tahir, A.; Heys, S.D.; Richardson, S.; Eremin, O. Distal procto-colitis and n-3 polyunsaturated fatty acids: the mechanism(s) of natural cytotoxicity inhibition. Eur. J. Clin. Invest. 2000, 30, 58–65.

Anderson, M.J.; Fritsche, K.L. Dietary polyunsaturated fatty acids modulate in vivo, antigen-driven CD4+ T-cell proliferation in mice. J. Nutr. 2004, 134, 1978–1983.

Anderson, R.E.; Sperling, L. Lipids of ocular tissues. VII. Positional distribution of the fatty acids in the phospholipids of bovine retina rod outer segments. Arch. Biochem. Biophys.1971, 144, 673–677.

Anti, M.; Armelao, F.; Marra, G.; Percesepe, A.; Bartoli, G.M.; Palozza, P.; Parrella, P.; Canetta, C.; Gentiloni, N.; De Vitis, I.; et al. Effects of different doses of fish oil on rectal cell proliferation in patients with sporadic colonic adenomas. Gastroenterology 1994, 107, 1709–1718.

Anti, M.; Marra, G.; Armelao, F.; Bartoli, G.M.; Ficarelli, R.; Percesepe, A.; De Vitis, I.; Maria, G.; Sofo, L.; Rapaccini, G.L.; et al. Effect of omega-3 fatty acids on rectal mucosal cell proliferation in subjects at risk for colon cancer. Gastroenterology 1992, 103, 883–891.

Arrington, J.L.; Chapkin, R.S.; Switzer, K.C.; Morris, J.S.; McMurray, D.N. Dietary n-3 poly-unsaturated fatty acids modulate purified murine T-cell subset activation. Clin. Exp. Immunol. 2001,125, 499–507.

Arrington, J.L.; McMurray, D.N.; Switzer, K.C.; Fan, Y.Y.; Chapkin, R.S. Docosahexaenoic acid suppresses function of the CD28 costimulatory membrane receptor in primary murine and Jurkat T cells. J. Nutr. 2001b, 131, 1147–1153.

Bagga, D.; Wang, L.; Farias-Eisner, R.; Glaspy, J.A.; Reddy, S.T. Differential effects of prosta-glandin derived from omega-6 and omega-3 polyunsaturated fatty acids on COX-2 expression and IL-6 secretion. Proc. Natl. Acad. Sci. U S A 2004, 100, 1751–1756.

Bang, H.O.; Dyerberg, J.; Hjoorne, N. The composition of food consumed by Greenland Eskimos. Acta. Med. Scand. 1976, 200, 69–73.

Bastard, J.P.; Jardel, C.; Bruckert, E.; Blondy, P.; Capeau, J.; Laville, M.; Vidal, H.; Hainque, B. Elevated levels of interleukin 6 are reduced in serum and subcutaneous adipose tissue of obese women after weight loss. J. Clin. Endocrinol. Metab. 200, 85, 3338–3342.

Becuwe, P.; Bianchi, A.; Didelot, C.; Barberi-Heyob, M.; Dauca, M. Arachidonic acid activates a functional AP-1 and an inactive NF-kappaB complex in human HepG2 hepatoma cells. Free Radic. Biol. Med. 2003, 35, 636–647.

Belluzzi, A. N-3 fatty acids for the treatment of inflammatory bowel diseases. Proc. Nutr. Soc. 2002, 61, 391–395.

Belluzzi, A.; Boschi, S.; Brignola, C.; Munarini, A.; Cariani, G.; Miglio, F. Polyunsaturated fatty acids and inflammatory bowel disease. Am. J. Clin. Nutr. 2000, 71, 339S–342S.

Belluzzi, A.; Brignola, C.; Campieri, M.; Pera, A.; Boschi, S.; Miglioli, M. Effect of an enteric-coated fish-oil preparation on relapses in Crohn's disease. N. Engl. J. Med. 1996, 334, 1557–1560.

Boniface, K.; Bak-Jensen, K.S.; Li, Y.; Blumenschein, W.M.; McGeachy, M.J.; McClanahan, T.K.; McKenzie, B.S.; Kastelein, R.A.; Cua, D.J.; de Waal Malefyt, R. Prostaglandin E2 regulates Th17 cell differentiation and function through cyclic AMP and EP2/EP4 receptor signaling. J. Exp. Med. 2009, 206, 535–548.

Bonilla, D.L.; Fan, Y.Y.; Chapkin, R.S.; McMurray, D.N. Transgenic mice enriched in omega-3 fatty acids are more susceptible to pulmonary tuberculosis: impaired resistance to tuberculosis in fat-1 mice. J. Infect. Dis. 2010, 201, 399–408.

Bonilla, D.L.; Ly, L.H.; Fan, Y.Y.; Chapkin, R.S.; McMurray, D.N. Incorporation of a dietary omega-3 fatty acid impairs murine macrophage responses to Mycobacterium tuberculosis. PLoS One 2010b, 5, e10878.

Browning, L.M.; Krebs, J.D.; Moore, C.S.; Mishra, G.D.; O'Connell, M.A.; Jebb, S.A. The impact of long chain n-3 polyunsaturated fatty acid supplementation on inflammation, insulin sensitivity and CVD risk in a group of overweight women with an inflammatory phenotype. Diabetes Obes. Metab. 2007, 9, 70–80.

Brunborg, L.A.; Madland, T.M.; Lind, R.A.; Arslan, G.; Berstad, A.; Froyland, L. Effects of short-term oral administration of dietary marine oils in patients with inflammatory bowel disease and joint pain: a pilot study comparing seal oil and cod liver oil. Clin. Nutr. 2008, 27, 614–622.

Buchanan, F.G.; DuBois, R.N. Connecting COX-2 and Wnt in cancer. Cancer Cell. 2006, 9, 6–8.

Bucher, H.C.; Hengstler, P.; Schindler, C.; Meier, G. N-3 polyunsaturated fatty acids in coronary heart disease: a meta-analysis of randomized controlled trials. Am. J. Med. 2002, 112, 298–304.

Calder, P.C. Immunomodulation by omega-3 fatty acids. Prostaglandins Leukot. Essent. Fatty Acids 2007, 77, 327–335

Calder, P.C. n-3 polyunsaturated fatty acids, inflammation, and inflammatory diseases. Am. J. Clin. Nutr. 2006, 83, 1505S–1519S.

Calle, E.E.; Kaaks, R. Overweight, obesity and cancer: epidemiological evidence and proposed mechanisms. Nat. Rev. Cancer 2004, 4, 579–591.

Camandola, S.; Leonarduzzi, G.; Musso, T.; Varesio, L.; Carini, R.; Scavazza, A.; Chiarpotto, E.; Baeuerle, P.A.; Poli, G. Nuclear factor kB is activated by arachidonic acid but not by eicosapentaenoic acid. Biochem. Biophys. Res. Commun. 1996, 229, 643–647.

Cappelli, P.; Di Liberato, L.; Stuard, S.; Ballone, E.; Albertazzi, A. N-3 polyunsaturated fatty acid supplementation in chronic progressive renal disease. J. Nephrol. 1997, 10, 157–162.

Caterson, B.; Flannery, C.R.; Hughes, C.E.; Little, C.B. Mechanisms involved in cartilage proteoglycan catabolism Matrix Biol. 2000, 19, 333–344.

Chan, D.C.; Watts, G.F.; Barrett, P.H.; Beilin, L.J.; Mori, T.A. Effect of atorvastatin and fish oil on plasma high-sensitivity C-reactive protein concentrations in individuals with visceral obesity. Clin. Chem. 2002, 48, 877–883.

Chang, W.C.; Chapkin, R.S.; Lupton, J.R. Predictive value of proliferation, differentiation and apoptosis as intermediate markers for colon tumorigenesis. Carcinogenesis 1997, 18, 721–730.

Chapkin, R.S.; Kim, W.; Lupton, J.R.; McMurray, D.N. Dietary docosahexaenoic and eicosapentaenoic acid: emerging mediators of inflammation. Prostaglandins Leukot. Essent. Fatty Acids 2009, 81, 187–191.

Chapkin, R.S.; McMurray, D.N.; Lupton, J.R. Colon cancer, fatty acids and anti-inflammatory compounds. Curr. Opin. Gastroenterol.2007, 23, 48–54.

Chapman-Kiddell, C.A.; Davies, P.S.; Gillen, L.; Radford-Smith, G.L. Role of diet in the development of inflammatory bowel disease. Inflamm. Bowel Dis. 2010, 16, 137–151.

Cheng, J.; Ogawa, K.; Kuriki, K.; Yokoyama, Y.; Kamiya, T.; Seno, K.; Okuyama, H.; Wang, J.; Luo, C.; Fujii, T.; Ichikawa, H.; et al. Increased intake of n-3 polyunsaturated fatty acids elevates the level of apoptosis in the normal sigmoid colon of patients polypectomized for adenomas/tumors. Cancer Lett. 2003, 193, 17–24.

Chin, J.P.; Gust, A.P.; Nestel, P.J.; Dart, A.M. Marine oils dose-dependently inhibit vasoconstriction of forearm resistance vessels in humans. Hypertension 1993, 21, 22–28.

Chizzolini, C.; Chicheportiche, R.; Alvarez, M.; de Rham, C.; Roux-Lombard, P.; Ferrari-Lacraz, S.; Dayer, J.M. Prostaglandin E2 synergistically with interleukin-23 favors human Th17 expansion. Blood 2008, 112, 3696–3703.

Chow, W.H.; Blot, W.J.; Vaughan, T.L.; Risch, H.A.; Gammon, M.D.; Stanford, J.L.; Dubrow, R.; Schoenberg, J.B.; Mayne, S.T.; Farrow, D.C.; et al. Body mass index and risk of adenocarcinomas of the esophagus and gastric cardia. J. Natl. Cancer Inst. 1998, 90, 150–155.

Chung, Y.C.; Chang, Y.F. Serum interleukin-6 levels reflect the disease status of colorectal cancer. J. Surg. Oncol. 2003, 83, 222–226.

Conquer, J. A.; Holub, B.J. Effect of supplementation with different doses of DHA on the levels of circulating DHA as non-esterified fatty acid in subjects of Asian Indian background. J. Lipid. Res. 1998, 39, 286–292.

Courtney, E.D.; Matthews, S.; Finlayson, C.; Di Pierro, D.; Belluzzi, A.; Roda, E.; Kang, J.Y.; Leicester, R.J. Eicosapentaenoic acid (EPA) reduces crypt cell proliferation and increases apoptosis in normal colonic mucosa in subjects with a history of colorectal adenomas. Int. J. Colorectal. Dis. 2007, 22, 765–776.

Damsgaard, C.T.; Frokiaer, H.; Lauritzen, L. The effects of fish oil and high or low linoleic acid intake on fatty acid composition of human peripheral blood mononuclear cells. Br. J. Nutr. 2008, 99, 147–154.

Davidson, L.A.; Nguyen, D.V.; Hokanson, R.M.; Callaway, E.S.; Isett, R.B.; Turner, N.D.; Dougherty, E.R.; Wang, N.; Lupton, J.R.; Carroll, R.J.; et al. Chemopreventive n-3 polyunsaturated fatty acids reprogram genetic signatures during colon cancer initiation and progression in the rat. Cancer Res. 2004, 64, 6797–6804.

Daviglus, M.L.; Stamler, J.; Orencia, A.J.; Dyer, A.R.; Liu, K.; Greenland, P.; Walsh, M.K.; Morris, D.; Shekelle, R.B. Fish consumption and the 30-year risk of fatal myocardial infarction. N. Engl. J. Med. 1997, 336, 1046–1053.

Deckelbaum, R.J.; Calder, P.C. Dietary n-3 and n-6 fatty acids: are there 'bad' polyunsaturated fatty acids? Curr. Opin. Clin. Nutr. Metab. Care 2010, 13, 123–124.

Dietrich, M.; Jialal, I. The effect of weight loss on a stable biomarker of inflammation, C-reactive protein. Nutr. Rev. 2005, 63, 22–28.

Erlinger, T.P.; Platz, E.A.; Rifai, N.; Helzlsouer, K.J. C-reactive protein and the risk of incident colorectal cancer. JAMA 2004, 291, 585–590.

Eslick, G.D.; Howe, P.R.; Smith, C.; Priest, R.; Bensoussan, A. Benefits of fish oil supplementation in hyperlipidemia: a systematic review and meta-analysis. Int. J. Cardiol. 2009, 136, 4–16.

Esposito, K.; Pontillo, A.; Di Palo, C.; Giugliano, G.; Masella, M.; Marfella, R.; Giugliano, D. Effect of weight loss and lifestyle changes on vascular inflammatory markers in obese women: a randomized trial. JAMA 2003, 289, 1799–1804.

Fain, J.N. Release of interleukins and other inflammatory cytokines by human adipose tissue is enhanced in obesity and primarily due to the nonfat cells. Vitam. Horm. 2006, 74, 443–477.

Fan, Y.Y.; Ly, L.H.; Barhoumi, R.; McMurray, D.N.; Chapkin, R.S. Dietary docosahexaenoic acid suppresses T cell protein kinase C theta lipid raft recruitment and IL-2 production. J. Immunol. 2004, 173, 6151–6160.

Fan, Y.Y.; Spencer, T.E.; Wang, N.; Moyer, M.P.; Chapkin, R.S. Chemopreventive n-3 fatty acids activate RXRalpha in colonocytes. Carcinogenesis 2003, 24, 1541–1548.

Fantuzzi, G. Adipose tissue, adipokines, and inflammation. J. Allergy Clin. Immunol. 2005, 115, 911–919; quiz 920.

Fasshauer, M.; Kralisch, S.; Klier, M.; Lossner, U.; Bluher, M.; Klein, J.; Paschke, R. Adiponectin gene expression and secretion is inhibited by interleukin-6 in 3T3-L1 adipocytes. Biochem. Biophys. Res. Commun. 2003, 301, 1045–1050.

Feagan, B.G.; Sandborn, W.J.; Mittmann, U.; Bar-Meir, S.; D'Haens, G.; Bradette, M.; Cohen, A.; Dallaire, C.; Ponich, T.P.; McDonald, J.W.; et al. Omega-3 free fatty acids for the maintenance of remission in Crohn disease: the EPIC Randomized Controlled Trials. JAMA 2008, 299, 1690–1697.

Feldmann, M.; Maini, R.N. The role of cytokines in the pathogenesis of rheumatoid arthritis. Rheumatology (Oxford) 1999, 38 Suppl 2, 3–7.

Feskens, E.J.; Kromhout, D. Epidemiologic studies on Eskimos and fish intake. Ann. N Y Acad. Sci.1993, 683, 9–15.

Feuerer, M.; Herrero, L.; Cipolletta, D.; Naaz, A.; Wong, J.; Nayer, A.; Lee, J.; Goldfine, A.B.; Benoist, C.; Shoelson, S.; Mathis, D. Lean, but not obese, fat is enriched for a unique population of regulatory T cells that affect metabolic parameters. Nat. Med. 2009, 15, 930–939.

Filion, K.B.; El Khoury, F.; Bielinski, M.; Schiller, I.; Dendukuri, N.; Brophy, J.M. Omega-3 fatty acids in high-risk cardiovascular patients: a meta-analysis of randomized controlled trials. BMC Cardiovasc. Disord. 2010, 10, 24.

Flachs, P.; Mohamed-Ali, V.; Horakova, O.; Rossmeisl, M.; Hosseinzadeh-Attar, M.J.; Hensler, M.; Ruzickova, J.; Kopecky, J. Polyunsaturated fatty acids of marine origin induce adiponectin in mice fed a high-fat diet. Diabetologia 2006, 49, 394–397.

Folsom, A.R.; Kaye, S.A.; Prineas, R.J.; Potter, J.D.; Gapstur, S.M.; Wallace, R.B. Increased incidence of carcinoma of the breast associated with abdominal adiposity in postmenopausal women. Am. J. Epidemiol. 1990, 131, 794–803.

Gallai, V.; Sarchielli, P.; Trequattrini, A.; Franceschini, M.; Floridi, A.; Firenze, C.; Alberti, A.; Di Benedetto, D.; Stragliotto, E. Cytokine secretion and eicosanoid production in the peripheral blood mononuclear cells of MS patients undergoing dietary supplementation with n-3 poly-unsaturated fatty acids. J. Neuroimmunol. 1995, 56, 143–153.

Gapinski, J.P.; VanRuiswyk, J.V.; Heudebert, G.R.; Schectman, G.S. Preventing restenosis with fish oils following coronary angioplasty. A meta-analysis. Arch. Intern. Med. 1993, 153, 1595–1601.

Geleijnse, J.M.; Giltay, E.J.; Grobbee, D.E.; Donders, A.R.; Kok, F.J. Blood pressure response to fish oil supplementation: metaregression analysis of randomized trials. J. Hypertens. 2002, 20, 1493–1499.

Geusens, P.; Wouters, C.; Nijs, J.; Jiang, Y.; Dequeker, J. Long-term effect of omega-3 fatty acid supplementation in active rheumatoid arthritis. A 12-month, double-blind, controlled study. Arthritis. Rheum. 1994, 37, 824–829.

Gil, A. Polyunsaturated fatty acids and inflammatory diseases. Biomed. Pharmacother. 2002, 56, 388–396.

GISSI-Prevenzione Investigators (Gruppo Italiano per lo Studio della Sopravvivenza nell'Infarto miocardico). Dietary supplementation with n-3 polyunsaturated fatty acids and vitamin E after myocardial infarction: results of the GISSI-Prevenzione trial. Lancet 1999, 354, 447–455.

Gonzalez-Periz, A.; Horrillo, R.; Ferre, N.; Gronert, K.; Dong, B.; Moran-Salvador, E.; Titos, E.; Martinez-Clemente, M.; Lopez-Parra, M.; Arroyo, V.; et al. Obesity-induced insulin resistance and hepatic steatosis are alleviated by omega-3 fatty acids: a role for resolvins and protectins. FASEB J. 2009, 23, 1946–1957.

Groeger, A.L.; Cipollina, C.; Cole, M.P.; Woodcock, S.R.; Bonacci, G.; Rudolph, T.K.; Rudolph, V.; Freeman, B.A.; Schopfer, F.J. Cyclooxygenase-2 generates anti-inflammatory mediators from omega-3 fatty acids. Nat. Chem. Biol. 2010, 6, 433–441.

Gunter, M.J.; Stolzenberg-Solomon, R.; Cross, A.J.; Leitzmann, M.F.; Weinstein, S.; Wood, R.J.; Virtamo, J.; Taylor, P.R.; Albanes, D.; Sinha, R. A prospective study of serum C-reactive protein and colorectal cancer risk in men Cancer Res. 2006, 66, 2483–2487.

Hallgren, C.G.; Hallmans, G.; Jansson, J.H.; Marklund, S.L.; Huhtasaari, F.; Schutz, A.; Stromberg, U.; Vessby, B.; Skerfving, S. Markers of high fish intake are associated with decreased risk of a first myocardial infarction. Br. J. Nutr. 2001, 86, 397–404.

Hansson, L.E.; Nyren, O.; Bergstrom, R.; Wolk, A.; Lindgren, A.; Baron, J.; Adami, H.O. Nutrients and gastric cancer risk. A population-based case-control study in Sweden. Int. J. Cancer 1994, 57, 638–644.

Harrington, L.E.; Hatton, R.D.; Mangan, P.R.; Turner, H.; Murphy, T.L.; Murphy, K.M.; Weaver, C.T. Interleukin 17-producing CD4+ effector T cells develop via a lineage distinct from the T helper type 1 and 2 lineages. Nat. Immunol. 2005, 6, 1123–1132.

Hibbeln, J.R. Fish consumption and major depression. Lancet 1998, 351, 1213.

Hibbeln, J.R. Seafood consumption, the DHA content of mothers' milk and prevalence rates of postpartum depression: a cross-national, ecological analysis. J. Affect. Disord. 2002, 69, 15–29.

Holla, V.R.; Mann, J.R.; Shi, Q.; DuBois, R.N. Prostaglandin E2 regulates the nuclear receptor NR4A2 in colorectal cancer. J. Biol. Chem. 2006, 281, 2676–2682.

Hu, F.B.; Cho, E.; Rexrode, K.M.; Albert, C.M.; Manson, J.E. Fish and long-chain omega-3 fatty acid intake and risk of coronary heart disease and total mortality in diabetic women. Circulation 2003, 107, 1852–1857.

Hudert, C.A.; Weylandt, K.H.; Lu, Y.; Wang, J.; Hong, S.; Dignass, A.; Serhan, C.N.; Kang, J.X. Transgenic mice rich in endogenous omega-3 fatty acids are protected from colitis. Proc. Natl. Acad. Sci. U S A 2006, 103, 11276–11281.

Hughes, D.A.; Pinder, A.C. n-3 polyunsaturated fatty acids inhibit the antigen-presenting function of human monocytes. Am. J. Clin. Nutr. 2000, 71, 357S–360S.

Hurst, S.; Rees, S.G.; Randerson, P.F.; Caterson, B.; Harwood, J.L. Contrasting effects of n-3 and n-6 fatty acids on cyclooxygenase-2 in model systems for arthritis. Lipids 2009, 44, 889–896.

Hurst, S.; Zainal, Z.; Caterson, B.; Hughes, C.E.; Harwood, J.L. Dietary fatty acids and arthritis. Prostaglandins Leukot. Essent. Fatty Acids. 2010, 82, 315–318.

Il'yasova, D.; Colbert, L.H.; Harris, T.B.; Newman, A.B.; Bauer, D.C.; Satterfield, S.; Kritchevsky, S.B. Circulating levels of inflammatory markers and cancer risk in the health aging and body composition cohort. Cancer Epidemiol. Biomarkers Prev. 2005, 14, 2413–2418.

Ito, Y.; Suzuki, K.; Tamakoshi, K.; Wakai, K.; Kojima, M.; Ozasa, K.; Watanabe, Y.; Kawado, M.; Hashimoto, S.; Suzuki, S.; Tokudome, S.; Toyoshima, H.; et al. Colorectal cancer and serum C-reactive protein levels: a case-control study nested in the JACC Study. J. Epidemiol. 2005, 15 Suppl 2, S185–189.

Itzkowitz, S.H.; Yio, X. Inflammation and cancer IV. Colorectal cancer in inflammatory bowel disease: the role of inflammation. Am. J. Physiol. Gastrointest. Liver Physiol. 2004, 287, G7–17.

Jellema, A.; Plat, J.; Mensink, R.P. Weight reduction, but not a moderate intake of fish oil, lowers concentrations of inflammatory markers and PAI-1 antigen in obese men during the fasting and postprandial state. Eur. J. Clin. Invest. 2004, 34, 766–773.

Jia, Q.; Lupton, J.R.; Smith, R.; Weeks, B.R.; Callaway, E.; Davidson, L.A.; Kim, W.; Fan, Y.Y.; Yang, P.; Newman, R.A.; et al. Reduced colitis-associated colon cancer in Fat-1 (n-3 fatty acid desaturase) transgenic mice. Cancer Res. 2008, 68, 3985–3991.

Jump, D.B.; Botolin, D.; Wang, Y.; Xu, J.; Christian, B.; Demeure, O. Fatty acid regulation of hepatic gene transcription. J. Nutr. 2005, 135, 2503–2506.

Kelley, D.S.; Taylor, P.C.; Nelson, G.J.; Mackey, B.E. Dietary docosahexaenoic acid and immuno-competence in young healthy men. Lipids 1998, 33, 559–566.

Kelley, D.S.; Taylor, P.C.; Nelson, G.J.; Schmidt, P.C.; Ferretti, A.; Erickson, K.L.; Yu, R.; Chandra, R.K.; Mackey, B.E. Docosahexaenoic acid ingestion inhibits natural killer cell activity and production of inflammatory mediators in young healthy men. Lipids 1999, 34, 317–324.

Kim, W.; Fan, Y.Y.; Barhoumi, R.; Smith, R.; McMurray, D.N.; Chapkin, R.S. n-3 polyunsaturated fatty acids suppress the localization and activation of signaling proteins at the immunological synapse in murine CD4+ T cells by affecting lipid raft formation. J. Immunol. 2008, 181, 6236–6243.

Kim, S.; Keku, T.O.; Martin, C.; Galanko, J.; Woosley, J.T.; Schroeder, J.C.; Satia, J.A.; Halabi, S.; Sandler, R.S. Circulating levels of inflammatory cytokines and risk of colorectal adenomas. Cancer Res. 2008b, 68, 323–328.

Kim, W.; Khan, N.A.; McMurray, D.N.; Prior, I.A.; Wang, N.; Chapkin, R.S. Regulatory activity of polyunsaturated fatty acids in T-cell signaling. Prog. Lipid Res. 2010, 49, 250–261.

Kim, W.; McMurray, D.N.; Chapkin, R.S. n-3 polyunsaturated fatty acids—physiological relevance of dose. Prostaglandins Leukot. Essent. Fatty Acids 2010b, 82, 155–158.

Kitayama, J.; Tabuchi, M.; Tsurita, G.; Ishikawa, M.; Otani, K.; Nagawa, H. Adiposity and gastro-intestinal malignancy. Digestion 2009, 79 Suppl 1, 26–32.

Kolonel, L.N. Nutrition and prostate cancer. Cancer Causes Control 1996, 7, 83–44.

Kremer, J.M. n-3 fatty acid supplements in rheumatoid arthritis. Am. J. Clin. Nutr. 2000, 71, 349S–351S.

Kremer, J.M.; Jubiz, W.; Michalek, A.; Rynes, R.I.; Bartholomew, L.E.; Bigaouette, J.; Timchalk, M.; Beeler, D.; Lininger, L. Fish-oil fatty acid supplementation in active rheumatoid arthritis. A double-blinded, controlled, crossover study. Ann. Intern. Med. 1987, 106, 497–503.

Kremer, J.M.; Lawrence, D.A.; Jubiz, W.; DiGiacomo, R.; Rynes, R.; Bartholomew, L.E.; Sherman, M. Dietary fish oil and olive oil supplementation in patients with rheumatoid arthritis. Clinical and immunologic effects. Arthritis. Rheum. 1990, 33, 810–820.

Kremer, J.M.; Lawrence, D.A.; Petrillo, G.F.; Litts, L.L.; Mullaly, P.M.; Rynes, R.I.; Stocker, R.P.; Parhami, N.; Greenstein, N.S.; Fuchs, B.R. Effects of high-dose fish oil on rheumatoid arthritis after stopping nonsteroidal antiinflammatory drugs. Clinical and immune correlates. Arthritis Rheum. 1995, 38, 1107–1114.

Kris-Etherton, P.M.; Taylor, D.S.; Yu-Poth, S.; Huth, P.; Moriarty, K.; Fishell, V.; Hargrove, R.L.; Zhao, G.; Etherton, T.D. Polyunsaturated fatty acids in the food chain in the United States. Am. J. Clin. Nutr. 2000, 71, 179S–188S.

Kristensen, S.D.; Schmidt, E.B.; Dyerberg, J. Dietary supplementation with n-3 polyunsaturated fatty acids and human platelet function: a review with particular emphasis on implications for cardiovascular disease. J. Intern. Med. Suppl. 1989, 731, 141–150.

Lagergren, J.; Bergstrom, R.; Nyren, O. Association between body mass and adenocarcinoma of the esophagus and gastric cardia. Ann. Intern. Med. 1999, 130, 883–890.

Larsson, S.C.; Wolk, A. Obesity and colon and rectal cancer risk: a meta-analysis of prospective studies. Am. J. Clin. Nutr. 2007, 86, 556–565.

Lee, J.Y.; Zhao, L.; Hwang, D.H. Modulation of pattern recognition receptor-mediated inflammation and risk of chronic diseases by dietary fatty acids. Nutr. Rev. 2010, 68, 38–61.

Leon, H.; Shibata, M.C.; Sivakumaran, S.; Dorgan, M.; Chatterley, T.; Tsuyuki, R.T. Effect of fish oil on arrhythmias and mortality: systematic review. BMJ 2008, 337, a2931.

Little, C.B.; Hughes, C.E.; Curtis, C.L.; Janusz, M.J.; Bohne, R.; Wang-Weigand, S.; Taiwo, Y.O.; Mitchell, P.G.; Otterness, I.G.; Flannery, C.R.; Caterson, B. Matrix metalloproteinases are involved in C-terminal and interglobular domain processing of cartilage aggrecan in late stage cartilage degradation. Matrix Biol. 2202, 21, 271–288.

Lopez-Garcia, E.; Schulze, M.B.; Manson, J.E.; Meigs, J.B.; Albert, C.M.; Rifai, N.; Willett, W.C.; Hu, F.B. Consumption of (n-3) fatty acids is related to plasma biomarkers of inflammation and endothelial activation in women. J. Nutr. 2004, 134, 1806–1811.

Lupton, J.R.; Chapkin, R.S. Chemopreventive effects of omega-3 fatty acids. Cancer chemoprevention I, Kelloff, G., Hawk, E.T., Sigman, C.C., Eds.; Humana Press: Totowa, N.J., 2004; 591–608.

Ma, D.W.; Seo, J.; Davidson, L.A.; Callaway, E.S.; Fan, Y.Y.; Lupton, J.R.; Chapkin, R.S. n-3 PUFA alter caveolae lipid composition and resident protein localization in mouse colon. FASEB J. 2004, 18, 1040–1042.

Ma, D.W.; Seo, J.; Switzer, K.C.; Fan, Y.Y.; McMurray, D.N.; Lupton, J.R.; Chapkin, R.S. n-3 PUFA and membrane microdomains: a new frontier in bioactive lipid research. J. Nutr. Biochem. 2004b, 15, 700–706.

MacInnis, R.J.; English, D.R.; Hopper, J.L.; Gertig, D.M.; Haydon, A.M.; Giles, G.G. Body size and composition and colon cancer risk in women. Int. J. Cancer 2006, 118, 1496–1500.

Marcheselli, V.L.; Hong, S.; Lukiw, W.J.; Tian, X.H.; Gronert, K.; Musto, A.; Hardy, M.; Gimenez, J.M.; Chiang, N.; Serhan, C.N.; Bazan, N.G. Novel docosanoids inhibit brain ischemia-reperfusion-mediated leukocyte infiltration and pro-inflammatory gene expression. J. Biol. Chem. 2003, 278, 43807–43817.

Massaro, M.; Scoditti, E.; Carluccio, M.A.; Montinari, M.R.; De Caterina, R. Omega-3 fatty acids, inflammation and angiogenesis: nutrigenomic effects as an explanation for anti-atherogenic and anti-inflammatory effects of fish and fish oils. J. Nutrigenet. Nutrigenomics 2008, 1, 4–23.

Matsuyama, W.; Mitsuyama, H.; Watanabe, M.; Oonakahara, K.; Higashimoto, I.; Osame, M.; Arimura, K. Effects of omega-3 polyunsaturated fatty acids on inflammatory markers in COPD. Chest. 2005, 128, 3817–3827.

McFarland, C.T.; Fan, Y.Y.; Chapkin, R.S.; Weeks, B.R.; McMurray, D.N. Dietary polyunsaturated fatty acids modulate resistance to Mycobacterium tuberculosis in guinea pigs. J. Nutr. 2008, 138, 2123–2128.

Mickleborough, T.D.; Lindley, M.R.; Ionescu, A.A.; Fly, A.D. Protective effect of fish oil supplementation on exercise-induced bronchoconstriction in asthma. Chest. 2006, 129, 39–49.

Mizoguchi, A.; Mizoguchi, E. Animal models of IBD: linkage to human disease. Curr. Opin. Pharmacol. 2010, 10, 578–587.

Moghaddam, A.A.; Woodward, M.; Huxley, R. Obesity and risk of colorectal cancer: a meta-analysis of 31 studies with 70,000 events. Cancer Epidemiol. Biomarkers Prev. 2007, 16, 2533–2547.

Moreno-Aliaga, M.J.; Lorente-Cebrian, S.; Martinez, J.A. Regulation of adipokine secretion by n-3 fatty acids. Proc. Nutr. Soc. 2010, 69, 324–332.

Morris, M.C.; Evans, D.A.; Bienias, J.L.; Tangney, C.C.; Bennett, D.A.; Wilson, R.S.; Aggarwal, N.; Schneider, J. Consumption of fish and n-3 fatty acids and risk of incident Alzheimer disease. Arch. Neurol. 2003, 60, 940–946.

Mozaffarian, D.; Geelen, A.; Brouwer, I.A.; Geleijnse, J.M.; Zock, P.L.; Katan, M.B. Effect of fish oil on heart rate in humans: a meta-analysis of randomized controlled trials Circulation 2005, 112, 1945–1952.

Mozaffarian, D.; Stein, P.K.; Prineas, R.J.; Siscovick, D.S. Dietary fish and omega-3 fatty acid consumption and heart rate variability in US adults. Circulation 2008, 117, 1130–1137.

Nagata, C.; Takatsuka, N.; Shimizu, H. Soy and fish oil intake and mortality in a Japanese community. Am. J. Epidemiol.2002, 156, 824–831.

Napolitani, G.; Acosta-Rodriguez, E.V.; Lanzavecchia, A.; Sallusto, F. Prostaglandin E2 enhances Th17 responses via modulation of IL-17 and IFN-gamma production by memory CD4+ T cells. Eur. J. Immunol. 2009, 39, 1301–1312.

Naugler, W.E.; Karin, M. NF-kappaB and cancer-identifying targets and mechanisms. Curr. Opin. Genet. Dev. 2008, 18, 19–26.

Nielsen, G.L.; Faarvang, K.L.; Thomsen, B.S.; Teglbjaerg, K.L.; Jensen, L.T.; Hansen, T.M.; Lervang, H.H.; Schmidt, E.B.; Dyerberg, J.; Ernst, E. The effects of dietary supplementation with n-3 polyunsaturated fatty acids in patients with rheumatoid arthritis: a randomized, double blind trial. Eur. J. Clin. Invest. 1992, 22, 687–691.

Nishimura, S.; Manabe, I.; Nagasaki, M.; Eto, K.; Yamashita, H.; Ohsugi, M.; Otsu, M.; Hara, K.; Ueki, K.; Sugiura, S.; Yoshimura, K.; Kadowaki, T.; Nagai, R. CD8+ effector T cells contribute to macrophage recruitment and adipose tissue inflammation in obesity. Nat. Med. 2009, 15, 914–920.

Noel, S.E.; Newby, P.K.; Ordovas, J.M.; Tucker, K.L. Adherence to an (n-3) fatty acid/fish intake pattern is inversely associated with metabolic syndrome among Puerto Rican adults in the Greater Boston area. J. Nutr. 2010, 140, 1846–1854.

Novak, T.E.; Babcock, T.A.; Jho, D.H.; Helton, W.S.; Espat, N.J. NF-kappa B inhibition by omega-3 fatty acids modulates LPS-stimulated macrophage TNF-alpha transcription. Am. J. Physiol. Lung Cell. Mol. Physiol. 2003, 284, L84–89.

Oh da, Y.; Talukdar, S.; Bae, E.J.; Imamura, T.; Morinaga, H.; Fan, W.; Li, P.; Lu, W.J.; Watkins, S.M.; Olefsky, J.M. GPR120 is an omega-3 fatty acid receptor mediating potent anti-inflammatory and insulin-sensitizing effects. Cell 2010, 142, 687–698.

Okuyama, H.; Kobayashi, T.; Watanabe, S. Dietary fatty acids—the N-6/N-3 balance and chronic elderly diseases. Excess linoleic acid and relative N-3 deficiency syndrome seen in Japan. Prog. Lipid Res.1996, 35, 409–457.

Olufadi, R.; Byrne, C.D. Clinical and laboratory diagnosis of the metabolic syndrome. J. Clin. Pathol. 2008, 61, 697–706.

Oomen, C.M.; Feskens, E.J.; Rasanen, L.; Fidanza, F.; Nissinen, A.M.; Menotti, A.; Kok, F.J.; Kromhout, D. Fish consumption and coronary heart disease mortality in Finland, Italy, and The Netherlands. Am. J. Epidemiol. 2000, 151, 999–1006.

Otani, T.; Iwasaki, M.; Sasazuki, S.; Inoue, M.; Tsugane, S. Plasma C-reactive protein and risk of colorectal cancer in a nested case-control study: Japan Public Health Center-based prospective study. Cancer Epidemiol. Biomarkers Prev. 2006, 15, 690–695.

Ouchi, N.; Kihara, S.; Arita, Y.; Okamoto, Y.; Maeda, K.; Kuriyama, H.; Hotta, K.; Nishida, M.; Takahashi, M.; Muraguchi, M. Adiponectin, an adipocyte-derived plasma protein, inhibits endothelial NF-kappaB signaling through a cAMP-dependent pathway. Circulation 2000, 102, 1296–1301.

Ouyang, W.; Kolls, J.K.; Zheng, Y. The biological functions of T helper 17 cell effector cytokines in inflammation. Immunity 2008, 28, 454–467.

Pais, R.; Silaghi, H.; Silaghi, A.C.; Rusu, M.L.; Dumitrascu, D.L. Metabolic syndrome and risk of subsequent colorectal cancer. World J. Gastroenterol. 2009, 15, 5141–5148.

Palacios-Pelaez, R.; Lukiw, W.J.; Bazan, N.G. Omega-3 essential fatty acids modulate initiation and progression of neurodegenerative disease. Mol. Neurobiol. 2010, 41, 367–374.

Park, H.; Li, Z.; Yang, X.O.; Chang, S.H.; Nurieva, R.; Wang, Y.H.; Wang, Y.; Hood, L.; Zhu, Z.; Tian, Q.; et al. A distinct lineage of CD4 T cells regulates tissue inflammation by producing interleukin 17. Nat. Immunol. 2005, 6, 1133–1141.

Park, K.; Mozaffarian, D. Omega-3 Fatty acids, mercury, and selenium in fish and the risk of cardiovascular diseases. Curr. Atheroscler. Rep. 2010, 12, 414–422.

Parr, C.L.; Batty, G.D.; Lam, T.H.; Barzi, F.; Fang, X.; Ho, S.C.; Jee, S.H.; Ansary-Moghaddam, A.; Jamrozik, K.; Ueshima, H.; et al. Body-mass index and cancer mortality in the Asia-Pacific Cohort Studies Collaboration: pooled analyses of 424,519 participants. Lancet Oncol. 2010, 11, 741–752.

Podolsky, D.K. Inflammatory bowel disease. N. Engl. J. Med. 2002, 347, 417–429.

Potter, J.D. Nutrition and colorectal cancer. Cancer Causes Control 1996, 7, 127–146.

Prescott, S.M. The effect of eicosapentaenoic acid on leukotriene B production by human neutrophils. J. Biol. Chem. 1984, 259, 7615–7621.

Prescott, S.M.; Stenson, W.F. Fish oil fix. Nat. Med. 2005, 11, 596–598.

Rakoff-Nahoum, S.; Medzhitov, R. Toll-like receptors and cancer. Nat. Rev. Cancer 2009, 9, 57–63.

Reddy, B.S. Chemoprevention of colon cancer by dietary fatty acids. Cancer Metastasis Rev. 1994, 13, 285–302.

Rees, D.; Miles, E.A.; Banerjee, T.; Wells, S.J.; Roynette, C.E.; Wahle, K.W.; Calder, P.C. Dose-related effects of eicosapentaenoic acid on innate immune function in healthy humans: a comparison of young and older men. Am. J. Clin. Nutr. 2006, 83, 331–342.

Rodriguez-Vita, J.; Lawrence, T. The resolution of inflammation and cancer. Cytokine Growth Factor Rev. 2010, 21, 61–65.

Rossi, A.S.; Lombardo, Y.B.; Lacorte, J.M.; Chicco, A.G.; Rouault, C.; Slama, G.; Rizkalla, S.W. Dietary fish oil positively regulates plasma leptin and adiponectin levels in sucrose-fed, insulin-resistant rats. Am. J. Physiol. Regul. Integr. Comp. Physiol. 2005, 289, R486–R494.

Rubin, D.T.; Kavitt, R.T. Surveillance for cancer and dysplasia in inflammatory bowel disease. Gastroenterol. Clin. North Am. 2006, 35, 581–604.

Sano, H.; Hla, T.; Maier, J.A.; Crofford, L.J.; Case, J.P.; Maciag, T.; Wilder, R.L. In vivo cyclooxygenase expression in synovial tissues of patients with rheumatoid arthritis and osteoarthritis and rats with adjuvant and streptococcal cell wall arthritis. J. Clin. Invest. 1992, 89, 97–108.

Saravanan, P.; Davidson, N.C.; Schmidt, E.B.; Calder, P.C. Cardiovascular effects of marine omega-3 fatty acids. Lancet 2010, 376, 540–550.

Sarra, M.; Pallone, F.; Macdonald, T.T.; Monteleone, G. IL-23/IL-17 axis in IBD. Inflamm. Bowel Dis. 2010, 16, 1808–1813.

Schapira, D.V.; Kumar, N.B.; Lyman, G.H.; Cavanagh, D.; Roberts, W.S.; LaPolla, J. Upper-body fat distribution and endometrial cancer risk. JAMA 1991, 266, 1808–1811.

Schmitz, G.; Ecker, J. The opposing effects of n-3 and n-6 fatty acids. Prog. Lipid Res. 2008, 47, 147–155.

Schottelius, A.J.;Dinter, H. Cytokines, NF-kappaB, microenvironment, intestinal inflammation and cancer. Cancer Treat. Res. 2006, 130, 67–87.

Seki, H.; Tani, Y.; Arita, M. Omega-3 PUFA derived anti-inflammatory lipid mediator resolvin E1. Prostaglandins Other Lipid Mediat.2009, 89, 126–130.

Seo, J.; Barhoumi, R.; Johnson, A.E.; Lupton, J.R.; Chapkin, R.S. Docosahexaenoic acid selectively inhibits plasma membrane targeting of lipidated proteins. FASEB J. 2006, 20, 770–772.

Shao, J.; Jung, C.; Liu, C.; Sheng, H. Prostaglandin E2 Stimulates the beta-catenin/T cell factor-dependent transcription in colon cancer. J. Biol. Chem. 2005, 280, 26565–26572.

Sheibanie, A.F.; Khayrullina, T.; Safadi, F.F.; Ganea, D. Prostaglandin E2 exacerbates collagen-induced arthritis in mice through the inflammatory interleukin-23/interleukin-17 axis. Arthritis Rheum. 2007, 56, 2608–2619.

Sheibanie, A.F.; Yen, J.H.; Khayrullina, T.; Emig, F.; Zhang, M.; Tuma, R.; Ganea, D. The proinflammatory effect of prostaglandin E2 in experimental inflammatory bowel disease is mediated through the IL-23→IL-17 axis. J. Immunol. 2007b, 178, 8138–8147.

Shoda, R.; Matsueda, K.; Yamato, S.; Umeda, N. Epidemiologic analysis of Crohn disease in Japan: increased dietary intake of n-6 polyunsaturated fatty acids and animal protein relates to the increased incidence of Crohn disease in Japan. Am. J. Clin. Nutr. 1996, 63, 741–745.

Sijben, J.W.; Calder, P.C. Differential immunomodulation with long-chain n-3 PUFA in health and chronic disease. Proc. Nutr. Soc. 2007, 66, 237–259.

Smith, W.L. Cyclooxygenases, peroxide tone and the allure of fish oil. Curr. Opin. Cell Biol. 2005, 17, 174–182.

Sonnenberg, G.E.; Krakower, G.R.; Kissebah, A.H. A novel pathway to the manifestations of metabolic syndrome. Obes. Res. 2004, 12, 180–186.

Sperling, R.I. Eicosanoids in rheumatoid arthritis. Rheum. Dis. Clin. North Am. 1995, 21, 741–758.

Sperling, R.I.; Weinblatt, M.; Robin, J.L.; Ravalese, J., 3rd; Hoover, R.L.; House, F.; Coblyn, J.S.; Fraser, P.A.; Spur, B.W.; Robinson, D.R.; et al. Effects of dietary supplementation with marine fish oil on leukocyte lipid mediator generation and function in rheumatoid arthritis. Arthritis Rheum. 1987, 30, 988–997.

Stenson, W.F.; Cort, D.; Rodgers, J.; Burakoff, R.; DeSchryver-Kecskemeti, K.; Gramlich, T.L.; Beeken, W. Dietary supplementation with fish oil in ulcerative colitis. Ann. Intern. Med. 1992, 116, 609–614.

Stillwell, W.; Wassall, S.R. Docosahexaenoic acid: membrane properties of a unique fatty acid. Chem. Phys. Lipids 2003, 126, 1–27.

Streppel, M.T.; Ocke, M.C.; Boshuizen, H.C.; Kok, F.J.; Kromhout, D. Long-term fish consumption and n-3 fatty acid intake in relation to (sudden) coronary heart disease death: the Zutphen study. Eur. Heart J. 2008, 29, 2024–2030.

Stulnig, T.M. Immunomodulation by polyunsaturated fatty acids: mechanisms and effects. Int. Arch. Allergy Immunol. 2003, 132, 310–321.

Sumarac-Dumanovic, M.; Stevanovic, D.; Ljubic, A.; Jorga, J. Simic, M.; Stamenkovic-Pejkovic, D.; Starcevic, V.; Trajkovic, V.; Micic, D. Increased activity of interleukin-23/interleukin-17 proinflammatory axis in obese women. Int. J. Obes. (Lond) 2009, 33, 151–156.

Svensson, M.; Schmidt, E.B.; Jorgensen, K.A.; Christensen, J.H. N-3 fatty acids as secondary prevention against cardiovascular events in patients who undergo chronic hemodialysis: a randomized, placebo-controlled intervention trial. Clin. J. Am. Soc. Nephrol. 2006, 1, 780–786.

Switzer, K.C.; Fan, Y.Y.; Wang, N.; McMurray, D.N.; Chapkin, R.S. Dietary n-3 polyunsaturated fatty acids promote activation-induced cell death in Th1-polarized murine CD4+ T-cells. J. Lipid Res. 2004, 45, 1482–1492.

Tashiro, T.; Yamamori, H.; Takagi, K.; Hayashi, N.; Furukawa, K.; Nakajima, N. n-3 versus n-6 polyunsaturated fatty acids in critical illness. Nutrition 1998, 14, 551–553.

Thies, F.; Nebe-von-Caron, G.; Powell, J.R.; Yaqoob, P.; Newsholme, E.A.; Calder, P.C. Dietary supplementation with gamma-linolenic acid or fish oil decreases T lymphocyte proliferation in healthy older humans. J. Nutr. 2001, 131, 1918–1927.

Tierney, A.C.; McMonagle, J.; Shaw, D.I.; Gulseth, H.L.; Helal, O.; Saris, W.H.; Paniagua, J.A.; Golabek-Leszczynska, I.; Defoort, C.; Williams, C.M.; et al. Effects of dietary fat modification on insulin sensitivity and on other risk factors of the metabolic syndrome-LIPGENE: a European randomized dietary intervention study. Int. J. Obes. (Lond). 2010, 1–10.

Tjonneland, A.; Overvad, K.; Bergmann, M.M.; Nagel, G.; Linseisen, J.; Hallmans, G.; Palmqvist, R.; Sjodin, H.; Hagglund, G.; Berglund, G.; et al. Linoleic acid, a dietary n-6 polyunsaturated fatty acid, and the aetiology of ulcerative colitis: a nested case-control study within a European prospective cohort study. Gut 2009, 58, 1606–1611.

Trayhurn, P.; Wood, I.S. Adipokines: inflammation and the pleiotropic role of white adipose tissue. Br. J. Nutr. 2004, 92, 347–355.

Trayhurn, P.; Wood, I.S. Signalling role of adipose tissue: adipokines and inflammation in obesity. Biochem. Soc. Trans. 2005, 33, 1078–1081.

Trebble, T.M.: Arden, N.K.; Wootton, S.A.; Calder, P.C.; Mullee, M.A.; Fine, D.R.; Stroud, M.A. Fish oil and antioxidants alter the composition and function of circulating mononuclear cells in Crohn disease. Am. J. Clin. Nutr. 2004, 80, 1137–1144.

Turner, D.; Shah, P.S.; Steinhart, A.H.; Zlotkin, S.; Griffiths, A.M. Maintenance of remission in inflammatory bowel disease using omega-3 fatty acids (fish oil): A systematic review and meta-analyses. Inflamm. Bowel Dis. 2010.

van der Tempel, H.; Tulleken, J.E.; Limburg, P.C.; Muskiet, F.A.; van Rijswijk, M.H. Effects of fish oil supplementation in rheumatoid arthritis. Ann. Rheum. Dis. 1990, 49, 76–80.

Virgili, N.; Farriol, M.; Castellanos, J.M.; Giro, M.; Podzamczer, D.; A, M.P. Evaluation of immune markers in asymptomatic AIDS patients receiving fish oil supplementation. Clin. Nutr. 1997, 16, 257–261.

Wang, D.; Wang, H.; Brown, J.; Daikoku, T.; Ning, W.; Shi, Q.; Richmond, A.; Strieter, R.; Dey, S.K.; DuBois, R.N. CXCL1 induced by prostaglandin E2 promotes angiogenesis in colorectal cancer. J. Exp. Med. 2006, 203, 941–951.

Warensjo, E.; Sundstrom, J.; Lind, L.; Vessby, B. Factor analysis of fatty acids in serum lipids as a measure of dietary fat quality in relation to the metabolic syndrome in men. Am. J. Clin. Nutr. 2006, 84, 442–448.

Wei, E.K.; Giovannucci, E.; Fuchs, C.S.; Willett, W.C.; Mantzoros, C.S. Low plasma adiponectin levels and risk of colorectal cancer in men: a prospective studyJ. Natl. Cancer Inst. 2005, 97, 1688–1694.

Weisberg, S.P.; Hunter, D.; Huber, R.; Lemieux, J.; Slaymaker, S.; Vaddi, K.; Charo, I.; Leibel, R.L.; Ferrante, A.W., Jr. CCR2 modulates inflammatory and metabolic effects of high-fat feeding. J. Clin. Invest. 2006, 116, 115–124.

Weisberg, S.P.; McCann, D.; Desai, M.; Rosenbaum, M.; Leibel, R.L.; Ferrante, A.W., Jr. Obesity is associated with macrophage accumulation in adipose tissue. J. Clin. Invest. 2003, 112, 1796–1808.

West, N.J.; Clark, S.K.; Phillips, R.K.; Hutchinson, J.M.; Leicester, R.J.; Belluzzi, A.; Hull, M.A. Eicosapentaenoic acid reduces rectal polyp number and size in familial adenomatous polyposis. Gut 2010, 59, 918–925.

Whelan, J.; McEntee, M.F. Dietary (n-6) PUFA and intestinal tumorigenesis. J. Nutr. 2004, 134, 3421S–3426S.

White, P.J.; Marette, A. Is omega-3 key to unlocking inflammation in obesity? Diabetologia 2006, 49, 1999–2001.

Wigmore, S.J.; Fearon, K.C.; Maingay, J.P.; Ross, J.A. Down-regulation of the acute-phase response in patients with pancreatic cancer cachexia receiving oral eicosapentaenoic acid is mediated via suppression of interleukin-6. Clin. Sci. (Lond) 1997, 92, 215–221.

Winer, S.; Paltser, G.; Chan, Y.; Tsui, H.; Engleman, E.; Winer, D.; Dosch, H.M. Obesity predisposes to Th17 bias. Eur. J. Immunol. 2009, 39, 2629–2635.

Wolf, A.M.; Wolf, D.; Rumpold, H.; Enrich, B.; Tilg, H. Adiponectin induces the anti-inflammatory cytokines IL-10 and IL-1RA in human leukocytes. Biochem. Biophys. Res. Commun. 2004, 323, 630–635.

Yan, S.R.; Joseph, R.R.; Rosen, K.; Reginato, M.J.; Jackson, A.; Allaire, N.; Brugge, J.S.; Jobin, C.; Stadnyk, A.W. Activation of NF-kappaB following detachment delays apoptosis in intestinal epithelial cells. Oncogene. 2005, 24, 6482–6491.

Yang, P.; Chan, D.; Felix, E.; Cartwright, C.; Menter, D.G.; Madden, T.; Klein, R.D.; Fischer, S.M.; Newman, R.A. Formation and antiproliferative effect of prostaglandin E(3) from eicosapentaenoic acid in human lung cancer cells. J. Lipid Res. 2004, 45, 1030–1039.

Yog, R.; Barhoumi, R.; McMurray, D.N.; Chapkin, R.S. n-3 polyunsaturated fatty acids suppress mitochondrial translocation to the immunologic synapse and modulate calcium signaling in T cells. J. Immunol. 2010, 184, 5865–5873.

Yokota, T.; Oritani, K.; Takahashi, I.; Ishikawa, J.; Matsuyama, A.; Ouchi, N.; Kihara, S.; Funahashi, T.; Tenner, A. J.; Tomiyama, Y.; et al. Adiponectin, a new member of the family of soluble defense collagens, negatively regulates the growth of myelomonocytic progenitors and the functions of macrophages. Blood 2000, 96, 1723–1732.

Yokoyama, M.; Origasa, H.; Matsuzaki, M.; Matsuzawa, Y.; Saito, Y.; Ishikawa, Y.; Oikawa, S.; Sasaki, J.; Hishida, H.; Itakura, H.; Kita, T.; Kitabatake, A.; Nakaya, N.; Sakata, T.; Shimada, K.; Shirato, K. Effects of eicosapentaenoic acid on major coronary events in hypercholesterolaemic patients (JELIS): a randomised open-label, blinded endpoint analysis. Lancet 2007, 369, 1090–1098.

Yu, H.; Pardoll, D.; Jove, R. STATs in cancer inflammation and immunity: a leading role for STAT3. Nat. Rev. Cancer 2009, 9, 798–809.

Yzebe, D.; Lievre, M. Fish oils in the care of coronary heart disease patients: a meta-analysis of randomized controlled trials. Fundam. Clin. Pharmacol. 2004, 18, 581–592.

Zainal, Z.; Longman, A.J.; Hurst, S.; Duggan, K.; Caterson, B.; Hughes, C.E.; Harwood, J.L. Relative efficacies of omega-3 polyunsaturated fatty acids in reducing expression of key proteins in a model system for studying osteoarthritis. Osteoarthritis Cartilage 2009, 17, 896–905.

Zatonski, W.A.; Lowenfels, A.B.; Boyle, P.; Maisonneuve, P.; Bueno de Mesquita, H.B.; Ghadirian, P.; Jain, M.; Przewozniak, K.; Baghurst, P.; Moerman, C.J.; et al. Epidemiologic aspects of gallbladder cancer: a case-control study of the SEARCH Program of the International Agency for Research on Cancer. J. Natl. Cancer Inst. 1997, 89, 1132–1138.

Zeyda, M.; Farmer, D.; Todoric, J.; Aszmann, O.; Speiser, M.; Gyori, G.; Zlabinger, G.J.; Stulnig, T.M. Human adipose tissue macrophages are of an anti-inflammatory phenotype but capable of excessive pro-inflammatory mediator production. Int. J. Obes. (Lond) 2007, 31, 1420–1428.

Zeyda, M.; Saemann, M.D.; Stuhlmeier, K.M.; Mascher, D.G.; Nowotny, P.N.; Zlabinger, G.J.; Waldhausl, W.; Stulnig, T.M. Polyunsaturated fatty acids block dendritic cell activation and function independently of NF-kappaB activation. J. Biol. Chem. 2005, 280, 14293–14301.

Zhang, S.M.; Buring, J.E.; Lee, I.M.; Cook, N.R.; Ridker, P.M. C-reactive protein levels are not associated with increased risk for colorectal cancer in women. Ann. Intern. Med. 2005, 142, 425–432.

Zhang, P.; Kim, W.; Zhou, L.; Wang, N.; Ly, L.H.; McMurray, D.N.; Chapkin, R.S. Dietary fish oil inhibits antigen-specific murine Th1 cell development by suppression of clonal expansion. J. Nutr. 2006, 136, 2391–2398.

Zhang, P.; Smith, R.; Chapkin, R.S.; McMurray, D.N. Dietary (n-3) polyunsaturated fatty acids modulate murine Th1/Th2 balance toward the Th2 pole by suppression of Th1 development. J. Nutr. 2005, 135, 1745–1751.

Zhao, Y.T.; Shao, L.; Teng, L.L.; Hu, B.; Luo, Y.; Yu, X.; Zhang, D.F.; Zhang, H. Effects of n-3 polyunsaturated fatty acid therapy on plasma inflammatory markers and N-terminal pro-brain natriuretic peptide in elderly patients with chronic heart failure. J. Int. Med. Res. 2009, 37, 1831–1841.

Zuniga, L.A.; Shen, W.J.; Joyce-Shaikh, B.; Pyatnova, E.A.; Richards, A.G.; Thom, C.; Andrade, S.M.; Cua, D.J.; Kraemer, F.B.; Butcher, E.C. IL-17 regulates adipogenesis, glucose homeostasis, and obesity. J. Immunol. 2010, 185, 6947–6959.

3

Fish Sources of Various Lipids Including n-3 Polyunsaturated Fatty Acids and Their Dietary Effects

Nobuya Shirai
National Food Research Institute, Kannondai, Tsukuba, Ibaraki, Japan

Introduction

All fish, regardless of whether they are caught from the sea or fresh water sources, contain n-3 polyunsaturated fatty acids (n-3 PUFAs). Thus, consumption of fish results in a natural intake of n-3 PUFAs. However, the proportion of n-3 PUFAs in fish oil varies with factors such as the species, season, and location. Further, the documented beneficial effects of n-3 PUFAs, such as lowering plasma lipids and improvement of brain function, may be modified according to other components in the fish oils. This chapter describes the characteristics of various fish oils and the effects of an intake of these oils on animal lipid profiles and behavior.

Sardine Oil

The Japanese sardine, *Sardinops melanostictus,* has been an important species for commercial fisheries, as it was widely distributed in the sea around Japan. Various reports have shown that sardine oil is rich in the n-3 PUFAs eicosapentaenoic acid (20:5n-3, EPA) and docosahexaenoic acid (22:6n-3, DHA) (Linko et al., 1985; Hayashi & Takagi, 1977; Bandarra et al., 1997; Gámez-Meza et al., 1999; Shirai et al., 2002a). In general the DHA content in sardines is higher than that of EPA, but this ratio can be inverted by changes in the phytoplankton on which they feed. Variations in the concentrations of EPA and DHA in the phytoplankton are caused by differences in their birth rate which, in turn, is influenced by location and the season. Table 3.A shows the seasonal variation of the fatty acid composition of the sardines caught from the sea of Hyga-Nada (Fig. 3.1) over a period of one year (1994 to 1995). For the majority of the year the DHA percentage was higher than the EPA percentage, but this was reversed during July.

Table 3.A. Seasonal Variation in the Lipid Content and the Fatty Acid Composition of Total Lipids from Sardine, *Sardinops melanostictus*.

	1994				1995				
	Aug.	Sep.	Nov.	Dec.	Feb.	Mar.	May	Jun.	Jul.
SFA	41.4 ±2.4	40.7 ±1.1	37.5 ±1.1	41.2 ±1.1	34.3 ±2.6	34.7 ±1.8	35.8 ±0.5	37.2 ±1.5	37.7 ±0.9
MUFA	19.9 ±2.1	18.9 ±0.8	19.4 ±2.2	19.3 ±1.4	15.1 ±1.5	16.4 ±1.3	19.6 ±2.0	19.0 ±1.1	22.4 ±2.6
PUFA	37.3 ±3.7	39.2 ±1.2	42.5 ±1.9	38.4 ±1.3	49.4 ±3.5	47.9 ±2.6	43.4 ±2.2	42.5 ±2.1	38.5 ±2.8
18:2n-6	1.3 ±0.1	1.4 ±0.1	1.4 ±0.1	1.2 ±0.1	1.1 ±0.1	1.1 ±0.2	0.9 ±0.1	1.1 ±0.1	0.7 ±0.1
20:4n-6	2.2 ±0.4	2.5 ±0.2	2.5 ±0.2	2.0 ±0.2	2.0 ±0.3	2.1 ±0.3	2.2 ±0.6	1.9 ±0.2	1.7 ±0.2
20:5n-3	8.8 ±1.9	10.4 ±1.6	9.5 ±1.7	8.6 ±1.1	6.9 ±0.6	9.4 ±1.3	10.5 ±1.1	11.6 ±1.5	18.9 ±1.1
22:6n-3	16.6 ±3.0	16.4 ±1.5	19.7 ±3.7	18.3 ±2.9	32.5 ±3.7	27.9 ±4.1	23.5 ±3.7	20.7 ±2.3	10.7 ±3.6
Unknown	1.1 ±0.5	1.0 ±0.2	0.3 ±0.5	0.9 ±0.4	1.0 ±0.1	0.9 ±0.2	1.2 ±0.2	1.3 ±0.5	1.4 ±0.4

Shiria et al., 2002a.

Fig. 3.1. Location of the sea of Hyuga-nada.

The total lipid content in the whole body of the sardine catches also varied with the time of the year and was low in February (1.8%) and high from July to September (7.0%–7.2%). This period of high lipid content correlated with increases in free histidine in the fish meat and the temperature of the sea (Fig. 3.2). July to September was also the time when the fatty acid composition of sardines reflected that of the phytoplankton (Fig. 3.3). Suyama et al. (1986) suggested that imidazole containing

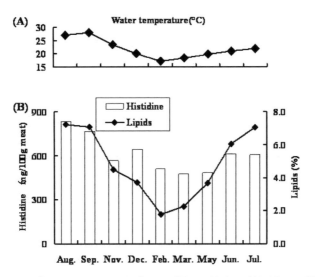

Fig. 3.2. The variation of water temperature in the sea of Hyuga-Nada and histidine and lipids content of dorsal meat in the sardine catch from the sea of Hyga-Nada. Source: Shirai et al., 2002a.

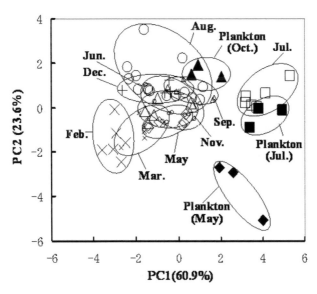

Fig. 3.3. The score plot of the first and second principal components after PCA, based on the fatty acid composition of sardine and plankton samples caught from the sea of Hyuga-Nada. Painted black markers indicate plankton collected from the sea of Hyuga-Nada. Source: Shirai et al., 2002a.

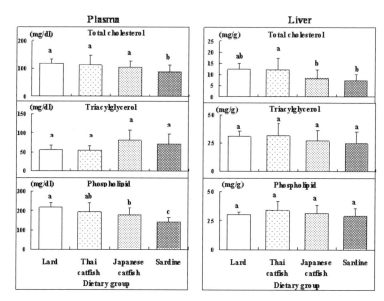

Fig. 3.4. Plasma and liver lipids concentrations of mice fed a diet containing lard, Thai catfish, Japanese catfish, or sardine oil for 4 weeks. Source: Shirai et al., 2001.

components, such as histidine, may play an important role in buffering the lactic acid that is produced during increased feeding activity in high water temperatures, indicating a relationship between the variation of lipid content in the sardines and the extent of their feeding activity.

In animal experiments, an intake of EPA- and DHA-rich sardine oil has been shown to significantly decrease plasma and liver lipid cholesterol and plasma phospholipid levels (Fig. 3.4). A similar effect in humans could make an important contribution to the maintenance of lipid balance and cardiovascular integrity. However, sardines are rapidly becoming less useful as a fish oil source because in recent years the catch has seriously declined due to decreasing resources.

Catfish Oil

Catfish are distributed all over the world and are often a component of human diets. Catfish oil also contains n-3 PUFAs, but the percentage in catfish is low compared with that of oil extracted from marine fish (Blank et al., 1992; Innis et al, 1995). Table 3.B shows the fatty acid composition of lipids extracted from dorsal meat of Cannel catfish (*Ictalurus punctatus*), Japanese catfish (*Silurus asotus*), and Thai catfish (*Clarias macrocephalus*). According to some reports, Cannel catfish, which is mainly consumed in America, has the lowest n-3 PUFA percentage (Chanmugam et al., 1986; Mustafa & Medeiros., 1985; Morris et al., 1995; Nettleton et al., 1990). However, Cannel catfish of Japanese origin can have a higher percentage of n-3 PUFA

Table 3.B. Fatty Acid and Dimethylacetal Composition in Dorsal Meat of Cultured Catfish.

Country	Japan		America	Thailand
		Ictalurus	*Ictalurus*	*Clarias*
Sample	*Silurus asotus*	*punctatus*	*punctatus*	*macrocephalus*
Fatty acid (%)				
14:0	0.6	1.3	1.0	0.8
16:0	19.3	18.8	20.9	27.8
18:0	7.5	5.1	7.1	8.2
Others	1.6	-	0.1	0.5
Σ SFA	28.9	25.2	29.1	37.4
16:1n-9	0.5	0.5	0.5	0.4
16:1n-7	2.2	2.5	1.7	1.9
18:1n-9	6.7	31.5	37.9	35.8
18:1n-7	5.9	2.9	1.8	2.6
20:1	0.8	1.7	1.0	0.5
22:1	-	0.2	-	-
Others	0.1	-	-	-
Σ MUFA	16.3	39.3	42.9	41.2
18:2n-6	4.4	9.2	13.3	11.3
18:3n-3	1.1	0.6	0.6	0.3
20:3n-9	-	-	0.6	0.1
20:4n-6	10.2	1.7	2.5	2.3
20:5n-3	3.2	3.9	0.6	-
22:5n-3	3.4	1.4	0.6	0.4
22:6n-3	20.6	13.6	3.5	2.2
Others	6.0	1.9	4.3	2.5
Σ PUFA	48.9	32.3	26.0	19.2
16:0DMA	1.9	0.7	0.5	1.0
18:0DMA	0.7	0.4	0.3	0.5
18:1DMA	1.7	1.7	0.9	0.5
Σ DMA	4.3	2.9	1.7	2.0
unknown	1.6	0.1	0.2	0.2

-:not found or trace, SFA:saturated fatty acid, MUFA:Monounsaturated fatty acid, PUFA:polyunsaturated fatty acid, DMA:dimethylacetal

Shiria et al., 2000.

than those of American origin. Cannel catfish cultured in Japan were fed on a rich n-3 PUFA diet had higher levels of fish oil comparing with those cultured in America on a diet containing less n-3 PUFA (Table 3.B). Thus, the percentage of n-3 PUFA in Cannel catfish may be related to both species and their diet. The n-3 PUFA of catfish cultured in Thailand was very low, and this may reflect a relationship with the water temperature of the habitat.

In a study mice were fed oil that was extracted from Japanese and Thai catfish and from sardines for 1 month, and the intake of Thai catfish oil had little influence on levels of plasma and liver lipids. In contrast, the intake of Japanese catfish oil resulted in a decrease in plasma phospholipids and liver cholesterol that was similar to that observed with an intake of sardine oil (Fig. 3.4). Not to mention, the effect on plasma and liver total cholesterol was less pronounced with a sardine oil intake. These results suggest that the source and species of catfish as a dietary component is important and an intake of catfish oil that is rich in n-3 PUFAs may be useful in the control of hyperlipemia.

Kazunoko Lipids

Kazunoko is one of a number of Japanese salted fish roe food products and is produced from herring. The other most frequently consumed salted roe products are Ikura, Tarako, and Tobiko, produced from salmon, pollock, and flying fish, respectively. Detailed analyses of the lipid composition of these fish roe has shown that they contain large amounts of EPA, DHA, phospholipids, and cholesterol (Shirai et al., 2006a; Kaitaranta, 1980; Tocher & Sargent, 1984; Bledsoe et al., 2003) (Table 3.C). Generally, large intakes of fish roe tend to be avoided because of their high cholesterol content. Despite this, Kazunoko lipids have been shown to decrease the plasma and liver cholesterol levels in mice, but the effect was less pronounced than that obtained with fish oils (Shirai et al., 2008; Higuchi et al., 2006). Thus, the positive benefits of n-3 PUFA in Kazunoko lipids in lowering cholesterol levels in the mice appear to be counteracted by the high cholesterol content of Kazunoko. The data confirm that an intake of Kazunoko should not be regarded as an aid in maintaining healthy lipid levels.

Some reports have indicated that a phospholipid intake improves the learning ability of mice (Lim and Suzuki, 2000; 2008; Leathwood et al., 1982; Carrié et al., 2002). We compared the learning ability of mice fed a 5% Kazunoko lipid or fish oil

Table 3.C. Lipid Components and Fatty Acid Composition of Total Lipids in the Ikura, Tarako, Tobiko, and Kazunoko.

	Ikura	Tarako	Tobiko	Kazunoko
Total lipid content (g/100g tissue)	14.5 ±0.7	3.7 ±0.4	3.2 ±0.2	3.0 ±0.3
Cholesterol content	0.59 ±0.10	0.35 ±0.02	0.43 ±0.01	0.26 ±0.04
Triacylglycerol content	8.09 ±0.68	0.64 ±0.06	0.42 ±0.01	0.54 ±0.06
Phospholipid content	5.51 ±0.12	2.28 ±0.23	2.39 ±0.19	2.14 ±0.26
Fatty acid composition (%)				
SFA	21.6 ±0.2	26.9 ±0.6	39.6 ±0.3	32.0 ±1.3
MUFA	33.1 ±0.4	25.0 ±0.3	14.4 ±0.1	25.0 ±0.9
PUFA	44.6 ±0.5	47.5 ±0.9	45.5 ±0.4	42.7 ±0.3
18:2n-6	1.0 ±0.0	1.0 ±0.0	1.1 ±0.0	0.7 ±0.1
20:4n-6	1.0 ±0.0	1.3 ±0.0	3.0 ±0.1	1.1 ±0.1
20:5n-3	13.6 ±0.1	18.8 ±0.3	7.0 ±0.1	15.0 ±0.6
22:6n-3	17.4 ±0.2	22.2 ±0.4	27.9 ±0.3	22.6 ±1.0

Ikura: salted salmon roe Tarako: salted Pollock roe
Tobiko: salted flyingfish roe Kazunoko: salted herring roe
SFA: saturated fatty acids MUFA: monounsaturated fatty acids
PUFA: polyunsaturated fatty acids
Shiria et al., 2006a.

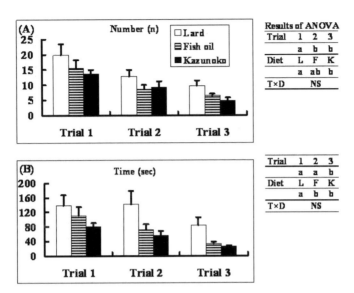

Fig. 3.5. The time taken to reach the maze exit, the number of times strayed into blind alleys, and the speed of mice fed lard, fish oil, or Kazunoko lipids diets (mean ± SE). Small different letters in the results of ANOVA indicate significant difference between each trial or between each dietary group at p<0.05 by Spjotroll/Stoline test. Source: Shirai et al., 2006b.

containing diet for 4 months in a study of maze behavior. Animals receiving Kazunoko lipids had a slightly improved learning ability compared with those receiving fish oil (Fig. 3.5). It is conceivable that the higher phospholipid component of the Kazunoko lipids contributed to this improvement. Thus, at the present time there is conflicting evidence for the possible beneficial effects of a Kazunoko lipid intake. Currently, a high consumption of salted fish roe cannot be recommended due not only to the high cholesterol levels but also to the large amount of salt and purine containing uric acid present in these roe. The latter components may contribute to hypertension and gout. Further studies are required to ascertain the impact and safety of intakes Kazunoko lipids on human health.

Sea Snake Oil

Sea snakes are marine reptiles that consume fish, and as a consequence their oil contains large amounts of n-3 PUFA compared with the oil from land snakes. Erabu sea snake, *Laticauda semifasciata*, is distributed throughout the Pacific Ocean and along the coast of the South China Sea. It has been used as a folk medicine for generations in China and Okinawa, Japan. Table 3.D shows that the fatty acids of oil from sea snakes caught from these areas include DHA (10.7 to 13.4%) and EPA (2.3 to 2.9%). From previous studies of oils containing high levels of these fatty acids it was proposed that an intake of sea snake oil might decrease plasma and liver lipids and

Table 3.D. Major Fatty Acid Composition of Oils from Fat Sacks of Erabu Sea Snakes Captured Near Okinawa, the Philippines, China, and Vietnam.

Fatty acid	Okinawa	Philippines	China	Vietnam
14:0	2.1 ±0.0	3.3 ±1.3	2.0 ±0.1	2.3 ±0.2
16:0	27.2 ±0.4	27.1 ±0.3	27.4 ±0.8	25.9 ±0.1
16:1n-7	5.2 ±0.1	5.3 ±0.3	4.8 ±0.0	5.8 ±0.2
18:0	9.1 ±0.0	8.9 ±0.8	9.3 ±0.4	8.9 ±0.1
18:1n-9	20.3 ±0.1	16.9 ±1.5	20.8 ±0.8	20.3 ±0.3
18:1n-7	2.8 ±0.0	3.7 ±0.4	2.8 ±0.0	2.5 ±0.3
20:4n-6	1.9 ±0.1	2.5 ±0.3	1.9 ±0.0	1.9 ±0.0
20:5n-3	2.2 ±0.1	2.9 ±0.7	2.3 ±0.0	2.7 ±0.3
22:4	1.3 ±0.0	1.1 ±0.2	1.4 ±0.0	1.3 ±0.1
22:5	2.3 ±0.0	3.1 ±0.3	2.2 ±0.0	2.2 ±0.0
22:6n-3	12.8 ±0.1	10.7 ±1.7	13.4 ±0.7	12.8 ±0.4

*Values were means ± S.D., n=10.
Shiria et al., 2002c.

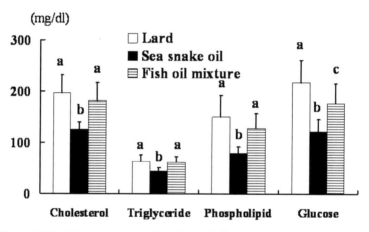

Fig. 3.6. Plasma lipid and glucose concentrations of mice fed lard, Erabu sea snake oil, or fish oil diets for 4 months (n = 8). Values for each sample with different italic letters in the same fraction are significantly different at p<0.05 by Duncan's pairwise comparisions. Source: Shirai et al., 2002b

improve learning ability. In mice fed a diet containing 5% of this oil for 4 months, levels of plasma and liver lipids and plasma glucose decreased to a greater extent than those from mice fed fish oil containing the same percentage of DHA (Fig. 3.6). Mice fed the sea snake oil had similar improvements in learning ability to those of mice with a fish oil intake (Shirai et al., 2004). Thus, an intake of sea snake oil might also make a positive contribution to lowering plasma lipids and the maintenance of brain function in humans.

The analeptic effects of sea snake have been known for a very long time. To study this further, we compared the effects of sea snake and fish oil on the swimming endurance of aged mice. It was of considerable interest that an intake of sea snake oil

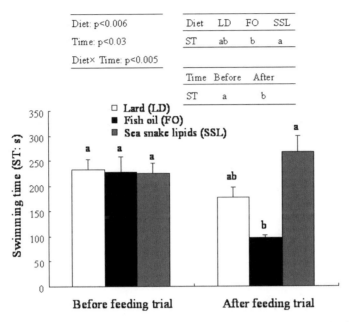

Fig. 3.7. Changes of swimming times to exhaustion of aged mice fed diet containing lard, fish oil, and sea snake oil before and after feeding trials. Each value represents the mean ± SE. Significant differences between the three diet groups before and after feeding trials are denoted by a, b, and ab (p<0.05). Source: Zhang et al., 2007.

for 4 months significantly extended the swimming endurance of aged mice whereas an intake of fish oil slightly decreased the time of swimming endurance when compared with the initial time recorded from these mice (Fig. 3.7). These results suggest that the improvement of swimming endurance, and possibly of plasma and liver lipid parameters, following an intake of sea snake oil may be associated with the components other than n-3 PUFA. It remains to be fully established whether sea snake oil is more effective in the maintenance of human health than fish oil that contains the same proportion of n-3 PUFAs.

Conclusion

The results of studies presented in this paper suggest that intakes of various oils containing n-3 PUFA, such as those obtained from sardine, catfish, Kazunoko, and sea snake, may be effective in maintaining healthy lipid profiles that contribute to human well-being. Although these effects appear to be primarily related to the n-3 PUFA contained in the oil, components other than these fatty acids present in fish oils, such as those obtained from Kazunoko and sea snake, could exert an influence by inhibitory or complementary mechanisms. Currently, the question is still open as to whether the intake of n-3 PUFA, even in high concentrations, is beneficial for human health. Further studies, particularly in humans, are required to address this question.

References

Bandarra, N.M.; Batista, I.; Nunes, M.L.; et al. Seasonal changes in lipid composition of sardine (Sardina pilchardus) J. Food Sci. 1997, 62, 40–42.

Blank, M.L., Cress, E.A.; Smith, Z.L.; et al. Meats and fish consumed in the American diet contain substantial amounts of ether-linked phospholipids. J. Nutr. 1992, 122, 1656–1661.

Bledsoe, G.E.; Bledsoe, C.D.; Rasco, B. Caviars and fish roe products. Crit. Rev. Food Sci. Nutr. 2003, 43, 317–356.

Carrié, I; Smirnova, M.; Clément M.; et al. Docosahexaenoic acid-rich phospholipid supplementation: effect on behavior, learning ability, and retinal function in control and n-3 polyunsaturated fatty acid deficient old mice. Nutr. Neurosci. 2002, 5, 43–52.

Chanmugam, P.; Boudreau, M.; Hwang, D.H. Differences in the n-3 fatty acid contents in pond-reared and wild fish and shellfish. J. Food Sci. 1986, 51, 1556–1557.

Gámez-Meza, N.; Higuera-Ciapara, I.; Calderon de la Barca, A.M.; et al. Seasonal variation in the fatty acid composition and quality of sardine oil from Sardinops sagax caeruleus of the gulf of California. Lipids 1999, 34, 639–642.

Hayashi, K.; Takagi, T. Seasonal variations in lipids and fatty acids of Sardine, Sardinops melanosticta. Bull. Fac. Fish Hokkaido Univ. 1977, 28, 89–94.

Higuchi, T.; Shirai, N.; Suzuki, H. Effect of dietary herring roe lipids on plasma lipids, glucose, insulin and adiponectin concentrations in mice. J. Agric. Food Chem. 2006, 54, 3750–3755.

Innis, S.M.; Rioux, F.M.; Auestad, N.; et al. Marine and freshwater fish oil varying in arachidonic, eicosapentaenoic and docosahexaenoic acids differ in their effects on organ lipids and fatty acids in growing rats. J. Nutr. 1995, 125, 2286–2293.

Kaitaranta, J.K. Lipid and fatty acids of a whitefish (Coregonus albula) flesh and roe. J. Sci. Food Agric. 1980, 31, 1303–1308.

Leathwood, P.D.; Heck, E.; Mauron, J. Phosphatidyl choline and avoidance performance in 17 month-old SEC/1ReJ mice. Life Sci. 1982. 30, 1065–1071.

Lim, S.Y.; Suzuki, H. Dietary phosphatidylcholine improves maze-learning performance in adult mice. J. Med. Food 2008, 11, 86–90.

Lim, S.Y.; Suzuki, H. Effect of dietary docosahexaenoic acid and phosphatidylcholine on maze behavior and fatty acid composition of plasma and brain lipids in mice. Int. J. Vitam Nutr. Res. 2000, 70, 251–259.

Linko, R.P.; Kaitaranta, J.K.; Vuorela, R. Comparison of the fatty acid in Baltic herring and available plankton feed. Comp. Biochem. Physiol. 1985. 82B, 699–705.

Morris, C.A.; Haynes, K.C.; Keeton, J.T.; et al. Fish oil dietary effects on fatty acid composition and flavor of channel catfish. J. Food Sci. 1995, 60, 1225–1227.

Mustafa, F.A.; Medeiros, D.M. Proximate composition, mineral content, and fatty acids of catfish (Icutalurus punctatus, Rafinesque) for different seasons and cooking methods. J. Food Sci. 1985, 50, 585–588.

Nettleton, J.A.; Allen, W.H.; Klatt, L.V.; et al. Nutrients and chemical residues in one- to two-pound Mississippi farm-raised channel catfish (Ictalurus punctatus), J. Food Sci. 1990, 55, 954–958.

Shirai, N.; Miyakawa, M.; Tokairin, S.; et al. The fatty acid composition in edible portion of wild and cultured catfish. Nippon Suisan Gakkaishi. 2000, 66, 859–868.

Shirai, N.; Suzuki, H.; Tokairin, S.; et al. Effect of Japanese catfish lipid intake on the lipid components of plasma and liver in adult mice. Fisheries Sci. 2001, 67, 321–327.

Shirai, N.; Terayama, M.; Takeda, H. Effect of season on the fatty acid composition and free amino acid content of the sardine Sardinops melanostictus. Comp. Biochem. Physiol. 2002a, 131B, 387–393.

Shirai, N.; Hayashi, K.; Suzuki, H.; et al. Effects of Erabu sea snake oil on the plasma and liver lipids in mice. Nutr. Res. 2002b, 22, 1197–1207.

Shirai, N.; Suzuki, H.; Shimizu, R. Fatty acid composition of oil extracted from fat sack of Erabu sea snake Laticauda Semifasciata in the Pacific Ocean and South China Sea Fisheries Sci. 2002c, 68, 239–240

Shirai, N.; Suzuki, H.; Shimizu, R. Effect of Erabu sea snake Laticauda Semifasciata oil intake on maze-learning ability in mice. Fisheries Sci. 2004, 70, 314–318.

Shirai, N.; Higuchi, T.; Suzuki, H. Analyses of lipid classes and the fatty acid composition of Japanese fish roe products Ikura, Tarako, Tobiko, and Kazunoko Food Chem. 2006a, 94, 61–67.

Shirai, N.; Higuchi, T.; Suzuki, H. Effect of lipids extracted from a salted herring roe food product on maze-behavior in mice. J. Nutr. Sci. Vitaminol. 2006b, 52, 451–456.

Shirai, N.; Higuchi, T.; Suzuki, H. A comparative study of lipids extracted from herring roe products and fish oil on glucose and adipocytokine levels in ICR mice. FSRT 2008, 14, 25–31.

Suyama, M.; Hirano, T.; Suzuki, T.; Buffering Capacity of free histidine and its related dipeptides in white and dark muscles of yellowfin tuna. Bull. Japan Soc. Sci. Fish 1986, 52, 2171–2175.

Tocher, D.R.; Sargent, J.R. Analyses of lipids and fatty acids in ripe roes of some northwest european marine fish. Lipids 1984, 19, 492–499.

Zhang, G.; Shirai, N.; Higuchi, T.; et al. Effect of Erabu Sea Snake (Laticauda semifasciata) Lipids on the Swimming Endurance of Aged. J. Nutr. Sci. Vitaminol. 2007, 53, 476–481.

4

Production of Marine Oils

Anthony P. Bimbo

International Fisheries Technology, Kilmarnock, Virginia

Introduction

The scientific and news media are filled with articles declaring the impending doom to the oceans and the creatures that live in them. Some articles have even predicted that this extinction of sea life will occur around 2048 (Worm et al., 2006). Once published, these statements take on a life of their own, independent of what might actually be happening. The statements result in confusion, sometimes panic in the markets, and usually raise many questions. The usual question: is there or will there be a shortage of raw materials for Omega-3 products? The market is concerned and wants to know if there will be any fish and will there be a sufficient supply of fish oil from sustainable sources for the Omega-3 market. Hopefully this chapter will answer those questions and concerns. The chapter is divided into 2 parts: production of crude fish oil and the further processing of the crude oil through the various refining and concentration steps.

Historical Background

When you look at the global growth in fisheries over the period of 1950–2008 (2008 is the latest year for these statistics) you find that the fisheries have been growing at a rate of about 11.5% per year. However, closer examination of the data shows that since 1988 most of the growth has come from aquaculture, and the landings of ocean fish have plateaued in the 80–90 million metric ton (mmt) range. This can be seen in Fig. 4.1.

In fact, extrapolating the available data shows that sometime between 2035 and 2040 aquaculture raised fish will probably reach the same level as the capture fisheries. This is shown in Fig. 4.2. Of course, this date will change depending on fishing quotas and biological and environmental issues in the aquaculture industry. This is important since aquaculture currently represents the major global consumer of fish oil production. The data does not include plants, mammals or mollusks.

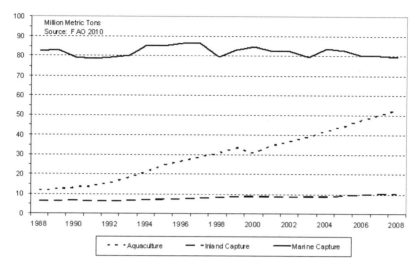

Fig. 4.1. Global landings of fish and shellfish by industry sector, marine mammals and aquatic plants excluded. Source: FAO 2010.

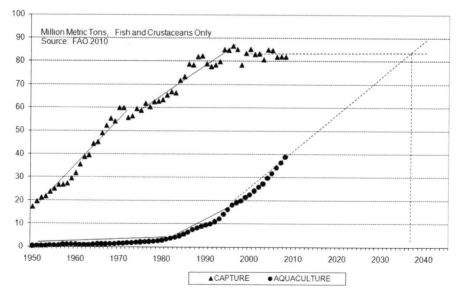

Fig. 4.2. Global fisheries capture vs. aquaculture production, fish and crustaceans only. Source: FAO 2010.

Marine Oils in Perspective

A number of countries and/or geographical entities are engaged in the production of fish oil. For this chapter the European Union (EU) will be represented by 25 countries with Denmark and Sweden grouped with the Scandinavian countries. Data showing production, imports, exports and apparent consumption for the major entities are shown in Table 4.A. Apparent consumption is defined as production plus imports minus exports.

Fish oil production has been relatively stable from 2001 to 2009 after a drastic decline during the major El Nino event in 1998 followed by 2 record recovery years in 1999 and 2000. The stability is probably due to the incorporation of quota systems in many of the producing countries and a general awareness that these stocks of fish must be managed. Aquaculture is a major industry in Chile, Norway, Canada, Asia, and the United Kingdom (UK), so those areas will be the major fish oil consumers. Norway is included in Scandinavia while the UK is in the EU25 group.

The consumption of fish oil actually can be defined by 3 periods: pre-1996, 1996–present, and the start of a transitional period that began in 2003 and continues. The pre-1996 period might be defined as the industrial (technical), animal feed and edible (hydrogenation) period. Fish oil was always a cheap fat, selling at a

Table 4.A. Five (5) Year Average Crude Fish Oil Production, Imports, Exports, and Apparent Consumption.

	Production	Imports	Exports	Apparent Consumption
		Metric Tons		
EU 25	48	103	10	140
Scandinavia	187	335	200	322
Peru	289		282	6
Chile	170	71	67	173
USA	74	23	60	28
Canada	5	38	6	37
Mexico	18	2	15	9
Asia	105	123	10	217
Morocco	32		28	9
Others	80	54	63	72
Total	1007	748	741	1014

EU members Denmark and Sweden included in Scandinavia.

Source: Fish Oil & Meal World, ISTA Mielke, Hamburg, Internet: www.oilworld.de.

discount to most of the edible fats and oils. Its primary edible use was in hydrogenated fats for margarine and baking, with industries in the UK, Germany and Netherlands accounting for most of the consumption. When the EU began legislating
against *trans* fatty acids, the market for fish oil started to decline, and this led into
the second period defined as the aquaculture period. Fish oil and the omega-3 fatty
acids in the oil are required by cold water carnivorous fish such as salmon and trout,
and as that industry developed so did the demand for fish oil. The El Nino in 1998
caused a major disruption in the availability of fish oil, and this caused the price to
rise significantly, which forced the aquaculture industry to lock in and secure sources
of oil that were so vital to their growth. This period continues to the present, however,
because fish oil prices have been on a steep upward curve and spiked in 2007–2008
when the biofuel industry started to grow and put pressure on all agricultural commodities. The aquaculture industry has begun to look for alternative sources of oil
with omega-3 fatty acids. This upward slope in the price of fish oil has also brought
us into the beginning of another transition to the pharmaceutical period. This third
transitional period can be seen in the following figure (4.3), which shows the his-
torical price increase of crude fish oil delivered into NW Europe from 1986 through
November 2010.

Sometimes, from a conservative perspective, it is better to use a 5-year running
average to depict the price of fish oil. This can be seen in Figure 4.4.

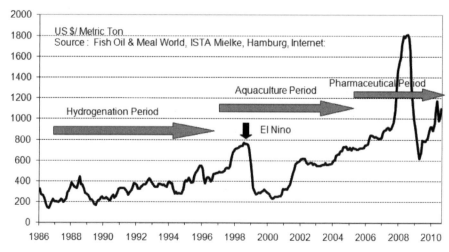

Fig. 4.3. Historic prices for crude fish oil of NW Europe, any origin. Source: Fish Oil & Meal World, ISTA
Mielke, Hamburg, www.oilworld.de.

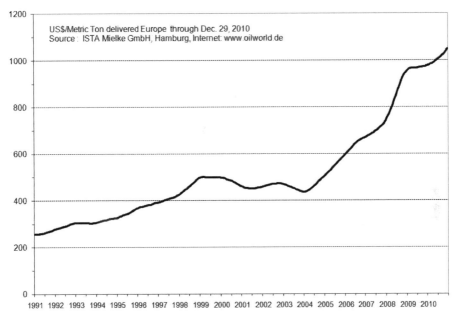

Fig. 4.4. Five (5) year running average price for fish oil delivered into Europe. Source: Fish Oil & Meal World, ISTA Mielke, Hamburg, www.oilworld.de.

Production of Crude Marine Oils

Raw Material

Fish landings can be characterized by their end use—either for edible use or for reduction to fishmeal and oil. The fish landings directed to the reduction industry arc utilized 100%. The edible-use fisheries can be characterized by their market use: frozen, fresh, canned, and cured. In this case, the edible portions are separated from the inedible portions. On average, the waste from edible-use fisheries is about 50% and breaks out as 15% heads, 14% frames, 4% skin, and 17% viscera. The utilization of the global landings by market use is shown in Table 4.B.

The market for fresh fish has been increasing while the market for fishmeal and oil production has been decreasing. However, this trend refers to captured fish only; about 25% of the raw material used for fishmeal and oil production comes from trimmings from the edible fisheries (which would include the fresh, frozen, canned and cured categories). The other category normally includes bait, fur, zoo animal food, and pet foods.

Table 4.B. Percent Distribution of the World Fish Catch by End Use from 1986–2008.

	Fresh	Frozen	Canned	Cured	Fishmeal And Oil	Other
1986	20	24	13	14	29	1
1987	21	24	12	15	27	1
1988	21	24	12	11	30	2
1989	24	24	13	10	28	2
1990	23	25	13	11	27	2
1991	23	25	13	11	27	2
1992	26	24	12	10	26	2
1993	25	25	12	10	26	2
1994	24	24	12	11	28	2
1995	34	21	8	9	23	4
1996	36	21	9	9	23	4
1997	38	21	8	9	21	4
1998	40	21	9	9	17	4
1999	39	20	8	9	20	5
2000	38	19	8	9	21	5
2001	39	20	9	9	18	5
2002	38	20	10	8	19	6
2003	40	21	11	9	16	4
2004	38	20	11	9	18	4
2005	38	21	12	9	17	4
2006	39	21	13	9	15	4
2007	39	21	12	9	15	5
2008	40	21	12	9	15	5

Source: NMFS 2010.

The main sources of raw material for fish oil production normally come from 3 or perhaps 4 areas (FAO 1986):

- Industrial fish caught specifically for fishmeal and oil production, which represents about 15–20% of the landings, or 20–26 million tons of fish using the 1999–2008 average.
- By-catch, or incidental catch, from other fisheries. There have been conservative estimates that the by-catch from edible fish operations is about 7 million tons, and almost 2 million of that comes from the shrimp fishery.

- By-products (cuttings or trimmings) from the edible fisheries that represent about 50% of the fish destined for food use. This represents a potential of about 53–56 million tons using the 1999–2008 average. Only a portion of that potential resource is currently used.

In terms of the omega-3 fatty acid market, there is a fourth category. While not specifically of fish origin, this includes other sources, such as marine and freshwater algae; krill; squid; and genetically modified oilseed plants, yeasts, and fungi designed to produce omega-3 in the lipid fraction.

There are a number of species of fish caught and used specifically for fish oil, and fishmeal production and these are shown in Table 4.C.

Trimmings from the edible fisheries represent about 25% of the raw material used for fishmeal and oil production. These species include those shown in Table 4.D.

The general classification of tuna actually includes a number of tuna and bonito species. These are shown in the following Figure 4.5.

Over 150 countries are engaged in the catching and/or processing of tuna for edible use, and 15 of them represent 75% of the production with major landings in Indonesia, Japan, Philippines, Taiwan, and South Korea. Most of the tuna oil entering the omega-3 market, however, appears to be coming from the remaining 135 countries, as well as from Japan and South Korea.

Not all fish landed for fish oil and fishmeal production or the trimmings from edible fish have a sufficient amount or the right ratio of the omega-3 fatty acids in the lipid. Table 4.E shows the main omega-3 fatty acid content of the fish oils of commerce. Included in this table are limited figures for some of the single-cell oils. The data for the single cell oils, krill oil, and SDA soybean oil are available from the respective GRAS submissions to the USFDA and can be accessed at http://www.accessdata. fda.gov/scripts/fcn/fcnNavigation.cfm?rpt=grasListing. The data for the squid oil can

Table 4.C. Species of Fish Caught for Fish Oil and Fishmeal Production.

Species	Country
Anchovy	Peru, Chile, So. Africa, Namibia, Mexico, Morocco
Jack (Horse) Mackerel	Peru, Chile
Capelin	Norway, Iceland, Russian Federation
Menhaden	USA, Atlantic and Gulf of Mexico
Blue Whiting	Norway, UK, Russian Federation, Ireland
Sand eel	Denmark, Norway, Faeroe Islands
Norway Pout	Denmark, Norway, Faeroe Islands
Sprat	Denmark, Russian Federation

Table 4.D. Fish Trimmings Used for Fish Oil and Fishmeal Production.

Species	Country
Catfish spp.	USA, Vietnam
Tuna spp.	Thailand, Japan, USA, Australia, South Korea, China, France, Ecuador, Maldive Islands and many others
Salmon	Norway, USA- Alaska (wild), UK, Ireland, Canada, Chile, Japan (wild)
Sardine/Pilchard	Peru, Chile, South Africa, Namibia, Japan, Spain, Mexico
White Fish spp.	UK, USA-Alaska, Canada, Chile
Dogfish	Canada, USA
Pollock	USA-Alaska, Russia
Horse Mackerel	Ireland, Norway, Denmark, Spain, Peru, Chile
Atlantic Herring	Iceland, Norway, Denmark, UK, Faeroe Islands, Sweden, Ireland, Canada
Mackerel spp.	UK, Peru, Chile, South Africa
Hoki (Blue Grenadier)	Australia, New Zealand
Non- fish species	
Krill	Norway, Poland, Ukraine, Japan, South Korea
Squid	Argentina, Chile, Peru, USA, Japan, China, South Korea, Russian Federation, France, Portugal, Spain, UK, Morocco, Mexico, Hong Kong, Taiwan, Ghana, Mauritania, South Africa, Senegal, Tunisia, Falkland Islands, Indonesia, Malaysia, Philippines, Thailand, New Zealand
Marine and Freshwater Algae	USA, Japan, Australia, Canada, USA (Hawaii), Israel

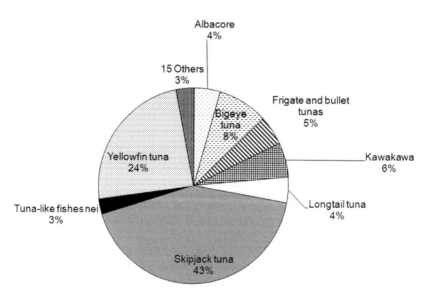

Fig. 4.5. Ten (10) year average global tuna landings. Source: FAO 2010.

Table 4.E. Omega-3 Fatty Acid Content of the Major Fish Oils of Commerce.

Species	C18:3	C18:4	C20:5	C22:5	C22:6	EPA+DHA	Total Omega-3
			% Total fatty acids				
EPA>DHA							
Anchovy	1	2	22	2	9	31	36
Atlantic Menhaden	1	3	14	3	12	26	33
Sardine/Pilchard	1	3	16	2	9	25	31
Gulf Menhaden	2	3	13	3	8	21	29
Pollock		2	16		4	20	22
Capelin	1	3	8		6	14	18
EPA≈DHA							
Sand Eel	1	5	11	1	11	22	29
Mackerel sp.	1	4	7	1	8	15	21
Blue whiting	1	3	7	1	8	15	20
Herring	2	3	6	1	6	12	18
DHA>EPA							
Tuna spp.	1	1	6	2	22	28	32
Norway Pout	1	3	9	1	14	23	28
Whitefish sp.	1	2	9	2	13	22	27
Salmon, wild	2	1	8	4	11	19	26
Jack Mackerel	1	2	7	2	12	19	24
Salmon, farmed	1	3	9	2	11	18	24
Sprat	2		6	1	9	15	18
Cod Liver, Atlantic	3	3	11	1	13	24	31
Cod Liver, Pacific	3	1	10	3	13	23	30
Very Low EPA and DHA							
Tilapia, farmed	2			3	5	5	10
Catfish, farmed	1		1	1	3	4	6
OTHERS							
Krill	1	4	18		10	28	33
Squid	1	1	14		15	29	31
SDA Soybean Oil	9–12	15–30	0		0	0	25–40
Single Cell Oils							
Ulkenia sp.			2		29	31	34
Crypthecod-inium sp.			1	1	42	43	46
Schizochytrium sp.			3		33	36	38
Nannochloropsis sp	2		31			31	35
Yarrowia lipolytica	2		56			56	59

Source: Bimbo 2010.

be accessed at Utilization of Squid By M. Sugiyama, S. Kousu, M. Hanabe http://books.google.com/books?id=bVXHe9nL0ycC&printsec=frontcover&dq=Utilization+of+Squid&hl=en&ei=E443TYytFcX_lgf8752MBw&sa=X&oi=book_result&ct=result&resnum=1&ved=0CCcQ6AEwAA#v=onepage&q&f=false.

Sustainability and Certifications

Because of issues related to availability of raw materials for food, aquaculture and pharmaceutical use, the various fishing industry participants have pursued policies aimed toward sustainability certification. Pressure from environmental groups, through the consumer end of the market chain, has been successful in moving this forward. Participants within the global fisheries, whether in the capture or value-added segment, have found it necessary to pursue certification in order to compete in the marketplace. The first certification group was the Marine Stewardship Council (MSC), followed later by Friends of the Sea (FOS). While the MSC certifies fisheries, FOS certifies fisheries and products. A sampling of the certified fisheries and those fisheries in the assessment process by MSC and fisheries certified by FOS are shown in Tables 4.F–4.H. The tables only include fisheries related either directly or indirectly to the fish oil and omega-3 market.

Table 4.F. Marine Stewardship Council (MSC) Certified Fisheries.

Species	Country
Alaska Salmon (Pacific)	USA
Bering Sea /Aleutian Islands, Gulf of Alaska Pollock	USA
Bering Sea /Aleutian Islands, Gulf of Alaska Pacific Cod	USA
Hoki (Blue Grenadier)	New Zealand
Atlantic Herring	Denmark, Faeroe Islands
North Sea Herring	Denmark, Norway
Spring Spawning Herring	Norway
Sardine, Purse Seine	Portugal
Pole and Line Skipjack Tuna	Japan
Atlantic Mackerel, Pole, Hand line and Purse Seine	Denmark, Ireland, Norway
Albacore Tuna, North and South Pacific	USA
Antarctic Krill	Aker Biomarine

There are currently 102 certified MSC fisheries.

Source: MSC 2010.

Table 4.G. Marine Stewardship Council (MSC) Fisheries in Assessment.

Species	Assessment stage	Country
Spiny Dogfish	3	British Columbia, Canada,
USA Atlantic Coast		
Anchovy	1	Argentina
Sardine	5	Gulf of California, Mexico
Pole, Line and Hand line Tuna	5	Maldive Islands
Pole, Line and Hand line Yellowfin and Skipjack Tuna	4	Mexico
Albacore Tuna Troll Fishery	5	New Zealand
Bering Sea and Sea of Okhotsk Pollock	2	Russia
Peruvian Anchovy Purse Seine[1]	Unknown	Peru

[1]There were some indications that this fishery was undergoing MSC certification review but there is no information on the MSC website.

There are 132 fisheries currently under assessment by MSC.

Source: MSC 2010.

Table 4.H. Friends of the Sea (FOS) Certified Fisheries.

Species	Catch method	Location
Anchovies	Purse Seine	Croatia , Morocco , Peru, Argentina
Mackerel	Purse Seine	Morocco , Peru , Portugal
Menhaden	Purse Seine	USA -North Atlantic and Gulf of Mexico
Sardines	Purse Seine	Morocco, Portugal
Tuna	Handline, Pole and Line	Frabelle Fishing Corp., Brazil, Maldives, Namibia, Senegal , Angola, Namibia, South Africa, Lee Fisheries, Philippines, Sri Lanka, Azores
Tuna	Purse Seine	Frabelle Fishing Corp., ANBAC, OPAGAC
Tuna	Troll	Ireland
Salmon	Various	USA-Alaska, Canada-BC

39 fisheries have been approved, 8 rejected by FOS.

Source: FOS 2010.

Harvesting

Fish harvesting methods depend upon the targeted species. Pelagic fish, which school in large numbers and feed near the surface, are caught by purse seine. This involves locating the school of fish, and then encircling it with a net, closing the bottom of the net, tightening the net, and then sucking fish and water aboard the vessel. The water is discharged and fish are stored in refrigerated holds on the vessel.

Some fish are located midway in the water column or near the bottom, and these are generally caught by a trawler. A large net is dispatched from the rear of the vessel and towed through the water. The depth of the trawl is controlled by sensors and a computer, depending on where the fish are located. One end of the net is wide where the fish enter and the other end is tapered. When the net is full, it is pulled onto the stern of the vessel and dumped. The fish are separated by species and then stored in holding bins on the vessel. Most nets are equipped with excluder devices to prevent the accidental catch of endangered or protected species. Since krill tend to deteriorate rapidly, Aker Biomarine has developed a technique where the krill are actually continuously extracted out of the net while the vessel is fishing, thus always processing essentially live krill.

Other harvesting methods involve long lines of baited hooks, sometimes several miles long. Trap nets are stationary nets on poles in rivers and bays. They are constructed in such a way that the fish swimming in the river or bay are directed along the net into a holding area or trap. The fishermen then visit the trap and dip the fish out of the trap and into the vessel. Another fishing method uses pole and line when there is a large school of fish. The fishermen essentially throw live bait into the water to attract the fish, and then by pulling the hooked line in the water, they essentially snag the feeding fish, pull them into the vessel, and repeat the process. This is done quite effectively with tuna and guarantees dolphin-safe methods and products.

Unloading Methods

When the fishing vessel reaches the processing plant, the fish are unloaded by either wet or dry methods. Wet unloading involves the use of water to transport the fish from the vessel to the plant. The water and fish are separated; the fish enter the storage facility and the water is screened, processed to remove fish pieces and oil, and then returned to the sea. Dry unloading can be done with positive displacement pumps (this will not work for edible fish, however, since they are essentially crushed and compacted) or pressure vacuum pumps where a small amount of water is used to move the fish to the suction hose. The fish and water are sucked into the vacuum chamber, the water is separated and recycled, and then the chamber is pressurized to push the fish onto conveyor belts for movement to the plant storage. These pumps normally have two chambers, so while one is sucking the fish into the system the other is pumping the fish into the plant storage. There are also vacuum unloaders that simply suck the fish out of the vessel's hold and discharge them onto a conveyor belt for transport to the plant storage. Vacuum unloaders work when the vessel can dock at the plant for unloading.

In smaller plants and in developing countries fish are caught in the usual manner but are separated onboard the vessel. The food fish are stored in the vessel hold with ice. Non-food fish destined for fishmeal and oil production are placed into bags and the bags are stored in the same holds under ice. The bags allow for an easy separation of edible and inedible fish. The process is reversed if the ratio of inedible fish to edible fish changes.

Process Flow

The raw material for fishmeal and oil production contains about 80% liquid (water and lipid) and 20% solids (protein and minerals). In order to produce fishmeal and oil the liquid must be removed from the solids. If you take a whole fish and squeeze it, nothing happens. If you take that same fish, bake it in an oven for 30 minutes, and then squeeze it you will express a liquid. The cooking process essentially breaks down the cells or denatures the protein, allowing the water and lipid to separate or be pressed out. This process of cooking and pressing is the heart of the wet reduction process.

For the production of fish oil and fishmeal, the fish are first cooked in long cylindrical cookers, which have either hollow flights or a hollow outer shell through which steam is introduced. This allows the fish to cook. The fat cells are generally broken at about 50°C and the protein is denatured at about 75°C. Liquid draining from the cooked fish is screened and the solids are returned to the process. The cooked fish are then pressed to remove as much liquid as possible. The pressed liquid and liquid from the cooker drainings are heated and pumped to a series of centrifuges. The first centrifuge is usually a decanter (horizontal bowl) designed to separate solids from liquid. The solids are returned to the process, and the liquid then goes to an oil/water separator centrifuge. The water goes to a storage tank for further processing. The oil is pumped to another centrifuge where water is introduced to "wash" the oil. The resultant fish oil is then pumped to storage. In some plants, the oil is first treated with activated carbon to reduce the level of dioxins, furans, some polychlorinated biphenyl (pcb), and polyaromatic hydrocarbon (pah) compounds.

The pressed fish solids, called presscake, is mixed with the solids from the decanter and conveyed to dryers where the fishmeal is produced. The water phase from the oil/water separation is pumped to evaporators where it is concentrated and then added back to the presscake or sold as a separate product. A flow diagram for the wet reduction process is shown in Figure 4.6.

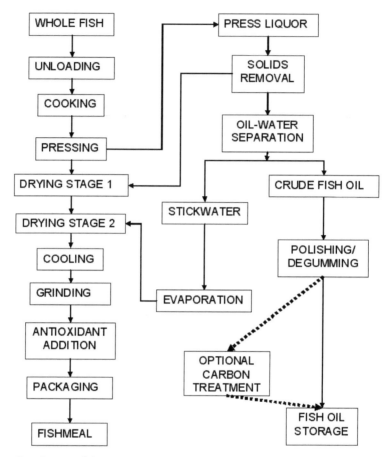

Fig. 4.6. Flow diagram of the wet reduction process. Source: Bimbo, 2007. Permission to reprint from Oily Press.

Other Production Methods

Solvent Extraction

Fish oil can also be produced by solvent extraction, which is the old fish protein concentrate process. In that process, raw fish are ground up either before or after deboning. The minced fish is mixed with solvent and extracted multiple times to remove the water, oil, and fishy tasting components. The solvent can be hexane, isopropyl or ethyl alcohol. The multiple solvent extraction steps are usually done countercurrently so that the second and third stage solvent can be reused as the first step of the next batch. The extract is separated from the solids by centrifuge and then dried. The liquid fraction is fractionally distilled to recover the solvent and produce the fish oil. Since some solvents produce azeotropes with water, the solvent recovery stage presents problems. Extracting wet fish with an azeotrope also extracts the water-soluble components. In the case of krill, where you may wish to produce two lipid fractions (neutral lipid and phospholipids), by choosing the correct solvent it is possible to obtain a high phospholipid fraction.

Hydrolysis

In some places, it does make economical sense to install fishmeal and oil plants. The volume of raw material may be too small for the normal fishmeal plant or the season may be too short for installing a large, expensive plant. In those cases, companies have found that the use of hydrolysis or autolysis can offer an economical alternative. Autolysis involves using the natural enzymes in the fish viscera along with organic acids to create an environment for the digestion of the raw fish material while preventing bacterial growth. The disadvantage of the autolysis method is that the natural enzymes have no specificity and the protein segments can vary quite dramatically. Hydrolysis, on the other hand, involves pasteurizing the fish raw material first to destroy the natural enzymes and kill any microbes that might be present. Extraneous enzymes are then added and the pH is adjusted. The enzymes that are added normally have some level of specificity so there is control over the reaction and the size of the protein segments formed.

Whether by autolysis or hydrolysis, the reaction results in a very diluted solution with non-digested bones, skin, and other portions of the fish. The digest must be separated from the undigested material by either decanter centrifuge or screening. The digest fraction must then be evaporated, and possibly dried, to finish the process.

Enzymatic Digestion

There were always 2 types of fish protein concentrate—concentrate produced by the chemical process or by the biological process. The chemical process involves solvent extraction, while the biological process involves the enzymatic digestion of the raw fish either before or after deboning. The process is performed at temperatures much lower than the other processes so the protein and oil fractions are not exposed to high temperatures. When the process is terminated, the digest is separated into three fractions: undigested solids (mostly bones, scales and skin), the digest, and the lipid fraction. The digest is then dried to produce a functional powder. In the solvent extraction process the final powder, while it possesses nutritional components, has no functionality. The enzymatic digest possesses nutritional components and is functional. It can be blended into liquids or whipped to produce foams and emulsions. The fish protein hydrolysis process is shown in Figure 4.7.

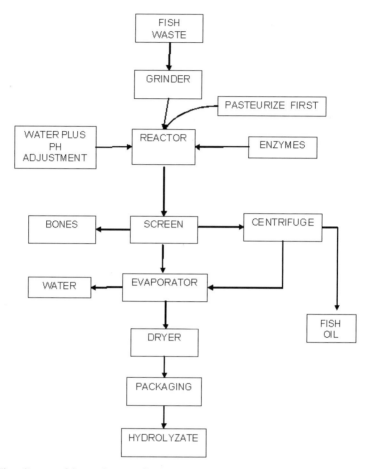

Fig. 4.7. Flow diagram of the production of hydrolyzates. Source: Bimbo, 2008.

Silage Production, Autolytic Digestion

Silage production is a form of autolysis in which the fish raw material is ground up and digested using the natural enzymes in the fish viscera plus an organic acid to control the pH. The acid suppresses microorganism growth without hindering the enzymes from digesting the proteins. The undigested residue must be separated from the digest before the process can be continued. Sometimes the raw material is first processed through a meat bone separator to separate the bones and skin from the meat. The meat is then added to the digester along with the acid, and this process results in less undigested residue. The digest is then centrifuged to separate the oil

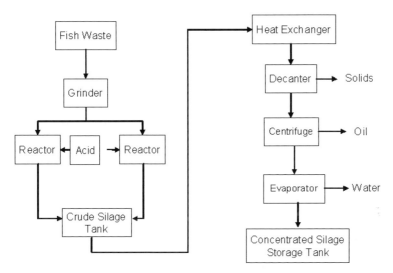

Fig. 4.8. Flow diagram for silage production. Source: Bimbo, 2008.

and water, and the water fraction can then be evaporated and dried as required by the market. If dried, a product with 75–80% protein can be produced. A typical flow diagram for the silage process is shown in Figure 4.8.

Dry Rendering

There are two rendering processes: wet and dry. The wet rendering process is routinely used in the fishmeal and oil industry, especially with fatty fish. The process allows the liquid fraction (oil and water) to be mechanically separated from the solid fraction (protein and oil). The solids are then dried while the oil and water are separated. The dry rendering process in normally used in the animal protein industry but has also been used with white fish (non-oily) to produce a fishmeal product with small volumes of oil. In this process, the raw material is "cooked" to remove the water (essentially the drying process in the fishmeal wet rendering process). The resultant dry cake is then pressed to remove any oil. Because the water has been removed, the lipid fraction can contain high levels of phospholipids (PL). The phospholipids normally hydrate in the wet rendering process and are recovered with the water fraction. In the dry rendering process, however, they are not hydrated and therefore remain dissolved in the lipid or oil fraction. Since there is interest in the fish phospholipids, it is possible to produce a PL fraction by hydrating the oil (also called degumming). A flow diagram for the dry rendering process is shown in Figure 4.9.

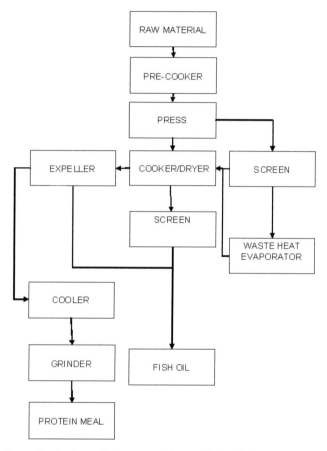

Fig. 4.9. Flow diagram for the dry rendering process. Source: Bimbo, 2008.

Other Marine Oils

Fish Liver Oils

Fish liver oils have been used as far back as the Middle Ages, and populations in Scandinavia have used them for thousands of years (Breivik, 1992). The most important raw material for the production of liver oils comes from the fisheries for cod, coalfish, and haddock. The livers of ling; tusk; several species of shark such as dogfish, Greenland shark and basking shark; and halibut have also been used in the production of liver oils. In order to obtain high quality, light colored oils with good flavor and odor containing a minimum of free fatty acids, it is important to eviscerate the fish and recover the livers so that they can be processed as quickly as possible.

The oil is contained in the liver protein and can normally be easily recovered by steam cooking the livers. This grade of oil is generally the medicinal grade liver oil. The residue can then be caustic treated to destroy the protein and release the residual oil. This grade of oil has higher vitamin content and is generally classified as a veterinary grade. In modern times, the cooked livers are processed through centrifuges that are more efficient at separating the liquid and solid fractions. The liquid fraction is then centrifuged to recover the oil. This centrifuge process recovers more primary grade oil than the old gravity separation processes. A typical process for the production of fish liver oils is shown in Figure 4.10.

The production of liver oils has varied over time and is dependent upon the availability of the raw materials used to supply the livers. In modern times, identified cod liver oil has been displaced by liver oils from various other fish species such as Alaska pollock, other gadoid species, and hake liver oils. Figure 4.11 shows the production of different liver oils from 1990 to 2008. The data show that defined cod liver oil has been replaced by other liver oils from species not easily identified (nei), according to FAO data (2010).

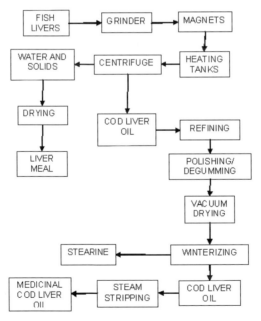

Fig. 4.10. Flow diagram for the production of liver oils. Source: Bimbo, 2007. Permission to reprint from Oily Press.

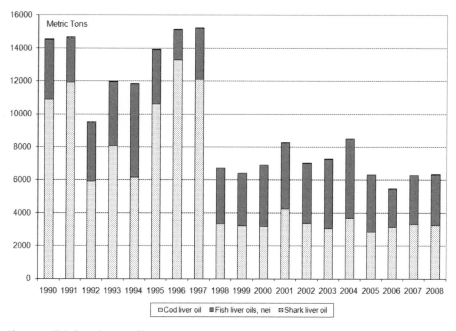

Fig. 4.11. Global production of liver oils. Source: FAO 2010.

Alaska, United States (US) is responsible for about 50% of the US landings of fish. The third largest species of fish landed in Alaska is the Pacific cod *Gadus macrocephalus*. According to US fisheries statistics, the Pacific cod landed in Alaska has ranged from 193,000 mt to 314,000 mt from 1990 to 2008. This can be seen in Figure 4.12.

According to some preliminary data, the Pacific cod liver contains about 50% lipid (personal communication Alaska 2010). On January 22, 2010, the Bering Sea and Aleutian Islands Pacific cod fishery was certified by the MSC. The fishery is regulated through a TAC (total available catch quota), which was 168,780 mt in 2010 and expected to be increased to 227,950 for 2011. If the liver were 5% of the weight of the fish, this would give a potential additional 5,700 metric tons of cod liver oil.

Krill Oil

While currently not a major source of oil or omega-3 fatty acids, krill oil has generated a great deal of interest in the last few years, possibly due to the unique form of the lipid in the krill—phospholipids. Krill is a term applied to describe over 80 species of open-ocean crustaceans known as *euphausiids,* most of which are planktonic. Of the seven species of *euphausiid* crustaceans commonly found in the Southern Ocean, only two of them regularly occur in dense swarms and are of particular

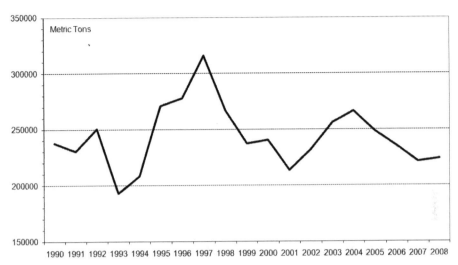

Fig. 4.12. Pacific cod landings in Alaska, US. Source: NMFS 2010.

interest to commercial fisheries: *E. superba* and *E. crystallorophias*. *E. superba* is the species commonly referred to as "Antarctic krill" and it is a widespread species that is subject to significant commercial fishing. Commercial fishing for krill is done in the Southern Ocean and in the waters around Japan. The total global production amounts to 150,000–200,000 metric tons annually. This is expected to increase as the market for krill oil increases. On May 25, 2010, Aker Biomarine's krill fishing operation was certified by the MSC.

If 1.5% of the total krill catch was recovered as oil after processing, the range in krill oil production would be 1,000–2,000 metric tons from 1993 to 2008. Figure 4.13 shows what the estimated krill oil production would have been over the period of 1973 to 2008. According to a recent Aker Biomarine financial report, their 2010 production of krill oil was 350 mt (FIS 2010).

As stated previously, the krill deteriorate very quickly after catching. There is very little published on how the krill are processed on-board the ships, and there are only a few possible options that could be employed:

- Option 1. The krill are caught and immediately frozen into blocks for transport to a shore-side facility. There the blocks can be processed into krill meal and oil using conventional rendering or solvent extraction.
- Option 2. The krill can be processed onboard the factory trawler, in which case simply drying the krill would remove a large part of the weight, namely the water. The dried krill meal would be high in fat.
- Option 3. The krill meal can be transported back to a shore-side facility where it can be solvent extracted to recover the oil and produce a high

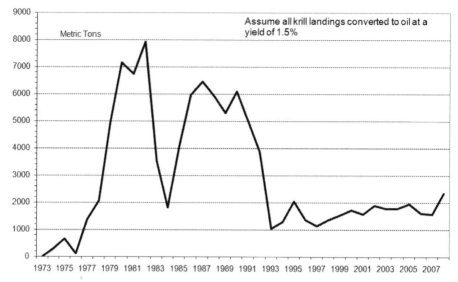

Fig. 4.13. Estimated global Antarctic krill oil production. Source: Based on data from FAO 2010.

protein meal. The extracted oil can then be further processed to separate the phospholipids fraction from the triglyceride fraction.

• Option 4. The krill can be enzymatically digested to release the oil and recover the protein fraction as a digest.

It would not be practical to solvent extract aboard the vessel because of safety concerns. Figure 4.14 outlines these four options for processing the krill.

Marine Mammal Oils

The oil from marine mammals once accounting for as much as 75% of the total global marine oil production, but today this oil accounts for less than 2% of the total aquatic animal oil production. This reduction in production is primarily due to conservation and a worldwide moratorium on whaling. There is some seal oil produced in Canada, Namibia, Chile, Norway, Greenland, and South Africa; however, no statistics are published on this production.

Cephalopod Oils

In recent years, interest in other sources of marine oil has led to the production of squid oil for the nutraceutical market (called calamari oil). Squid landings have ranged from 2.2 to 3.3 million metric tons (mmt) over the last 10 years (1999–2008). Argentina, China, Taiwan, South Korea, Japan, and Peru are the major countries catching squid, but there are 89 countries engaged in the squid fishery.

Figure 4.15 shows the relative amount of marine mammal and squid oil produced from 1990 to 2008. There is no explanation for the squid oil production peak

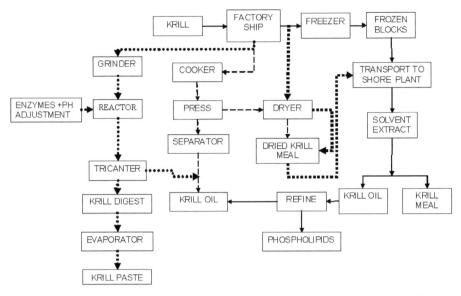

Fig. 4.14. Four (4) options for processing krill. Source: Bimbo, 2007. Permission to reprint from Oily Press.

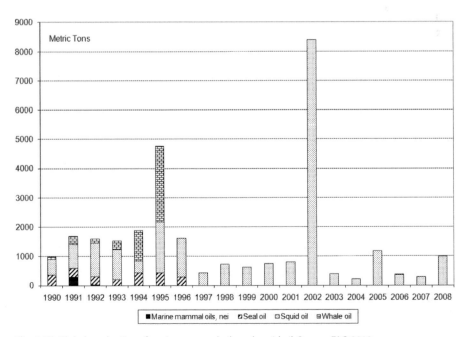

Fig. 4.15. Global production of marine mammal oils and squid oil. Source: FAO 2010.

in 2002; the data was checked for typographical errors by FAO, and they contacted all reporting countries to verify the data.

Processing Marine Oils

Marine oils like other edible fats and oils contain minor amounts of non-triglyceride substances. While some of these might be beneficial, such as tocopherols for stability and astaxanthin in salmon and krill oil, other impurities are objectionable because they render the oil dark colored, cause it to foam or smoke, or are precipitated when the oil is heated during subsequent processing steps. Other issues might be the stearine fraction, which is a triglyceride mixture that has a high melting point and precipitates at ambient temperature, causing the oil to cloud and form a residue in storage tanks and containers. Other impurities reduce acceptability because the flavors and odors they produce result in reduced stability and shelf life in the foods to which they are added. Fish oils and animal fats also contain cholesterol, which normally must be removed if a concentrate is produced. So the oil must be further processed to remove these substances while retaining their desirable features. With the new emphasis on "natural" products, certain compounds that were once considered undesirable such as the natural red color in salmon and krill oil are today considered highly desirable. To preserve the natural compounds the oil must be minimally processed (virgin). However, the processor must walk a very fine line; on one hand he must preserve the natural compounds, while on the other he must be assured that any environmental contaminants in the oil are removed. Often this requires sourcing oils from areas where the level of environmental contamination is minimal.

Pre-Refining and Degumming

The pre-refining step generally takes place in storage where the oil quality can be affected by the conditions under which it is stored. Free fatty acid increases, oxidation, color setting, and contamination by insoluble impurities can all take place under improper storage conditions. Simple procedures such as extending intake pipelines to the bottom of storage tanks; eliminating contact with iron, copper, and copper alloys; and constructing a sump in the tank at its lowest point so that moisture and other insoluble impurities can be drained away, thus preventing free fatty acid increases, are all beneficial. If the oils contain large quantities of stearine, equipping the tank with slow moving side mounted agitators and water-heated coils will reduce the collection of residue on the walls and bottom of the tank.

Degumming involves the treatment of crude oils with water, a salt solution, or dilute acid such as phosphoric to remove phospholipids, waxes, and other impurities. Degumming causes the phospholipids to hydrate and become insoluble in the oil. They can then be readily separated by gravity or centrifuge. Degumming can also take place in the crude oil storage tank if moisture is present. This results in foots settling out in the bottom of the storage tank. Arsenolipids are also hydrated, so this step can reduce the arsenic levels in the crude fish oil.

Refining Process

The addition of an alkali solution to crude oil results in the neutralization of the free fatty acids, degradation of some of the color bodies in the oil, and hydration of the phospholipids if the oil was not pre-degummed. The neutralized fatty acids plus any hydrated phospholipids and color bodies are removed from the oil as soap stock. Depending on the level of control of the process, some neutral oil is also lost in the refining process. Fish oil can be refined by either a continuous centrifugation process or a batch process, but the continuous process is generally the process of choice. Refining losses are generally 1.5–2.0 times the level of free fatty acids in the oil but can be as high as 5–10 times if there is a high level of phospholipids present in the oil. The refined oil is usually vacuum-dried before bleaching to prevent moisture from compromising the bleaching earth.

Treatment of the soapstock presents problems. Generally, the soap is mixed with water and then acidulated using a mineral acid. The soap is converted back to the free fatty acids and is separated from the acid/water phase by either gravity or centrifuge. In the past, this material was sold as a cheap fat for blending purposes or as a cheap rust inhibiting material. Today, because of the versatility of the biodiesel processes, it can be converted to methyl esters for use in biofuels. Disposal of the acidic water is still a major issue since it must be treated before disposal.

There are other methods of refining fish oils. Spinelli (1987) described a process utilizing supercritical carbon dioxide for purifying fish oils. The solvent is selected for the removal of odor bodies, pigments, and oxidation products that contribute to off flavors. By changing operating parameters and performing successive extraction steps, it is possible to remove contaminants, free fatty acids, oxidation products, and other unwanted compounds, including cholesterol. Krukonis (1989) described the process as producing a product that was too pure since it removed all the natural antioxidants, thus making the oil extremely unstable.

Adsorptive Bleaching

Bleaching is used to improve the color, flavor, and oxidative stability of the oil and to remove impurities, such as traces of soap, oxidation products, and heavy metals. Bleaching involves the heating of the oil under a vacuum after the addition of an activated bleaching earth. Sometimes the oil is first treated with activated carbon to remove contaminants and then bleached. The clay and carbon adsorb the contaminants, and then the oil is filtered. If water is present in the oil it will be adsorbed by the clay and this will inactivate the bleaching properties of the clay. Bleaching will generally reduce the peroxide value to extremely low levels and reduce the TOTOX number [(2 × Peroxide Value) + Anisidine Number] by as much as 50%. The bleaching process, whether atmospheric or vacuum, is usually a batch process, although some continuous vacuum bleaching plants are operational.

Winterization or Fractionation

Winterization is a form of fractionation and has been called cold pressing or cold clearing. It is a physical fractionation designed to separate the stearine and olein fractions of the oil. The process evolved from observations that when fats or oils are stored outside in tanks during the winter, the natural cooling of the oil precipitated the high-melting-point triglycerides in the oil. The result was two fractions—a top clear oil (olein) and a bottom waxy semi-solid (stearine). In terms of fish oil, specifically menhaden oil, the process does enhance the omega-3 content by separating out a stearine fraction with a higher saturated fatty acid content. However, since the oil is still in the triglyceride form, there is still some omega-3 fatty acids in the stearine fraction. Table 4.I compares the fatty acid profile of the starting menhaden oil with the olein and stearine fractions.

The stearine fraction is interesting. It contains omega-3 fatty acids with 17% EPA+DHA but has the solid fat index and melting point somewhat similar to lard and palm oil. In other words, it has a chemical composition similar to liquid omega-3 oils with a physical characteristic similar to hydrogenated fats but without the *trans* fatty acids. These unique characteristics are shown in Table 4.J.

Fractionation of the triglycerides can also be done with solvents. Solvent crystallization allows the oil temperature to be reduced further than what is possible

Table 4.I. Partial Fatty Acid Profile of Crude Oil, Winterized Oil and the Stearine Fraction from Menhaden Oil (% of total weight).

	Crude Oil	Winterized Oil	Stearine
C14:0	10	8	12
C16:0	20	17	29
C18:0	3	3	5
C16:1	12	12	9
C18:1	9	10	10
C18:3	2	2	2
C18:4	3	3	2
C20:1	1	1	1
C20:4	2	2	2
C20:5	14	14	10
C22:5	3	3	2
C22:6	9	12	7

Source: Omega Protein Corporation personal communication 2010.

Table 4.J. Melting Point and Solid Fat Index (SFI) for Menhaden Stearine Compared to Other Fats.

		Solid Fat Index				
	Melting Point °F	50	70	80	92	100
Butter	97	32	12	9	3	0
Cocoa Butter	85	62	48	8	0	0
Coconut Oil	79	55	27	0	0	0
Lard	110	25	20	12	4	2
Palm Oil	103	34	12	9	6	4
Palm Kernel Oil	84	49	33	13	0	0
Tallow	118	39	30	28	23	18
Menhaden Stearine	102	26	17	15	8	1

Source: Bimbo 1989. Permission from AOCS to reprint.

without the solvent. Hexane and acetone have been used as the solvents. With acetone, it is possible to reduce the temperature to –60°C, at which point the free cholesterol can be crystallized out. (Kohubu, 1984).

Deodorization

Deodorization is the last major step in the processing of refined edible oils. The finishing step is responsible for removing both undesirable compounds naturally present in fats and oils and those that might be produced by previous processing steps. This step initially establishes the oil characteristics of flavor and odor that are most readily recognized by the consumer. (Gavin, 1978).

The steam deodorization process is essentially steam distillation and takes advantage of the differences in volatility between the triglyccrides and the substances that give the oil and fats their natural flavors and odors. The volatile compounds are stripped from the non-volatile oil with the use of steam. The process also destroys some oxidation compounds and removes aldehydes formed during the previous processing steps.

Vacuum Stripping

Vacuum stripping started out as another less costly method for deodorization. The process utilizes a molecular still, short path vacuum distillation system, falling film evaporator, thin film distillation, or any other of a number of names for the equipment. Essentially, the system differs from a steam deodorizer because no steam is used. The system relies on high temperature, short retention time, and very high vacuum to "distill" out or strip the volatile materials from the oil. The process involves the pumping of the raw material oil into the system, spreading it out as a thin film by use of a wiper, and allowing the oil to flow down the walls while the volatiles are pulled to the center, which is a cold finger condenser. The two fractions then exit the system and are collected. All of this is done under vacuums as low as 0.05 mm. By setting up several systems in a series, it is possible to remove free cholesterol in the first stage and then other volatiles and contaminants in a second stage. Because volatiles are removed, the oil is also deodorized to some extent.

A combination of the two processing steps, steam deodorization and vacuum stripping, is also employed and is covered by a US patent (Marshner, 1989). In that case, the vacuum stripping is used to remove cholesterol and other compounds with the steam deodorization used as the finishing step. Steam deodorization and/or vacuum stripping are generally finishing steps in producing edible oil with good flavor and odor, such as the standard fish oils that you find in the soft gelatin capsules. However, these oils can also serve as the starting point for the production of esters, concentrates and purified omega-3 fatty acid products. The previously described steps are designed to purify the oil by removing a number of potential impurities and contaminants. A review of these steps and the compounds removed is shown in Table 4.K.

Table 4.K. Processing Steps Used to Purify Fats and Oils and Compounds Removed.

Carbon Treatment	Removal of dioxins, furans, and polyaromatic hydrocarbons (PAH)
Oil Storage	Insoluble Impurities
Degumming	Phospholipids, sugars, resins, proteinaceous compounds, trace metals and other materials.
Alkali Refining	Free fatty acids, pigments, phospholipids, oil insoluble material, water soluble material, trace metals
Water Washing/Silica Treatment	Soaps, oxidation products and trace metals
Drying	Moisture
Adsorptive Bleaching and Carbon Treatment	Pigments, oxidation products, trace metals, sulfur compounds, dioxins, furans, and PAH
Winterization	Higher melting triglycerides, waxes, enhancing unsaturated triglycerides
Deodorization	Free fatty acids, mono-di glycerides, aldehydes, ketones, chlorinated hydrocarbons and pigment decomposition products
Vacuum Stripping or Thin Film, Molecular or Short Path Distillation,	Removal of chlorinated hydrocarbons, fatty acids, oxidation products, PCB and free cholesterol

Source: Bimbo, 1998. Permission to reprint from AOCS.

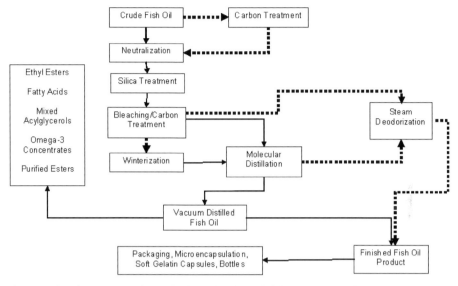

Fig. 4.16. Flow diagram to produce a deodorized triglyceride fish oil. Source: Bimbo, 1998. Permission to reprint from AOCS.

A flow diagram of the processing steps used to produce a vacuum-stripped or steam-deodorized fish oil ready for gelatin capsules or for further processing is shown in Figure 4.16. If the ultimate product is a concentrated ester or fatty acid, then the winterization step is omitted so that the maximum amount of omega-3 fatty acids can be recovered from the starting oil. This oil is then further processed to produce ethyl esters, fatty acids, or concentrates.

If the ultimate product or products are a concentrated or purified ester or fatty acid, then additional processing steps are needed to further purify or enhance the omega-3 fatty acid content of the oil. These additional processing steps are required to first split the triglyceride oil and convert it to the basic fatty acids, which are then converted to the ethyl esters. The esters are then fractionated to remove the undesirable fatty acids: the saturates, monos, and di-unsaturates. What remains are the fatty acids with 4, 5, or 6 double bonds. This concentrated omega-3 ethyl ester can then be further purified to produce relatively pure C20:5 and C22:6 (EPA and DHA). The processing steps employed are shown in Table 4.L.

The processing sequence to go from a vacuum distilled oil to the purified omega-3 fatty acids in shown in Figure 4.17.

While not shown in Figure 4.17, enzymatic processes are being used in the omega-3 fatty acid production. In addition to the use of enzymes for hydrolysis of the fish raw material to release the oil at lower temperatures, enzymes can be used to selectively remove saturates and monounsaturates from the triglyceride, thus

Table 4.L. Additional Processing Steps for the Production and Purification of Omega-3 Fractions.

Interesterification	Rearrangement of triglycerides to a more random distribution
Hydrolysis or Esterification	Splitting triglycerides, producing fatty acids or esters
Urea Complexing	Low temperature crystallization of the urea complex removes saturates and monounsaturates
Molecular Distillation	Concentrating esters or fatty acids
Supercritical Fluid Extraction (SCF)	Purification of esters or fatty acids to produce 85%+ pure compounds and to remove cholesterol
Preparative High Pressure Liquid Chromatography (HPLC)	Purification of esters or fatty acids to produce 95%+ pure compounds
Re-esterification	Conversion of concentrated fatty acids or ethyl esters back to the triglyceride form which is considered more natural

Source: Bimbo, 1998. Permission to reprint from AOCS.

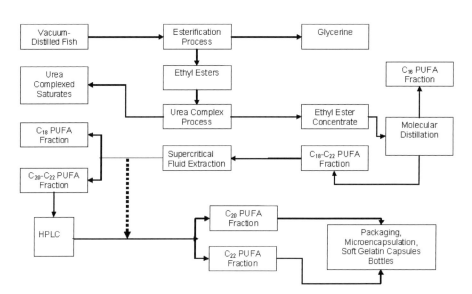

Fig.4.17. Flow diagram to produce, fractionate, and purify omega-3 fatty acid esters. Source: Bimbo, 1998. Permission to reprint from AOCS.

concentrating the omega-3 fatty acids. Enzymes can also be used in the alcoholysis reaction to produce and concentrate esters up to 50% enrichment. In this case, the enzyme acts as the catalyst in the reaction (Xu et al., 2007).

It becomes obvious from the processing steps that as you move from standard crude fish oil to an 85–90% pure omega-3 ethyl ester you will encounter major product loss. The processing required to produce concentrates ultimately results in a large volume of by-product material. The by-products must be utilized because disposal presents problems. In some areas, the by-product esters are used as a biodiesel fuel, but the product must meet the stringent ASTM specifications for this use.

Ernst (2010) states that the yield has a decisive effect on the cost effectiveness of ethyl ester concentration plants and that oils with low initial omega-3 content can be used to produce omega-3 concentrates in low volume (30%), with the majority of the by-product (the light phase) representing about 70% of the feed material. This light phase is then further distilled to separate a light fraction that can meet the US ASTM D6751 standards while the heavy fraction, which contains more of the long chain PUFA, can be returned to the omega-3 concentrate fraction. Starting with a 17% EPA, 6% DHA oil, they were able to produce a 33%/22% (20%) and a 45%/10% (10%) concentrate (EPA/DHA), with the remaining light phase representing 70% suitable for biodiesel production. Figure 4.18 shows the estimated yields (losses) of product as one proceeds from a crude fish oil to the Omega-3 concentrates. The losses are quite substantial.

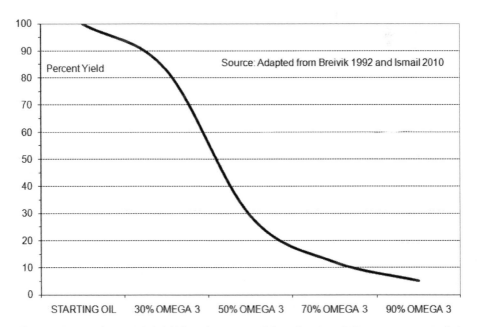

Fig. 4.18. Estimated percent (%) yield from the starting oil through a ninety (90) percent omega-3 ethyl ester. Source: Adapted from Breivik, 1992 and Ismail, 2010.

Table 4.M. Friends of the Sea (FOS) Certified Fish Oil Products.

Company	Product	Country
Austral Group SA[1]	Fishmeal and oil	Peru
EPAX AS	Fish oil	Norway
GC Reiber Oils	Fish oil	Norway
Omega Protein Inc.	Fish oil	USA
Pharma Marine	Calamari (squid) oil	Norway
Sovapec–Maromega Sarl	Fishmeal and oil	Morocco
Le Gousseant	Fishmeal from aquaculture	France
Dr. Loges	Fish oil	Germany
Nexgen Pharma	Fish oil	USA
Nordic Naturals	Fish oil	USA
Pharma Marine AS	Fish oil	Norway
Winterisation Europe	Fish oil	France
Biosym	Fish oil	Denmark
Polaris	Fish oil	France

[1]Austral Group is part of Austevoll Seafood ASA.

There are currently over 198 products certified by FOS.

Source: FOS 2010.

Sustainability and Certifications

Previously we listed the fisheries that were either MSC or FOS certified and mentioned that FOS also certifies products. Table 4.M lists the fish oil derived products certified by FOS as currently listed on their Web site.

Further details on the parameters used to process fish oil and generate derivatives such as ethyl esters concentrates and pharmaceutical grade oils and especially on their applications are reported in Chapter 5.

References

Bimbo, A. Raw material sources for the Omega-3 market: trends and sustainability. Paper presented at the 2010 International Society for Nutraceuticals and Functional Foods. St. John's Newfoundland, Canada, September 7, 2010.

Bimbo, A.P. Alaska Seafood By-Products: Potential Products and Markets and Competing Products 2008 Update. Report for the Alaska Fisheries Development Foundation, Fairbanks Alaska. Report available at http://www.afdf.org/past_research/2008_by_product_mkt_study.pdf (accessed November 24, 2010).

Bimbo, A.P. Fish Oils: Past and present food uses. JAOCS 1989, 66(12), 1717–1726.

Bimbo, A.P. Guidelines for characterizing food grade fish oil. Inform 1998, 9(5), 473–483.

Bimbo, A.P. Processing of marine oils. In Long Chain Omega-3 Specialty Oils; Breivik, H., Ed. The Oily Press: Bridgewater, England, 2007, pp 84–101.

Breivik, H. Concentrates: a Scandinavian Viewpoint. Paper presented at the AOCS Short Course, Modern Applications of Marine Oils, Toronto, Canada, May 7–8, 1992.

Ernst, M. Reprocessing of ethyl ester light phase. Inform 2010, 21(9), 535–536.

FAO 1986, Fishery Industries Division. The production of fishmeal and oil. FAO Fish. Technical Paper, 142, 1–63.

FIS 2010. Aker Biomarine's revenues reach EUR 16.5 mln in H1 http://fis/worldnews/worldnews.asp?l=e&ndb=1&id=37954 (accessed August 9, 2010).

Fisheries Global Information System (FAO-FIGIS). http://www.fao.org/fishery/topic/18044/en (accessed January 10, 2011).

Friends of the Sea 2010. http://www.friendofthesea.org/certified-products.asp (accessed December 2010).

Gavin, A.M. Edible Oil Deodorization. JAOCS 1978, 55, 783–791.

Ismail, A. Sustainability of fish oil. Paper presented at the Supply Side East Nutrition Conference, Meadowlands, NJ, April 26, 2010.

Kokubu, K.; Hayashia S.; Kodama K. Purification Method for marine oils, Japanese patent No. S58–883410. 1984.

Krukonis. Supercritical fluid processing of fish oils: Extraction of polychlorinated biphenyls. JAOCS 1989, 66, 818–821.

Marine Stewardship Council 2010. http://www.msc.org/track-a-fishery/all-fisheries (accessed December 2010).

Marshner, S.; Fine, J. Process for deodorizing and reducing cholesterol in fats and oils by employing flash vaporization and thin film stripping. US Patent No. 4,804,555. 1989.

NMFS Fisheries of the United States 2009. Current Fishery Statistics No. 2009. USDC, NOAA, NMFS Silver Spring, Maryland, 2010.

Oil World Annual 2010, Fish Oil & Meal World, T. Mielke, editor. www.oilworld.de. (accessed January 2011)

Omega Protein 2010. Personal communication, December 13, 2010.Spinelli, J.; Stout, V.F.; Nilsson, W.P. Purification of fish oils. US Patent No. 4,692,280. 1987.

Xu, X.; H-Kittikun, A.; Zhang, H.Enzymatic processing of omega-3 specialty oils. In Long Chain Omega-3 Specialty Oils Breivik, H., Ed. Oily Press: Bridgewater, England, 2007, pp 141–164.

Worm, B.; Barbier, E.B.; Beaumont, N.; Duffy, J.E.; Folke, C.; Halpern, B.S.; Jackson, J.B.C.; Lotze, H.K.; Micheli, F.; Palumbi, S.R.; et al. Impacts of Biodiversity Loss on Ocean Ecosystem Services. Science 2006, 314(5800), 787–790.

5

Processing of Omega-3 Oils

Ernesto M. Hernandez
Omega Protein Inc., Houston, Texas

Omega-3 oils, being highly polyunsaturated, are inherently unstable and prone to oxidation, regardless of the source. As a result, this can cause problems of rapid oil deterioration during extraction, processing, cooking, or storage. Therefore, special precautions have to be taken such minimizing exposure to high temperatures and air. Omega-3 oils that are not properly processed and handled very likely will result in unwanted fishy taste and aroma and, more importantly, in the deterioration of the essential omega-3 fatty acids. These in fact are the major obstacles effecting the incorporation of omega-3 fats in fortified food products and nutritional supplements.

Refined, Bleached, Winterized, and Deodorized Fish Oil

The main components in fish oil are basic triglycerides, or three fatty acids bound to a glycerol backbone. Phospholipids are also present in appreciable amounts but not as abundant in vegetable oils like soybean oil. The non-saponifiable components commonly found in fish oil are cholesterol and squalene. Other components found are vitamins A, D, and E. The vitamins are commonly removed for the most part during processing.

Fish oil has a wide variety of fatty acids, from 16-carbon chain length palmitic and palmitoleic acids to 20- and 22-carbon long fatty acids like eicosapentaenoic (EPA) and docosahexaenoic (DHA) acids, respectively. Vegetable oils, on the other hand, are mainly 18-carbon and 16-carbon chain length fatty acids. The most common fatty acids found in oilseeds are stearic, oleic, linoleic, linolenic, and palmitic acids (Erickson, 1995). Most naturally occurring fats and oils, including fish oils, have a *cis* geometric configuration. This means the hydrogens in the double bonds of the fatty acid chain are oriented in opposite directions. This has become a relevant issue for the food industry because processes such as hydrogenation orient the hydrogen into the same side of the molecule producing a *trans* geometry. There is reported work that shows that an excess of *trans* fats in the diet may induce cardiovascular disease (Mozaffarian, 2006).

The degree of unsaturation or number of double bonds in the fatty acids of triglyceride oils will determine the degree of susceptibility to oxidation. This is particularly true for fish oils. EPA and DHA in fish oils have 5 and 6 double bonds, respectively, making them very susceptible to several types of reactions, especially oxidation. Reactions of deterioration of fish start the moment the oil is extracted. Oxidation starts very quickly in the presence of metals, air, and heat, which induce the rapid incorporation of oxygen into the double bonds of triglyceride oils. The sequence of oxidation reaction usually starts with the incorporation of oxygen into double bonds, forming first peroxides and then the peroxides decomposing into carboxylic compounds such as ketones and aldehydes. These compounds are responsible for the development of rancidity and off flavors in the fish and other vegetable oils.

Fish oil's susceptibility to oxidization and deterioration is greatly exacerbated by the presence of iron, a well known pro-oxidant, naturally present in fish muscle. This metal has been identified as being the most responsible for the generation of fishy notes in omega-3 oils in both short and long chain poly unsaturated fatty acids (PUFAs). Fishy off-flavors have been identified as volatiles comprising alkenals, alkadienals, alkatrienals, and vinyl ketones. These compounds, generally identified as fishy, metallic, and rancid off-flavors, are considered one of the major deterrents to increased consumption of fish and fish oils.

The reaction of hydrolysis can also readily occur in the presence of heat and moisture. This is basically the breakdown of triglycerides into free fatty acids, mono, and di-glycerides. Hydrolysis also occurs from lipase activity, an enzyme, which is particularly active in fish, that is stored at ambient conditions. Polymerization of the fatty acids in fish oil can also occur due to the reactivity of the double bonds and is usually coupled with an oxidation reaction in the presence of heat and catalysts like metals. The presence of polymerized material in fish oil is an indication of quality control problems. In industrial applications, polyunsaturated oils are commonly used in paints and varnishes.

Another indication of quality control problems is the presence of isomerization. The most common isomerization reaction in fish and in vegetable oils is the change from *cis* to *trans* configuration in the fatty acids. This can occur in fish oil if deodorization is conducted at unusually high temperatures and for long periods of time. Other isomerization reactions occur when the double bond in fatty acids changes position due to excessive exposure to heat, resulting in the formation of conjugated double bonds (Hernandez, 2005).

In order to make it edible, fish oil is commonly refined using the conventional processing steps similar to vegetable oil processing (i.e., neutralization, bleaching, winterization, and deodorization). The step of winterization is sometimes added after bleaching to remove the more saturated fraction in fish oil and to enrich the polyunsaturated content of the final oil. Also after bleaching, fish oil is processed by molecular distillation in order to remove contaminants such as pesticides and chlorinated aromatic hydrocarbons (Breivik & Thorstad, 2005). Fig. 5.1 shows a diagram for fish

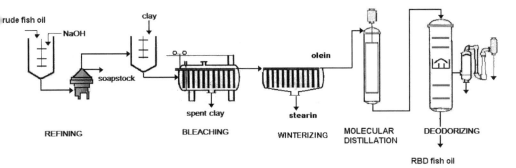

Fig. 5.1. Processing of Fish Oil.

oil refining process in general. First, the free fatty acids in the oil are neutralized by mixing in a caustic solution added in stoichiometric amounts. This produces a waste stream of saponified matter, or soapstock, that is usually removed by centrifugation. The residual saponified matter that remains in the oil after centrifugation is removed by water washing where the water is also removed by centrifugation. Instead of water some companies use silica gels that can adsorb the residual saponified matter and that then can be removed by filtration.

The formula to calculate the amount of caustic to be added is based on the principle that one mole of sodium hydroxide will neutralize one mole of free fatty acids in the oil plus an excess to ensure that most of the acidity in the oil is neutralized and the phosphatides are hydrated. The formula for the calculation of caustic addition, commonly used in vegetable oil refining and adapted for fish oil refining, is as follows:

$$\% \text{ Caustic Addition} = \frac{(\% \text{ FFA} \times 0.142) + \% \text{ Excess}}{\%\text{NaOH Sol}/100}$$

where 0.142 is the molecular weight ratio NaOH and average weight of fatty acids (as oleic acid) in the oil. Percent excess of caustic solution used will depend on the amount of gums in the crude oil or whether the oil was already degummed (Erickson, 1995). The concentration of the caustic solution used is typically 8%–15%. The oil is normally refined at 150°F.

After refining with caustic and water washing, the fish oil is bleached to remove color, decompose peroxides, and to remove metals and contaminants. This step is particularly important for fish oils to ensure the removal of harmful metals such as mercury and arsenic. The most common bleaching agent used in oil processing is acid activated clay. In the case of fish oil, activated carbon is also used in conjunction with the bleaching clay to help better remove environmental contaminants such as PCBs and dioxins (Oterhals, 2007).

Dosage of bleaching clay is dictated to achieve zero peroxides instead of a predetermined color. Normal conditions for vacuum bleaching of are clay dosage: 0.3%–0.6%; temperature: 100–110°C; vacuum pressure: 28–29 in Hg; processing time: 15–30 min.

The most common bleaching agent used in oil processing is acid-activated bleaching clays, which consist of bentonites or montmorrillonites treated with a hydrochloric or sulfuric acid to create active sites for color compounds. These clays are also the most active for decomposition of peroxides in edible oils. As mentioned above, clay dosages are designed to produce zero peroxides; however, an excess of bleaching clay is likely to increase the free fatty acids in the bleached oil. Neutral clays are hydrated aluminas, better known as fuller's earth, and are also commonly used to remove pigments from soybean oil; however, they are not as effective as acid-activated clays.

As mentioned in Chapter 4, fish oils also have a saturated fraction that crystallizes and forms a haze and a solid layer during storage, especially at cold temperatures. Winterization or chill fractionation is used to remove the more saturated fraction and, therefore, to increase the amount of omega-3 in the clear oil. The process is carried out on the bleached oil by chilling the oil to temperatures below 5°C. The solid fraction generated can then be removed by filtration.

Removal of fishy odor and taste from fish oil is done during deodorization and is usually the last processing step. This is done by bubbling steam through the oil at high vacuum and relatively high temperatures. In this process the steam strips off not just odors and flavors but other impurities, such as free fatty acids, peroxides, and contaminants such as organochlorine pesticides and polycyclic aromatic hydrocarbons.

The oil being fed to a deodorizer has to meet certain quality requirements from the bleaching operation in order to prevent damaging the oil at the higher processing temperatures found in the deodorizer. Typical conditions of deodorization of marine oils are pressure: 1-6 mmHg; temperature: 180–190°C; holding time: 15–120 min. Both batch and continuous deodorizer systems use multistage steam ejectors with barometric intercondensers to create and maintain vacuum. Usually the vacuum system consists of 4 stage steam ejectors. Some systems use 3 stages, but a fourth stage is needed if vacuum below 3 mmHg is desired. It is important that temperatures used not be over 190°C, as isomerization form *cis* to *trans* configuration of the polyunsaturated fatty acids may occur.

Most deodorizers are constructed of stainless steel 304. For physical refining/deodorization, 316 stainless steel is used in order to withstand the corrosive effect of free fatty acids in the oil at the higher temperatures.

After the oil is deodorized, packaging is done under nitrogen to prevent onset of oxidation. Natural and synthetic antioxidants such as tocopherols, TBHQ, and rosemary extracts are commonly used to help prevent deterioration of the oil.

Processing itself can result in the removal of nutrients like fat-soluble vitamins. Refining bleaching and deodorizing of menhaden oil removes more than 80% of vitamin A and vitamin D. Vitamin E levels are reported to decrease by about half as a result of processing. Bleaching the oil with Fuller's earth causes the major loss of reti-nols. Treating the fish oil with steam for several hours causes the major loss of vitamin D (Scolt & Latshaw, 1991).

While refining, bleaching, and deodorizing effectively removes impurities such as free fatty acids, metals, and some contaminants, it also has been reported that due to the removal of antioxidants such as tocopherols during processing, the crude oils can be more stable than their refined-bleached counterparts (Wasundra et al., 1998).

Removal of Contaminants

Because fish oils are currently sold as supplements for the health-oriented consumer, they are closely monitored for quality and harmful contaminants such as heavy met-als, dioxins, furans and PCBs. Dioxins (polychlorinated dibenzo-p-dioxins) and furans (polychlorinated dibenzofurans) are environmental contaminants generated in small amounts during combustion and as by-products in the manufacture of certain chemicals. Dioxins have been shown to be toxic to certain animals, and some of them are known to be carcinogenic in humans. Dioxins also have been implicated in dis-rupting the endocrine (hormone) systems in humans and wildlife. PCBs (Polychlori-nated Biphenyls) are a group of closely related chemicals, and some individual PCBs, namely dioxin-like PCBs, exhibit toxicity similar to those of toxic dioxins. Both diox-ins and PCBs are closely monitored in foods, nutritional supplements, and pharma-ceuticals, and very strict limits have been set by several regulatory organizations like the EPA and FDA. For fish oil the standards adopted by GOED (Global Organiza-tion for EPA and DHA) are based on European standards and are listed in Table 5.A. Dioxin limits are normally expressed in World Health Organization (WHO)

Table 5.A. Quality Parameters for Fish Oil (GOED, 2010).

acid value	3 mg KOH/g max
peroxide value	5 meq/kg max
anisidine value	20 max
TOTOX	26 max
PCDDs and PCDFs	2 ppt max
dioxin-like PCBs	3 ppt max
marker PCBs	90 ppb max
mercury	0.1 ppm max
cadmium	0.1 ppm max
lead	0.1 ppm max
arsenic	0.1 ppm max

toxic equivalents using WHO-toxic equivalent factors (TEFs). Marker PCBs include IUPAC congeners 28, 52, 101, 118, 138, 153, and 180. Dioxin-like PCBs include Non-Ortho PCBs (PCB, 77, 81, 126, 169) and Mono Ortho PCBs (PCB 105, 114, 118, 123, 156, 157, 167, 189).

Molecular or short path distillation is being more commonly utilized in fish oil processing to ensure the removal of contaminants. This step is normally conducted after bleaching and winterization (Breivik & Thorstad, 2005). This technology effectively eliminates contaminants such as PCBs, dioxins, furans, pesticides, and herbicides. Though not as effective as molecular distillation, activated carbon treatment can be used to remove environmental contaminants. It has been reported (Oterhals et al., 2007) that polychlorinated dibenzo-*p*dioxins and dibenzofurans (PCDD/F) showed a very rapid adsorption behavior, and the concentration and toxic equivalent (TEQ) level could be reduced by 99%. On the other hand, adsorption of dioxin-like polychlorinated biphenyls (DL-PCB) was less effective and depended on *ortho* substitution; non-*ortho* PCB were adsorbed more effectively than mono-*ortho* PCB with a maximum of 87 and 21% reduction. Activated carbon treatment had no effect on the level of polybrominated diphenyl ethers, found in flame retardants.

As mentioned before, the bleaching and activated carbon treatment steps efficiently remove toxic heavy metals such as mercury, cadmium, lead, and arsenic, and the deodorization step efficiently removes other environmental contaminants such as herbicides and pesticides.

Concentration of Omega-3 Fatty Acids

Consumption of omega-3 fats in the form of concentrates continues to increase in both supplement market and as ethyl ester concentrates of EPA/DHA in pharmaceutical applications. A blend of ethyl esters of EPA and DHA, in concentrations as high as 84%, is the only source of omega-3 approved as a pharmaceutical for the reduction of blood triglycerides (Breivik et al., 1997b). A concentration of omega-3 fatty acids has the advantage that it allows the manufacturer to offer higher amounts of EPA/DHA in smaller quantities of oil, making it more convenient for the consumer and bringing a better price for the producer and the retailer. Several methods of concentration have been practiced commercially. Molecular distillation, or short path distillation, is so far the most widely utilized.

Short Path Distillation

Short-path distillation is a process where volatile components can be rapidly vaporized at a wide range of temperatures in a very short time due to high vacuum used. Vacuum can be of the order of 10^{-4} to 10^{-6} bar at which volatility of most compounds becomes so high that it can allow operating at lower temperatures. Another feature of short path distillation is the efficient mass transfer rates that take place as the feed forms a thin film on the inner wall of the distillation unit. This is achieved through wiping brushes or rolling elements rotating inside the column while spreading out the film. The vaporized components are condensed immediately on a cold surface in the interior of the distillation unit. Since the diffusion path is only a few centimeters long, thermal deterioration is minimized.

In the case of lipid application, short path distillation has been used to prevent acyl migration when processing diacylglycerols in reaction mixtures (Xu et al., 2002), for the purification of monoacylglycerols (Szelag, 1983), for recovery of carotenoids from palm oil (Batistella & Wolf-Maciel, 1998), recovery of tocopherols (Bruegel et al., 1996), fractionation of squalene (Sun et al., 1997; Pietsch & Jaeger, 2007), and for the reduction of cholesterol in butter and lard (Fedeli & Barreteau, 1994).

Taking advantage of being able to operate at milder temperature conditions, short path distillation is now widely used in several of the processing steps for fish oil. It is now commonly used to concentrate EPA/DHA from marine oils and to remove volatile contaminants in both supplements and pharmaceutical applications.

The basic process to concentrate EPA and DHA consists in first converting the fatty acids in the oil into ethyl esters. The difference in volatility of the shorter chain 16- and 18-carbon fatty acids esters and the longer chain 20- and 22-carbon EPA/DHA ethyl ester allows the short path distillation process to concentrate EPA and DHA ethyl esters to levels of over 50% (Breivik et al., 1997a). After the EPA/DHA ethyl ester concentrates are generated, then they can be consumed directly in supplement and pharmaceutical applications or can be esterified into triglycerides by reacting them with glycerols, using chemical or enzymatic catalysts. Fig. 5.2 shows a general diagram for the concentration of EPA and DHA. Ethyl esters of EPA/DHA have not been approved yet for food applications.

Supercritical Fractionation

Extraction of omega-3s with supercritical carbon dioxide has been reported to offer major advantages over short path distillation in that the process temperature can be kept low, the solvent is easily removed from the final product, and the solvent can also be easily recycled. In this type of extraction it is also necessary to convert the fatty acids in the triglyceride into methyl or ethyl esters.

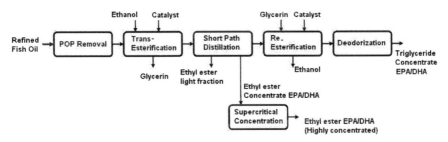

Fig. 5.2. Concentration of EPA/DHA as Ethyl Esters and Triglycerides.

Liang and Yeh (1991) studied the solubilities of ethyl palmitate, ethyl oleate, ethyl eicosapentaenoate (EPA), and ethyl docosahexaenoate (DHA) in supercritical carbon dioxide. Basically, the solubilities of fatty acid ethyl esters increased with pressure and decreased as the temperature was increased. They used an empirical equation, similar to Chrastir's equation, to describe the relationship between solute solubility and the density of carbon dioxide. They were able to qualitatively estimate the separation efficiency of isolating EPA and DHA ethyl esters from fatty acid esters. The operating conditions yielding high solubility gave a fast extraction rate but resulted in low separation efficiency. They demonstrated the use of their model in experiments conducted for the recovery of ethyl EPA and ethyl DHA from a model mixture containing 4 fatty acid ethyl esters and also from esterified squid visceral oil.

Systems for continuous extraction of omega-3 using supercritical CO2 have been demonstrated (Riha & Brunner, 2000; Borch-Jensen & Mollerup, 1997; Staby & Mollerup, 1993).

Urea Crystallization

Concentration of long chain omega-3s using urea complexation has been reported as a more simple technique for obtaining PUFA concentrates, either in the form of ethyl esters or free fatty acids. This is a widely studied technique for separation of saturated and monounsaturated fatty acids from longer chain polyunsaturated fatty acids. The basic process consists of hydrolyzing the triglycerides in the oil into their constituent fatty acids either by lipase activity or by chemical means using alcoholic KOH or NaOH. Then the free fatty acids or ethyl esters are mixed with an ethanol solution of urea for complex formation. The basic premise is that the saturated and mono-unsaturated fatty acids easily complex with urea and crystallize out on cooling; this allows their subsequent separation by simple filtration. The resulting liquid or non-urea complexed fraction is enriched with PUFA. Urea complexation has the advantage that complexed crystals are extremely stable, so the conditions of filtration don't have to be as cold as in the process of winterization (Liu, 2006; Udaya et al., 1999).

Gamez-Meza et al. (2003) surveyed five commercial lipase preparations from *Pseudomonas* sp. in the hydrolysis of sardine oil. They evaluated the efficiency of the combined process of lipase hydrolysis and the preparation of an EPA and DHA concentrate by urea complexation. Results showed that an immobilized lipase preparation (PSCI) produced the highest degree of hydrolysis for EPA and DHA. After complexation of saturated and less unsaturated free fatty acids in urea, the highest concentration of EPA and DHA in the liquid phase was 46.2% and 40.3%, respectively. Liu et al. (2006), using process optimization by response surface methodology (RSM), reported on the use of urea complexation of tuna head oil to concentrate DHA and EPA of oil. Variables such as urea-to-fatty acid ratio, crystallization temperature, and crystallization time were taken into consideration in order to optimize the conditions to obtain a maximum concentration of DHA and EPA. DHA and

EPA content in the concentrate was 85.02% at a urea-to-fatty acid ratio of 15 (mole/mole), a crystallization temperature of 5°C, and a crystallization time of 20 h.

The combined use of molecular distillation and urea complexation has been reported as a more efficient way to obtain EPA/DHA concentrates of ethyl esters from squid oil (Huang & Liang, 2001). Ethyl esters were treated with a urea solution first, and the ethyl esters concentrated of EPA/DHA were further fractionated by molecular distillation. Urea complexation has also been used in combination with supercritical extraction to concentrate ethyl esters of EPA and DHA from tuna cooking juice, reporting 80% yield of EPA and DHA; an increase of DHA ethyl esters from 9.23% to 30.6% after urea treatment and then to 46.20% after molecular distillation was reported (Hsieh et al., 2005).

Chromatography-Molecular Sieves

Chromatographic techniques have emerged recently as a viable method for the recovery of EPA and DHA. Silver resin chromatography has been used for preparation of omega-3 concentrates of EPA and DHA with different degrees of success. Early work on the use of an XN1010 resin column saturated with silver ions to concentrate EPA and DHA was reported by Adolf and Emken (1985). They reported fractionation of polyunsaturated fatty acid esters from fish oil concentrate based on the number of double bonds by using solvent programming (acetonitrile in methanol). Larger samples (4-9 g) of fish oil concentrate esters, menhaden oil fatty acids and esters were concentrated to up to 82% levels of omega-3 using a 100% silver resin column and isocratic elution with 30, 35, or 45% acetonitrile in acetone. Belarbi et al. (2000) reported on a process for large scale recovery of highly pure (95%) EPA esters from microalgal biomass and fish oil using argentated silica gel column chromatography. They describe the process in three main steps: 1) simultaneous extraction and transesterification of the algal biomass, 2) argentated silica gel column chromatography of the crude extract, and 3) removal of pigments by a second column chromatographic step. The process was scaled up by a factor of nearly 320 by increasing the diameter of the chromatography columns. Compared to the green alga *Monodus subterraneus,* the diatom *Phaeodactylum tricornutum* had important advantages as a potential commercial producer of EPA. The silver contamination in the final purified EPA was negligibly small (<210 ppb).

The use of molecular sieves combined with supercritical extraction has been reported to be a more efficient way to concentrate omega-3s (Hao et al., 2008; Cao & Hur, 2005). Cao and Hur (2005) used a modified molecular sieve 13X to adsorb methyl ester of EPA and DHA in hexane organic solvent and then desorbed them in SC-CO2 fluid. Hao et al. (2008) studied coupling molecular sieve 13X with supercritical fluid extraction in one step in order to obtain pure ethyl esters of EPA and DHA. They reported concentrations of 85%–96% EPA and 95%–98% DHA. This process has advantages such as enhancing the solubility and selectivity for PUFAs by adjusting the operation pressure and temperature of supercritical fluids. It also avoids the use of organic solvents and simplifies the conventional multi-step processes.

Stabilization for Omega-3 Oils

The tendency of polyunsaturated omega-3 oils to oxidize has given rise to a wide variety of new antioxidants and processes aimed at the protection of the PUFAs. The main objectives for the protection of omega-3s are to prevent development of rancidity and off flavors and also to preserve the bioactivity of the essential fatty acids.

It is widely recognized that when the more saturated fatty acids, such as omega-6, and monounsaturated fatty acids, such as oleic acid, oxidize slightly they produce more appealing flavors and odors than the products of oxidation from omega-3 fats. As mentioned before, the susceptibility of polyunsaturated fats to oxidized and deteriorate is greatly exacerbated by the presence of metals, especially copper and iron (Kamal-Eldin & Yanishlieva, 2002). Iron from fish muscle and other food ingredients has been identified as being the most responsible for the generation of fishy notes in omega-3 oils in both short chain PUFA in linolenic acids (C18:3) from vegetable oils and in longer chain PUFA from fish oils. While omega-3 oils generate fishy notes, other fatty acids such as oleic acid (18:1) have been identified with cooked beef fat flavor. Saturated and monounsaturated fatty acids have been positively correlated with pork taste. In one study where pigs were fed a diet high in linolenic acid (18:3) it was found that the fat from bacon was considered to have fishy flavor (Romans et al., 1995). The specific compounds identified as having a fishy off flavor after oxidation of fish oils were 2-trans, 4-cis, and 7-cis-decatrienal (Venkateshwarlu et al., 2004).

Research done on milk and mayonnaise fortified with fish oil identified over 60 different volatiles comprising alkenals, alkadienals, alkatrienals, and vinyl ketones as having strong fishy odor. The most potent odors found were from 1-penten-3-one, (Z)-4-heptenal, 1-octen-3-one, 1,5-octadien-3-one, (E,E)-2,4-heptadienal, and (E,Z)-2,6-nonadienal with 1-Penten-3-one reported as the major contributor to the unpleasant sharp fishy off-flavor in fish oil. It is generally recognized that the off flavors identified as fishy, metallic, and rancid off-flavors are considered major deterrents for the increased consumption of fish and fish oils (Donelly et al., 1998; Jacobsen et al., 1999). Fish oil has been added directly into some foods with various degrees of success (Warnants, 1999), as long as the oil was properly processed and protected with either the use of antioxidants or encapsulation.

Antioxidants

The most common way to protect omega-3 fats against oxidation is with the use of antioxidants. So in order to determine what types of antioxidants to use it is important to understand the main mechanisms that can trigger oxidation. Omega-3 oils typically oxidize by the free radical chain mechanism. This process of oxidation includes 4 distinct phases: initiation, propagation, chain branching, and termination. The reaction of oxidation is basically initiated by the abstraction of a hydrogen atom from a lipid molecule to form alkoxyl radicals. This triggers a new chain reaction where peroxyl radicals are formed. This reaction in turn then leads to hydro peroxides

(ROOH). These peroxides can themselves form several free radicals R•, RO•, and ROO• by branching. These reaction moieties can themselves then trigger another chain reaction, thus propagating the reaction of oxidation further. The rate of hydrogen abstraction generated in the propagation phase by RO• is much higher than by ROO• (Kamal-Eldin & Yanishlieva, 2002; Frankel, 1998).

It has also been reported that fatty acid molecules having methylene-interrupted double bonds, like linoleic, linolenic, arachidonic (AA), EPA, and DHA, *bis*-allylic methylene groups are the most prone to react. The tendency of these acids to oxidize was found to increase approximately twofold for each *bis*-allylic methylene group. The relative rates of autoxidation of stearate, oleate, linoleate, and linolenate have been reported as 1, 11, 114, and 179, respectively (Donnelly et al., 1998; Frankel, 1998, Frankel et al., 1996).

The main mechanisms by which the autoxidation reaction of PUFAs can be prevented may involve aromatic phenols that can break the oxidation chain reaction from the participation of aromatic compounds. The chain-reaction-breaking antioxidants act by competing with the substrate for the chain-carrying radical species normally present in the highest concentration. Other types of antioxidants that prevent oxidation include hydroperoxide decomposers, metal-chelating agents, and singlet oxygen quenchers (Yanishlieva-Maslarova, 2001).

The antioxidants commonly used to stabilize omega-3 oils can be natural and synthetic. Common synthetic antioxidants that are widely used in foods include butylated hydroxyanisole (BHA), butylated hydroxytoluene (BHT), propyl gallate (PG), and tertbutylhydroquinone (TBHQ). However, there are some concerns regarding their safety at high levels; therefore, their use in foods is limited, typically to less than 200 pmm. Until recently natural antioxidants had limited application in omega-3 oils. However, with the development of new plant and herb extracts, the use of natural antioxidants in vegetable oils, in general, and omega-3 fats, in particular, is now more common. Extracts from spices like rosemary and oregano have been found to be as effective antioxidants as some synthetic ones. Rosemary extract has been found to be particularly effective in combination with other antioxidants such as tocopherols (Xin & Shun, 1996; Lee & Shibamoto, 2002; Shahidi & Zhong, 2010). Antioxidant activities of volatile extracts isolated from thyme, basil, rosemary, chamomile, lavender, and cinnamon were evaluated by Lee and Shibamoto (2002). The antioxidant activity was evaluated by two independent assays: the aldehyde/carboxylic acid assay and the conjugated diene assay. It was reported that thyme extract was similar in antioxidant activity to BHT and alpha α-tocopherol in the conjugated diene assay.

The active compounds in rosemary extract are reported to be phenols such as carnosol, rosmanol, rosamaridiphenol, and phenolic acids such as carnosic acid and rosmarinic acid. This extract has been used effectively to prevent oxidation of fish oils like sardine. Rosemary compounds also have been reported to inhibit the formation of products of oxidation like conjugated dienes and pentenal, but not propanal in fish oil emulsions (Frankel et al., 1996). Kahkonen et al. (1999) reported on extensive study to examine the antioxidative activity of 92 phenolic extracts from edible and nonedible plant materials (berries, fruits, vegetables, herbs, cereals, tree materials, plant sprouts, and seeds), testing effect on autoxidation of methyl linoleate. The content of total phenolics in the extracts was determined spectrometrically according to the Folin-Ciocalteu procedure and calculated as gallic acid equivalents (GAE). Among edible plant materials, remarkable high antioxidant activity and high total phenolic content were reported in berries, especially aronia and crowberry.

Antioxidant systems work more efficiently when used in combination with other antioxidants and also other chelating agents such as ascorbyl palmitate and EDTA. A strong synergistic effect was reported between α-tocopherol (0.02%) and rosemary extract (0.02%) in sardine oil 9 (Wada & Fang, 1992). Dry oregano, at 0.5% in mackerel oil stored at 40°C in the dark, had an antioxidant effect that was comparable to the effects of 200 ppm BHT or 0.5% dry rosemary (Tsimidou, 1995). When oregano was tested at 1%, the activity was comparable to 200 ppm TBHQ. A mixture of α-tocopherol and rosemary extract (0.035% + 0.035%) was useful in inhibiting fish lipid oxidation catalyzed by Fe^{2+} or haemprotein.

Extracts of sesame seed have also been reported to be effective antioxidants. Extract from sesame seeds/cakes obtained by solvent extraction are reported to be effective antioxidants in common vegetable oils/emulsions/lipid systems in foods, cosmetics, and pharmaceuticals (Jayalekshmy et al., 2008). The antioxidant extract/concentrate reported is basically a mixture of compounds like sesamol, sesamin, episesamin, sesamolin, related derivatives, tocopherols, lygophenols/ferulic acid, denatured proteins, sugars, lipids, minerals, and browning products that result from the Maillard reaction. Also, extracts of marine red alga have been reported to have strong antioxidant activity when tested in linoleic acid and fish oil. This algal extract was shown to have comparable or better efficacy than BHA and BHT when tested at 0.01, 0.03, and 0.05% in fish oil. They show positive results in inhibiting oxidation of linoleic acid and fish oil at 0.05%. However, this extract had a poor DPPH radical scavenging ability. The extracts, therefore, may not be able to quench radicals from the medium, but may delay lipid peroxidation by suppressing radical initiation by metal chelation (Athukorala et al., 2003).

Depending on the method of application, some extracts can have contradictory effects. Green tea extracts have been reported to have both antioxidant and prooxidant activity. They were tested in seal blubber and in menhaden oils, and they were reported to have prooxidant activity possibly due to the catalytic effect of chlorophyll. One report also shows that after removal of chlorophyll by a chromatographic technique, the resultant extract showed an excellent antioxidant activity in both oils (Wanasundara & Shahidi, 1998). It was also reported that when individual green tea catechins were separated and applied at levels of 200 ppm it prevented the oxidation of seal blubber and menhaden oil under the *Schaal* oven test conditions at 65°C for 144 h (Ho et al., 1997).

Proteins, protein hydrolysates, and peptides and amino acids have been shown to provide significant antioxidant activities. Specific peptides with antioxidant activity have been identified (Chen et al., 1996). Other studies on whey protein reported on the relationship between the antioxidant properties of the whey peptides and the presence of aromatic amino acids (Pena-Ramos et al., 2004). The radical scavenging activity of peptides is likely due to the hydrogen atom donor activity of the phenolic and indolic groups and higher stability of the phenoxyl and indolyl radicals compared to that of simple peroxyl radical (Philanto, 2006).

Polysaccharides from plant, bacterial, and fungal sources have also been shown to be a potential source of antioxidants. The antioxidant activities of polysaccharides in these studies were attributed to their radical scavenging, reducing power, and metal ion chelation, as well induction of gene expression of antioxidant enzymes such as superoxide dismutase and glutathione peroxidise. It has also been suggested that degree of substitution (DS) has an effect on antioxidant activity of polysaccharides, possibly through interruption of inter- and intramolecular hydrogen bonding by the bulk moieties introduced. (Pasanphan et al., 2010; Shahidi & Zhong, 2010; Chen et al., 2005).

As mentioned, antioxidants are reported to be more effective when used in combination with chelating agents such as ascorbyl palmitate, which is probably the most widely used as a fat-soluble chelating agent. Other chelating agents, such as EDTA, have also been reported to be effective when used in combination with other antioxidants (Jacobsen et al., 1999). In general, the most effective antioxidant agents reported involve ternary systems and include free radical scavengers, chelating compounds, and synergistic agents like phospholipids. A mixture of natural antioxidants such as tocopherols, rosemary extract, ascorbic acid, and phosphatidyl choline was reported to be an effective antioxidant system in several types of fish oils and polyunsaturated vegetable oils (Chang et al., 1991). See Table 5.B.

Table 5.B. Main Antioxidants Used In Food and Supplements.

Antioxidants	Source
BHA (butylated hydroxyanisole), BHT (butylated hydroxytoluene), PG (propyl gallate), TBHQ (tertbutylhydroquinone)	synthetic
tocopherols, alpha, beeta, gamma, delta	seeds, cereal and legume grains, nuts, vegetable oils
tocotrienols alpha, beeta, gamma, delta	palm oil, rice bran oil
polyphenolics ferulic acid, quercetin, catechin, resveratrol, cyanidin	rosemary extracts fruits, vegetables, nuts, cereals
carotenoids b-carotene, lycopene, astaxanthin, fucoxanthin	carrots, tomato, fish/shellfish, marine algae
chelating agents: ascorbyl palmitate, EDTA	synthetic

Omega-3 Oils Blends

Polyunsaturated oils are blended with other oils and other food ingredients to increase their stability, facilitate delivery, modify functionality, and increase the nutritional value of foods and supplements. Simple blending of omega-3 oils with more stable oils like high oleic sunflower, palm, and partially hydrogenated oils, has been reported to appreciably increase the overall stability of polyunsaturated fats (Frankel & Huang, 1994; Neff et al., 1994; Chapman et al., 1998; Sundram et al., 1999). This blending can be a direct mix or blended chemical using interesterification techniques. The basic premise is that the more saturated fatty acids impart a stabilizing effect, as well as the antioxidants naturally present in the oil (i.e., tocopherols and tocotrienols). A mathematical model describing the relationship between oxidative stability, measured as oxidative stability index (OSI), and fatty acid composition of 21 oil blends has been reported (Chu & Kung, 1998). The order of positive influence of fatty acids on oxidative stability was palmitic acid followed by stearic acid and oleic acid. With regards to unsaturated fatty acids, linolenic acid had the most negative influence on OSI, followed by linoleic acid and oleic acid.

More recently, biotechnology has had an appreciable impact in the blending and transformation of several vegetable oils and animal fats, including omega-3 oils. These new technologies have given rise to a number of new biotech-based lipid products, such as non *trans* products and specialized delivery of bioactive lipids (Schörken & Kempers, 2009; Hernandez, 2008).

Immobilized lipases have now become accepted as a mainstream technology for fat modification and are widely used in commercial applications (Holm & Cowan, 2008). With regards to omega-3 fats, interesterification reactions lipases are used to chemically blend them with several types of oils and fats to improve nutritional and rheological properties of the final product (Osorio et al., 2009; Hernández-Martin & Otero, 2008; Halldorsson et al., 2004; Zu-yi & Ward, 1993).

Emulsions

Emulsions with edible oils are generally manufactured as delivery system, are an intrinsic part of particular foods, or are themselves the final food product such as margarines, salad dressings, and beverages. In the case of omega-3 oils, emulsions can serve as a delivery system and also can provide protection against oxidation. An emulsion consists basically of two immiscible liquids (usually oil and water), with one being dispersed in the other in the form of small spherical droplets. The size of the dispersed droplets typically will vary between 0.1 and 100 microns (McClements & Decker, 2000). Emulsions of omega-3s are usually in a system where the oil is in the form of oil droplets dispersed in an aqueous phase. These omega-3 emulsions with fish oil have been incorporated into different emulsified food products such as mayonnaise, yogurt, milk, and spreads with various levels of success (Let et al., 2004; Kolanowski et al., 2002).

Oxidative stability of fish oil in emulsion has been extensively studied, and a series of antioxidant systems active in both the continuous and dispersed phases have been developed. Most reports show that effectiveness of the antioxidants will depend on the partitioning of antioxidant molecules into the different emulsion phases (Kolanowski et al., 2007; Frankel et al. 2002; Coupland & McClements, 1997). The principle for the antioxidant effect is that non-polar antioxidants such as tocopherols are particularly effective in oil-in-water emulsion systems and that metal chelating agents and secondary antioxidants such as EDTA and citric acid provide synergistic effects by preventing dissolved metals from triggering any further oxidation. Blending omega-3 oils with less saturated and more stable oil in the emulsion is also an effective means to provide protection of the fish oil against oxidation. Reported results showed that addition of rapeseed oil to fish oil (1:1) prior to emulsification into milk significantly protected the emulsions against oxidative deterioration. Addition of propyl gallate and a citric acid ester to the fish oil prior to emulsification also protected the fish-oil-enriched milk during storage. It was suggested that the tocopherols in rapeseed oil may be the protective factor (Let at al., 2004).

Microencapsulation

Microencapsulation technology has been traditionally used in the food industry for flavor encapsulation where flavors are stabilized and their release controlled (Madene et al., 2006). Encapsulation is now widely used in the food and supplements industries to protect and enable the incorporation of bioactive component in fortified foods and nutritional supplements. This technology is used to ensure that the taste, aroma, or texture of food is not adversely affected (Kolanowski et al., 2007; Jónsdóttir et al., 2005; Pszczola, 1998; Brazel, 1999). Also, micro encapsulation is used to mask off-flavors contributed by certain vitamins and minerals, permit time-release of the nutrients, enhance stability at extremes storage conditions, and minimize undesirable chemical interactions with other ingredients (Augustin & Sanguansri, 2008)

Encapsulation has become a preferred method to protect fish oils against oxidation, and it is also used as an efficient delivery system for several applications in the food and nutritional supplements industries. The most basic encapsulation system consists of a dried product that was generated from an aqueous dispersion of an oil-in-water emulsion, where the water phase is removed by evaporation and the dispersed oil phase is converted into a dry powder. The water is normally removed by spraying the emulsion into a hot dry air stream, resulting in moisture evaporation. Spray drying is commonly conducted in the following stages: atomization, air contact, evaporation, and product recovery. The resulting powder then goes through a cooling phase or can go through other subsequent processing steps such as fluidized bed plating (Beindorff & Zuidam,2010).

Microencapsulation has been used successfully in many food and supplement applications, such as cereals, nutrition bars, and beverages. Milk is a particularly difficult product to fortify because of the tendency of off-flavors to readily appear. This makes it necessary to protect the fish oil through microenecapsulation to prevent deterioration and shorter shelf life (Keogh et al., 2001). In the case of fish oil, it is recommended that when encapsulating omega-3 oils relatively low spray-drying temperatures should be used in order to minimize the lipid oxidation (Baik et al., 2004; Drusch et al., 2007). The use of nitrogen instead of air to dry the dispersion can also be used, but it increases manufacturing costs appreciably. Examples of carriers used in microencapsulated omega-3 oils include maltodextrin, glucose syrup, proteins, sugars, gums, pectin, modified cellulose (e.g., hydroxypropyl methylcellulose or methylcellulose), and modified starch (Keogh et al., 2001; Kagami et al., 2003; Jónsdóttir et al., 2005; Drusch et al., 2006a; Kolanowski et al., 2004; 2006; 2007).

Silica powders and tricalciumphosphate can be added at the end of the microencapsulation process to facilitate the flowing properties of the powders (Drusch et al., 2006b).

Nano Particles

Emulsions are inherently unstable systems that require emulsifiers and homogenization forces to be applied to disperse an immiscible phase into another fluid. This instability is due to several factors: the fat content, the type and ratio of the emulsifier, the viscosity of the continuous phase, the volume fraction of the dispersed phase, and the droplet size distribution and temperature. Liposomes, because of their smaller particle size and other physicochemical factors, are more thermodynamically stable systems and are considered more stable carriers of bioactive compounds (Muller & Keck, 2004).

The ability of phospholipids to form liposomes has been utilized to manufacture thermodynamically stable dispersions of omega-3 fats. These liposomes which can enclose an aqueous phase can better preserve the oxidative stability of encapsulated molecules, including several classes of lipids and fatty acid ethyl esters (Aseki et al., 2002; Imai et al., 2008).

Liposomes are essentially closed bilayer membranes in the form of vesicles or sacs containing an entrapped aqueous phase. Liposomes having a single bilayer membrane are referred to as unilamellar vesicles; lipsomes having a number of concentric lipid bilayers separated by an aqueous phase are referred to as multilamellar vesicles. A stable liposome is typically prepared by dissolving a lipophilic material in a phospholipid followed by the addition of water or an aqueous solution and mixing by an emulsification method such as sonicating to produce a liposome having the lipophilic material encapsulated in the lipid bilayer. In the case of fish oil, the resulting liposomes have been reported to provide effective antioxidant protection (Haynes et al., 1991).

Nanoemulsions made from salmon oil and marine lecithin by high pressure homogenization showed that crude salmon oil was well protected by its own natural antioxidants (tocopherols and astaxanthin). It was also reported that marine lecithin generated the most stable liposomes. The use of marine phospholipids as emulsifiers in nanoemulsions preparation increases notably the stability of salmon oil against oxidation with droplet sizes ranging between 160 and 200 nm 160 nm (Belhaj et al., 2010). Nanoemulsions are also considered an efficient form of delivery of bioactive compounds in general and of omega-3 fatty acids in particular for applications in food and pharmaceutical industries.

The latest generation of lipid dispersions at nano scale levels are nanoencapsulation systems of solid lipid micro and nanoparticles (SLN) (Weiss et al., 2008; Cortesi, 2002). These systems show the advantages of liquid nanoemulsions (or microemulsions) of rapid dispersion and also of high permeability of the bioactive compounds through the walls of the digestive system. This has the added benefit of stabilizing the encapsulated compound for ease of handling and delivery. Solid lipid nano particles consist of a core of solid lipid with the bioactives being a part of the lipid matrix (Schubert & Muller-Goymann, 2005). The particle is stabilized by a surfactant layer, which may consist of a single surfactant or combination of emulsifying agents with different HLB (hydrophilic lipophilic balance) values. This allows for the manufacture of particles with distinctly different properties, such as targeted applications and time release effect. Particles from crystallized lipids instead of liquid lipids have been shown to have better release control and provide more stability of incorporated bioactives, especially in the case of omega-3s. This is explained by the fact that the mobility of bioactives can be manipulated by controlling the physical state of the lipid matrix. These properties have been used to develop new products such of drug carrier systems that can be water soluble, lipophilic, and target-specific (Muller & Keck, 2004; Sivaramakrishnana et al., 2004; Gupta et al., 2006).

References

Adolf, R.O.; Emken, E.A. The isolation of omega-3 polyunsaturated fatty acids and methyl esters of fish oils by silver resin chromatography. JAOCS 1985, 62, 1592–1595.

Aseki, M.; Yamamoto, K.; Miyashita, K. Oxidative stability of polyunsaturated fatty acid in phosphatidylcholine liposomes. Biosci. Biotech. Biochem. 2002,12, 25732577.

Athukorala, Y.; Lee, K.; Shahidi, F.; Heu, M.S.; Kim, H.T.; Lee, J.S.; Jeon, Y.J. Antioxidant efficacy of extracts of an edible res algae (Crateloupoia filicina) in linoleic acid and fish oil. J.Food Lipids 2003, 10, 313–327.

Augustin, M.A.; Sanguansri, L. Encapsulation of Bioactives. Food Material Science. SCIENCE Principles and Practic. J.M. Aguilera, Peter J. Lillford, Eds.; Springer: New York, 2008,577–601.

Baik, M.Y.; Suhendro, E.L.; Nawar, W.W.; McClements, D.J.; Decker, E.A.; Chinachoti, P. Effects of antioxidants and humidity on the oxidative stability of microencapsulated fish oil. J. Am. Oil Chem Soc. 2004, 81(4), 355–360.

Batistella, C.B.; Wolf-Maciel, M.R. Recovery of carotenoids from palm oil by molecular distillation. Computers Chem. Eng. 1998, 22, S53–S60.

Beindorff, C.; Zuidam, N.J. Microencapsulation of fish oil. Encapsulation technologies for active food ingredients and food processing, N.J. Zuidam, V.A. Nedović, Eds.; Springer: New York, 2010.

Belarbi, E.H.; Molina, E.; Chisti, Y. A process for high yield and scaleable recovery of high purity eicosapentaenoic acid esters from microalgae and fish oil. Enzyme Microb. Technol. 2000, 26, 516–529.

Belhaj, N.; Arab-Tehrany, E.; Linder, M. Oxidative kinetics of salmon oil in bulk and in nano-emulsion stabilized by marine lecithin. Process Biochem. 2010, 45, 87–195.

Borch-Jensen, C.; Mollerup. J. Phase equilibria of fish oil in sub- and supercritical carbon dioxide. Fluid Phase Equil. 1997, 138, 179–211.

Brazel, C.S. Microencapsulation: offering solutions for the food industry. Cereal Foods World 1999, 44, 388–393.

Breivik, H.; Haraldsson, G.G.; Kristinsson, B. Preparation of highly purified concentrates of eicosapentaenoic acid and docosahexaenoic acid. J. Am. Oil Chem. Soc. 1997a, 11, 1425–1429.

Breivik, H.; Boslashed, B.; Dahl,K.H.; Helk, K.; Krokan, H.E.; Kaare, H. Treatment and prevention of risk factors for cardiovascular diseases, 1997b, US Patent 5,698,594.

Breivik, H.; Thorstad, O. Removal of organic environmental pollutants from fish oil by short-path distillation. Lipid Technol. 2005, 17, 55–58.

Bruegel, B.; Johannisbauer, W.; Nitsche, M.; Schwarzer, J. 1996, German Patent 19652522.

Cao, X.J.; Hur, B.K. Separation of EPA and DHA from fish oil using modified zeolite 13X and Supercritical CO2. J. Ind. Eng. Chem. 2005, 11(5), 762–768.

Chang, S.S.; Bao, Y.; Pelura, T.J. Fish oil Antioxidants. 1991, US Patent 5,023,100.

Chapman, K.W.; Sagi, I.; Regenstein, J.M.; Bimbo, T.; Crowther, J.B.; Stauffer, C.E. Oxidative stability of hydrogenated menhaden oil shortening blends in cookies, crackers, and snacks. JAOCS 1998, 73, 167–172.

Chen, H.M.; Muramoto, K.; Yamauchi, F.; Nokihara, K. Antioxidant activity of designed peptides based on the antioxidative peptide isolated from digests of a soybean protein. J. Agric. Food Chem. 1996, 44, 2619–2623.

Chen, H.; Zhang, M.; Xie, B. Components and antioxidant activity of polysaccharide conjugate from green tea. Food Chem. 2005, 90, 17–21.

Chu, Y.H.; Kung, Y.L. A study on vegetable oil blends. Food Chem. 1998, 62, 191–195.

Cortesi, R.; Esposito, E.; Luca, G.; Nastruzzi, C. Production of lipospheres as carriers for bioactive compounds. Biomaterials 2002, 23, 2283-2294

Coupland, J.N.; McClements, D.J. Lipid oxidation in food emulsions. Trends Food Sci. Technol., 1996, 7, 83–91.

Donnelly, J.L.; Decker, E.A.; McClements, D.J. Iron-catalyzed oxidation of Menhaden oil as affected by emulsifiers. J. Food Sci. 1998, 63, 997–1000.

Drusch, S.; Serfert, Y.; Schwarz, K. Microencapsulation of fish oil with n-octenylsuccinate derivatised starch: flow properties and oxidative stability. Eur. J. Lipid Sci. Technol. 2006a, 108, 501–512.

Drusch, S.; Serfert, Y.; Van den Heuvel, A.; Schwarz, K. Physicochemical characterization and oxidative stability of fish oil encapsulated in an amorphous matrix containing trehalose. Food Res. Intern. 2006b, 39, 807–815.

Drusch, S.; Serfert, Y.; Scampicchio, M.; Schmidt-Hansberg, B.; Schwarz, K. Impact of physico-chemical characteristics on the oxidative stability of fish oil microencapsulated by spray-drying. J. Agric. Food Chem. 2007, 55, 11044–11051.

Erdweg, K.J. Molecular and short path distillation. Chem. Ind. 1983, 5, 342–345.

Erickson, D.R. Practical Handbook of Soybean and Utilization. AOCS: Champaign, IL, 1995, 184–202.

Frankel, E.N. Lipid Oxidation. Oily Press: Dundee, Scotland, UK, 1998.

Frankel, E.N.; Huang, S.W. Improving the oxidative stability of polyunsaturated vegetables oils by blending with high-oleic sunflower oil. J. Am. Oil Chem. 1994, 71, 255–259.

Frankel, E.N.; Huang, S.W.; Prior, E.; Aeschbach, R. Evaluation of antioxidant activity of rosemary extracts, carnosol and carnosic acid in bulk vegetable oils and fish oil and their emulsions. J. Agric. Food Chem. 1996, 72, 201–208.

Frankel, E.N.; Satue-Gracia, T.; Meyer, A.S.; German, B. Oxidative stability of fish and algae oils containing long-chain polyunsaturated fatty acids in bulk and in oil-in-water emulsions. J. Agric. Food Chem. 2002, 50, 2094–2099.

Gamez-Meza, N.; Noriega-Rodrıgueza, J.A.; Medina-Juareza, L.A; Ortega-Garcia, J.; Monroy-Rivera, J.; Toro-Vazquez, F.J.; Garcıa, H.S.; Angulo-Guerrero, O. Concentration of eicosapentaenoic acid and docosahexaenoic acid from fish oil by hydrolysis and urea complexation. Food Res. Intern. 2003, 36, 721–727.

Gupta, S.; Moulik, P.; Hazra, B.; Ghosh, R.; Sanyal, S.K.; Datta, S. New pharmaceutical microemulsion system for encapsulation and delivery of diospyrin, a plant-derived bioactive quinonoid compound drug deliver. 2006, 13, 193–199.

Halldorsson, A.; Kristinsson, B.; Haraldsson, G.G. Lipase selectivity toward fatty acids commonly found in fish oil. Eur. J. Lipid Sci. Technol. 2004, 106, 79–87.

Hao, L.P.; Cao, X.J.; Hur, B.K. Separation of single component of EPA and DHA from fish oil using silver ion modified molecular sieve 13X under supercritical condition. J.Ind.Eng. Chem. 2008, 14, 639–643.

Haynes, L.C.; Levine, H.; Finley, J.W. Liposome composition for the stabilization of oxidizable substances. US Patent 5,015,483.

Hernandez, E. Structured lipids as delivery systems. delivery and controlled release of bioactives in foods and nutraceuticals, N. Garti, Ed.; CRC Press: Boca Raton, FL, 2008.

Hernandez, E.; Quezada, N. Uses of Phospholipids as Functional Ingredients. Phospholipid Technology and Applications, F.D. Gunstone, Ed.; Oily Press: Bridgewater, England, 2008..

Hernandez, E. Production, Processing and Refining of Oils. Healthful Lipids, C. Akoh, O.M. Lai, Eds.; AOCS: Champaign, IL, 2005, 48–64.

Hernandez-Martín, E.; Otero, C. Selective enzymatic synthesis of lower acylglycerols rich in polyunsaturated fatty acids. Eur. J. Lipid Sci. Technol. 2008, 110, 325–333.

Ho, C.T.; Chen, C.W.; Wanasundara, U.N.; Shahidi, F. Natural antioxidants from tea. Natural Antioxidants. Chemistry, Health Effects, and Applications. F. Shahidi, Ed.; AOCS: Champaign, IL, 1997, 213-223.

Holm, H.C.; Cowan, D. The evolution of enzymatic interesterification in the oils and fats industry. Eur. J. Lipid Sci. Technol. 2008, 110, 679–691.

Hsieh, C.W.; Chang, C.J.; KO, W.C. Supercritical CO2 extraction and concentration of n-3 polyunsaturated fatty acid ethyl esters from tuna cooking juice. Fish. Sci. 2005, 71, 441–447.

Hwang, L.S.; Liang, J.H. Fractionation of urea-pretreated squid visceral oil ethyl esters. JAOCS 2001, 78(5), 473–476.

Jacobsen, C.; Schwarz, K.; Stöckmann, H.; Meyer, A.S.; Adler-Nissen, J. Partitioning of selected antioxidants in mayonnaise. J. Agric. Food Chem. 1999, 47, 3601–3610.

Jayalekshmy, A.; Arumughan, C.; Suja, K.P. Antioxidant sesame extract. US patent 7,396,554.

Jónsdóttir, R.; Bragadóttir, M.; Arnarson, G.O. Oxidatively derived volatile compounds in micro-encapsulated fish oil monitored by solid-phase micro-extraction. J. Food Sci. 2005, 70(7), 433–440.

Kagami, Y.; Sugimura, S.; Fujishima, N.; Matsuda, K.; Kometani, T.; Matsumura, Y. Oxidative stability, structure, and physical characteristics of microcapsules formed by spray drying of fish oil with protein and dextrin wall materials. J. Food Sci. 2003, 68(7), 2248–2255.

Kahkonen, M.P.; Hopia, A.I.; Vuorela, H.J.; Rauha, J.P.; Pihlaja, K.; Kujala, T.S.; Heinonen, M. Antioxidant activity of plant extracts containing phenolic compounds. J. Agric. Food Chem. 1999, 47, 3954–3962.

Kamal-Eldin, A.; Yanishlieva, N.V. N-3 fatty acids for human nutrition: stability considerations. Eur. J. Lipid Sci. Technol., 2002, 104, 825–836.

Keogh, M.K.; O'Kennedy, B.T.; Kelly, J.; Auty, M.A.; Kelly, P.M.; Fureby, A.; Haahr, A.M. Stability of oxidation of spray-dried fish oil powder microencapsulated using milk ingredients. J. Food Sci. 2001, 66(2), 217–224.

Kolanowski, W.; Laufenberg, G.; Kunz, B. Fish oil stabilisation by microencapsulation with modified cellulose. Int. J. Food Sci. Nutr. 2004, 55(4), 333–343.

Kolanowski, W.; Ziolkowski, M.; Weissbrodt, J.; Kunz, B.; Laufenberg, G. Micro encapsulation of fish oil by spray drying-impact on oxidative stability. Part I. Eur. Food Res. Technol. 2006, 222, 336–342.

Kolanowski, W.; Jaworska, D.; Weissbrodt, J.; Kunz, B. Sensory assessment of microencapsulated fish oil powder. J. Am. Oil Chem. Soc. 2007, 84(1), 37–45.

Kolanowski, W.; Swiderski, F.; Berger, S. Possibilities of fish oil application for food products enrichment with omega-3 PUFA. Int. J. Food Sci. Nutr. 2002, 50(1), 39–49.

Lanzani, A.; Bondioli, P.; Mariani, C.; Folegatti, L.; Venturinis, S.; Fedeli, E.; Barreteau, P. A new short path distillation system applied to the reduction of cholesterol in butter and lard. J. Am. Oil Chem. Soc. 1994, 71, 609–614.

Lee, K.G.; Shibamoto, T. Determination of antioxidant potential of volatile extracts isolated from various herbs and spices. J. Agric. Food Chem. 2002, 50, 4947–4952.

Let, M.; Jacobsen, C.; Meyer, A.S. Effects of fish oil type, lipid antioxidants and presence of rapeseed oil on oxidative flavour stability of fish oil enriched milk. Eur. J. Lipid Sci. Technol. 2004, 106, 170-182.

Liang, J.H.; Yeh, A. Process conditions for separating fatty acid esters by supercritical C02. J. Am. Oil Chem. Soc. 1991, 68, 687-692.

Liu, S.; Zhang, C.; Hong, P.; Ji, H. Concentration of docosahexaenoic acid (DHA) and eicosapentaenoic acid (EPA) of tuna oil by urea complexation: optimization of process parameters. J. Food Eng. 2006, 73, 203–209.

Madene, A.; Jacquot, M.; Scher, J.; Desobry, S. Flavour encapsulation and controlled release: a review. Int. J. Food Sci. Technol. 2006, 41, 1–21.

McClements, D.J.; Decker, E.A. Lipid oxidation in oil-in-water emulsions: impact of molecular environment on chemical reactions in heterogeneous food systems. J. Food Sci. 2000, 65, 1270–1282.

Mozaffarian, D.; Katan, M.B.; Ascherio, A.; Stampfer, M.J.; Willett, W.C. Trans fatty acids and cardiovascular disease. New Engl. J. Med. 2006, 354(15)m, 1601–1613.

Muller, R.H.; Keck, C.M. Challenges and solutions for the delivery of biotech drugs—a review of drug nanocrystal technology and lipid nanoparticles. J. Biotechnol. 2004, 113, 151-170.

Neff, W.E.; El-Agaimy, M.A.; Mounts, T.L. Oxidative stability of blends and interesterified blends of soybean oil and palm olein. J. Am. Oil Chem. 1994, 71, 1111–1116.

Osório, N.; Maeiro, I.; Luna, D.; Ferreira-Dias, S. Interesterification of fat blends rich in Omega-3 polyunsaturated fatty acids catalyzed by immobilized lipase on modified sepiolite. New Biotechnol. 2009, 25, S111–S112.

Oterhals, A.; Solvang, M.; Nortvedt, R.; Berntssenc, M. Optimization of activated carbon-based decontamination of fish oil by response surface methodology. Eur. J. Lipid Sci. Technol. 2007, 109, 691–705.

Pasanphan, W. ; Buettner, G.R. ; Chirachanchai, S. Chitosan gallate as a novel potential polysaccharide antioxidant: an EPR study. Carbohydr. Res. 2010, 345, 132–140.

Pena-Ramos, E.A.; Xiong, Y.L.; Arteaga, G.E. Fractionation and characterization forantioxidant activity of hydrolyzed whey protein. J. Sci. Food Agric. 2004, 84, 1908 -1918.

Pihlanto, A. Antioxidative peptides derived from milk proteins. Int. Dairy J. 2006, 16, 1306–1314.

Pietsch, A.; Jaeger, P. Concentration of squalene from shark liver oil by short-path distillation. Eur. J. Lipid Sci. Technol. 2007, 109, 1077–1082.

Pszczola, D.E. Encapsulated ingredients: providing the right fit. Food Technol. 1998, 52(12), 70–77.

Riha, V.; Brunner, G. Separation of fish oil ethyl esters with supercritical carbon dioxide. J. Supercrit. Fluids 2000, 17, 55–64.

Scolt, K.C.; Latshaw, J.D. Effects of commercial processing on the fat-soluble vitamin content of Menhaden fish oil. JAOCS 1991, 68(4), 234–236.

Schorken,U.; Kempers, P. Lipid biotechnology: Industrially relevant production processes. Eur. J. Lipid Sci. Technol. 2009, 111, 627–645.

Schubert, M.A.; Muller-Goymann, C.C. Characterisation of surface-modified solid lipid nanoparticles (SLN): Influence of lecithin and nonionic emulsifier. Eur. J. Pharm. Biopharm. 2005, 61, 77–86.

Shahidi, F.; Zhong, Y. Novel antioxidants in food quality preservation and health promotion. Eur. J. Lipid Sci. Technol. 2010, 112, 930–940.

Sivaramakrishnana, R.; Nakamurab, C.; Mehnertb, W.; Kortingc, H.C.; Kramera, K.D.; Scha¨fer-Korting, M. Glucocorticoid entrapment into lipid carriers-characterization by parelectric spectroscopy and influence on dermal uptake. J. Control. Rel. 2004, 97, 493–502.

Staby, A.; Mollerup, J. Solubility of fish oil fatty acid ethyl esters in sub- and supercritical carbon dioxide. J. Am. Oil Chem. Soc. 1993, 70, 583–588.

Sun, H.; Wiesenborn, G.; Tostenson, K.; Gillespie, J.; RayasDuarte, P. Fractionation of squalene from amaranth seed oil. J. Am. Oil Chem. Soc. 1997, 74, 413–418.

Sundram, K.; Perlman, D.; Hayes, K. Blends of palm fat and corn oil provide oxidation-resistant shortenings for baking and frying. US Patent 5,874,117.

Tsimidou, M.; Papavergou, E.; Boskou, D. Evaluation of oregano antioxidant activity in mackerel oil. Food Res. Int. 1995, 28, 431–433.

Venkateshwarlu, G.; Let, M.B.; Meyer, A.S.; Jacobsen, C. Modeling the sensory impact of defined combinations of volatile lipid oxidation products on fishy and metallic off-flavors. J. Agric. Food Chem. 2004, 52(2), 311–317.

Wada, S.; Fang, X. The synergistic antioxidant effect of rosemary extract and α-tocopherol in sardine oil model system and frozen-crushed fish meat. J. Food Process. Preserv. 1992, 16, 263–274.

Wanasundara, N.; Shahidi, F. Antioxidant and pro-oxidant activity of green tea extracts in marine oils. Food Chem. 1998a, 63, 335–342.

Wanasundara, U.N.; Shahidi, F.; Amarowicz, R. Effect of processing on constituents and oxidative stability of marine oils. J. Food Lipid. 1998b 5, 29–41.

Warnants, N.; Van Oeckel, M.J.; Boucque. C.V. Effect of incorporation of dietary polyunsaturated fatty acids in pork backfat on the quality of salami. Meat Sci. 1999, 49(4), 435–445.

Weiss, J.; Decker, E.A.; McClements, D.J.; Kristbergsson, K.; Helgason, T.; Awad, T. Solid lipid nanoparticles as delivery systems for bioactive food components. Food Biophys. 2008, 3, 146–154.

Xin, F.; Shun, W. Enhancing the antioxidant effect of α-tocopherol with rosemary in inhibiting catalyzed oxidation caused by iron (II) and hemoprotein. Food Res. Int. 1993, 26, 405–411.

Xu, X.; Jacobsen, C.; Nielsen, N.S.; Heinrich, M.T.; Zhoua, D. Purification and deodorization of structured lipids by short path distillation. Eur. J. Lipid Sci. Technol. 2002, 104, 745–755.

Yanishlieva-Maslarova, N.V. Inhibiting oxidation. Antioxidants in Food. Practical Applications. J. Pokorny, N. Yanishlieva, M. Gordon, Eds.; Woodhead Publishing: Cambridge, UK, 2001, 22–70.

Zu-yi, L.; Ward, O.P. Enzyme catalyzed production of vegetable oils containing omega-3 polyunsaturated fatty acid. Biotechnol. Letters 1993, 5, 185–188.

·•6•··

Synthesis and Properties of Structured Lipids with Omega-3s

Nathalie Quezada[1] and Ernesto M. Hernandez[2]

[1]*Oregon State University, Corvallis, Oregon and* [2]*Omega Protein Inc., Houston, Texas*

Structured Lipids

Lifestyle-related diseases, such as obesity, hyperlipidemia, arteriosclerosis, diabetes mellitus, cancer, and hypertension, are increasing in industrialized countries (Nagao & Yanagita, 2005). It has been reported that one-third of human cancers are associated with dietary habits and lifestyle (Doll, 1992). Modulation of dietary fat quality such as the balance omega-6/omega-3 has been related to immunologic diseases (Harbige, 2003) and the development of insulin resistance and the metabolic syndrome. Quality of dietary lipids could decrease the mortality of lifestyle-related diseases (Vessby, 2003). Furthermore, fat consumption in the Western diet has gone through a series of changes both on the amount consumed and the types of fats recommended. As a consequence, modifications have taken place in dietary recommendations from the scientific community, several fat-containing foods have been reformulated by the manufacturers, new labeling laws of fat-containing foods have been enforced, and new dietary and functional fats are being introduced through specialty fat products.

Structured lipids are tailor-made fats and oils with improved functional and nutritional properties through the incorporation of new fatty acids or the change of the position of existing fatty acids on the glycerol backbone (Akoh, 1995); these modifications are usually conducted by interesterification. The reaction of interesterification is basically the exchange of carbonyl groups of fatty acids within and between the triglyceride molecules.

There are three reactions associated with interesterification: acidolysis or ester exchange with other fatty acids, alcoholysis or ester exchange with alcohols, and ester exchange between triglycerides or transesterification (Willis et al., 1998; Hernandez, 2008). See Fig. 6.1.

In the case of alcoholysis, specific applications have been found with the development of ethyl esters of omega-3. This is basically the esterification of oils high in omega-3 such as fish oil with ethanol, which has found a wide application in the

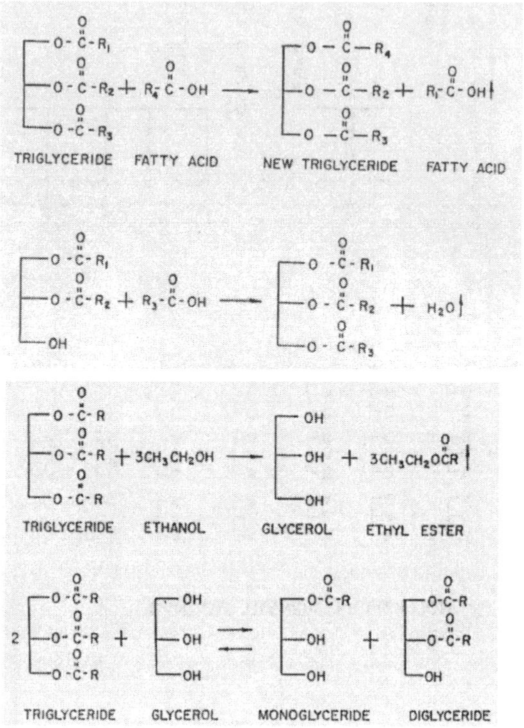

Fig. 6.1. Acidolysis and Alcoholysis Reactions.

pharmaceutical industry. Interesterification of reactions can be conducted using chemical-based catalysts or enzymes.

Chemical interesterification has some advantages over enzymatic modification, including lower catalyst cost, use of readily available industrial equipment, and shorter reaction times (Konishi et al., 1993). Enzymatic interesterification generally uses lipases and in some instances uses phospholipases. The introduction of immobilized lipases has allowed the use of this technology at industrial scale. A wide variety of immobilized lipases are now commercially available and widely used for industrial and research applications and offer several advantages over chemical interesterification, such as lower processing temperatures with less chemicals being used in the process and the ability to be specific to the position of the esterification in the glycerol molecule. The selectivity of these reactions can also be manipulated towards the production of triacylglycerols with specific fatty acids chains, as well (Ghazali et al., 1995). Their position selectivity and regiospecificity of lipases and phospholipases has allowed for the production of novel fats and oils (Akon & Min, 1998; Haman & Shahidi, 2005). Immobilized lipase can also be used for the production of ethyl esters of fatty acids like stearic acid, CLA, and DHA. The production and yields of these tailored triacylglycerols can be manipulated by changing enzyme loading, reaction times, and temperatures (Torres et al., 2003; Osorio et al., 2001).

Lipases used in interesterification are usually of microbial origin, but there are some products derived from plant sources such as oats and maize. Most lipases are 1,3 position specific but also can be non position specific. Table 6.A shows some examples of lipases and specificity.

Lipases usually tend to hydrolyze fatty acids from the triglyceride in a hydrophilic environment; however, if the amount of water in the reaction is restricted, hydrolysis is minimized and lipase catalyzed interesterification becomes the dominant reaction. A small amount of water needs to be present, however, to maintain the integrity of the enzyme. Lipases can catalyze interesterification with triglycerides and also with free fatty acids. Lipases can be immobilized in micro porous supports such as amberlite resins, diatomaceous earths, cellulose, silica gel, clay, carbon, or alumina.

As mentioned, lipids can be restructured with specific fatty acids to make oils healthier and also to modify the rheological and functional properties of certain fats. Studies have also reported on the effect a specific fatty acid and its position in the triacylglyceride have in the metabolic fate and health benefits of a structured lipid (Akoh & Min, 1998). More recently biotechnology has had an appreciable impact in the blending and transformation of several vegetable oils and animal fats, including omega-3 oils. These new technologies have given rise to a number of new biotech-based lipid products, from non-*trans* products to specialized delivery of bioactive lipids (Schörken & Kempers, 2009; Hernandez, 2008). Also, structured lipids have been developed for nutrition and medical applications such as infant formulas, low calorie fats, and enteral and parenteral nutrition, as well as food applications with specific functionalities such as plastic fats, cocoa butter alternatives, and frying oils (Osborn & Akoh, 2002).

Table 6.A. Regiospecificity and Fatty Acid Specificity.

	Regiospecificity	Fatty acid specificity
Porcine pancreas	1,3-	NFAs
Candida cylidracae	NPS	NFAS
Staphylococus aureus	NPS	NFAS
Aspergillus niger	1,3-	C:18
Candida rugosa	NPS	C:18
Mucor miehei	1,3-	—
Rhizopus arrhizus	1,3-	C:8, C:10

NPS, Non position specific; NFAS, Non fatty acid specific (Malcata, 1996)

Synthesis of Structured Lipids

As mentioned above, two main processes have been used for the production of structured lipids: chemical interesterification and enzymatic synthesis. More specifically, chemical synthesis involves hydrolysis of a mixture of triacylglycerides of different species and then re-esterification after random mixing (transesterification). The reaction is usually catalyzed by alkali metals or alkali metal alkylates (Akoh, 1995). This process requires high temperature and anhydrous conditions. Even though this process is inexpensive and easy to scale up, it lacks specificity and offers little or no control over the positional distribution of fatty acids in the final product (Willis et al., 1998).

As pointed out, enzymatic reactions usually involve lipases. These enzymes occur widely in nature and hydrolyze, in the presence of water, triacylglycerols to diacylglycerols, monoacylglycerols, free fatty acids and glycerol. In a water-starved environment, lipases catalyze reactions of interesterification and are essentially a combination of esterification and hydrolysis reactions (Akoh & Min, 1998). A continuous removal of water is important during the reaction in order to increase the esterification reactions, minimizing hydrolysis, and obtaining high yield products. However, it is important to keep some moisture in the system as it is necessary to maintain enzyme dynamics during non-covalent interactions. Thus, a balance between hydrolysis and esterification is important.

The production of structured lipids can also be carried out in organic solvents, where substrates are soluble. In this case the hydrolysis can be minimized, and it is less likely to cause enzyme inactivation in esterification reactions. However, the type of solvent for the reaction should be considered carefully, as it can dramatically affect the reaction kinetics and catalytic efficiency, as well as the stability of the enzyme. Hydrophilic or polar solvents can penetrate into the hydrophilic core of the proteins and alter their functional structure (Senanayake & Shahidi, 1999).

Optimum temperatures for active lipases range from 30 to 60°C. When the temperature increases, enzyme molecules unfold by destruction of the disulfide bonds, hydrolysis of peptide bonds, and deamidation of aspargine and glutamine residues. These processes can be avoided in a water-free environment or by immobilization of the enzyme, which result in great thermo-stability. Other factors that affect enzymatic activity and product yield include pH, substrate molar ratio, enzyme activity and load, incubation time, specificity of enzyme to substrate type and chain length, and region-specificity (Akoh & Min, 1998). As mentioned, the advantages of the enzymatic reactions over the chemical reactions are the low operation temperatures that result in appreciably energy savings and minimization of thermal degradation.

Due to the advantages of the enzymatic process and with the aim of producing SL at industrial level, many authors have been focused on the synthesis of SL using continuous reactors (Xu et al., 1998; Shimada et al., 1999; Gonzalez Moreno et al., 2004; Hamam & Budge, 2010) or super critical carbon dioxide systems (Liu & Shaw, 1997; Liu et al., 2007). Response Surface Methodology (RSM) can be effectively used to reduce experimental numbers and optimize reaction conditions and responses for production of structured lipids (Sahin et al., 2006; Shuang et al., 2009).

Structured Lipids with Omega-3s

A great deal of research has been reported on the methods of preparation and applications of structured lipids with omega-3s. Also, a wide variety of fatty acids have been used in the synthesis of these structured lipids to study their functionality and health properties of each to maximize the health benefits. Short and medium chain fatty acids have been used for the production of structured lipids due to the advantages associated with their digestibility, absorption, and metabolism (Forssel et al., 1993; Lepine et al., 1994; Fomuso & Akoh, 1997; Lee & Foglia, 2000; Mu & Hoy, 2000; Fomuso, 2002; Senayake & Shahidi, 2002a; Nielsen et al., 2005; Straarup et al., 2006; Bektas et al., 2008; Ding et al., 2009; Lin et al., 2009; Sengupta & Ghosh, 2010).

Structured lipids have been also synthesized with polyunsaturated fatty acids such as conjugated linoleic acid (Lee et al., 2004, 2006; Lumor & Akoh, 2005; Villeneuve et al., 2007; Guo & Sun, 2007; Goli et al., 2008; Rocha & Hernandez, 2008; Hossein et al., 2008; Hernandez-Martin et al., 2009), conjugated linolenic acid (Lumor & Akoh, 2005), and gamma linolenic acid (Shimada et al., 1999; Sahin et al., 2005).

Special attention has been given to the synthesis of structured lipids with omega-3 fatty acids due to the health benefits related with them, such as cardiovascular disease, inflammation, cancer, immune response, diabetes, hypertension, and renal disorders (Branden & Carroll, 1986; Sardesai, 1992; Conquer et al., 1997; Mori et al., 2004).

Incorporation of omega-3s into vegetable oils has been reported in sunflower oil, corn oil, peanut oil, and soybean oil (Li & Ward, 1993; Huang & Akoh, 1994). Also, Akoh et al. (1995) successfully modified trilinolein by incorporating DHA and EPA using two immobilized lipases, IM60 from Mucor miehei and SP435 from Candida antarctica, in a solvent and solvent-free environment. These authors reported on the effects of reaction parameters, such as type of solvent, enzyme load, time course, and molar ratio of substrates with regard to the n-3 PUFA incorporation using SP435 immobilized enzyme. High yields were reported, even in the absence of organic solvent. Huang et al. (1994) studied the incorporation of EPA in melon seed oil. This oil is used in cooking in African and Middle Eastern countries and contains large amounts of linoleic acid (up to 71%; Akoh & Nwosu, 1992). The aim of these authors was to improve the n-3 and n-6 fatty acid ratio of this oil through the incorporation of EPA for nutritional purposes.

Some authors have focused on the synthesis of structured lipids of omega-3s with medium chain fatty acids using fish oil and medium chain fatty acids. This would have the combined effect of providing high energy from the medium chain fatty acids and the cardiovascular health benefits provided by omega-3s (Lee & Akoh, 1998; Senanayake & Shahidi, 2002b). Lee and Akoh (1996) studied the incorporation of EPA in trilaurin and tricaprylin at specific positions. Also, these authors studied the effect of additives such as water and glycerol on the rate of the enzymatic reaction when using IM 60 and SP435 lipases. The rate of incorporation of EPA in the medium-chain triglycerols was increased by adding water when using IM 60, but the opposite effect was found when using SP435. The addition of glycerol to the reaction process had limited effects on the rate of incorporation of EPA for both enzymes. Lee and Akoh (1998) incorporated EPA and DHA into tricaprylin in hexane at 55°C using SP435 and studied the oxidative stability of the modified lipids. After purification, through short-path distillation, the structured lipid contained 46.9 mol % caprylic acid, 23.3 mol% EPA, and 21.7 mol % DHA as major fatty acids. Jennings and Akoh (1999) incorporated capric acid into fish oil by enzymatic acidolysis in a solvent, using hexane, and a solvent-free process. The optimal incorporation of capric acid was 41.2% after 48 h using hexane and 46.4% after 72 h for the solvent-free reaction. Akoh and Moussata (2001) synthesized structured lipids of menhaden fish oils with caprylic acids in a laboratory-scale pack bed reactor by enzymatic acidolysis using Lipozyme IM. The incorporation of caprylic acid in the fish oil was 29.5%. Xu et al. (2000) studied the incorporation of caprylic acid in menhaden oil by enzymatic acidolysis in a packed bed reactor and optimized the reaction using RSM. The optimal reaction temperature was 65°C, substrate molar ration ranged from 4 to5 and residence time was 180–220 min. Haman and Shahidi (2004) studied the incorporation of capric acid in Docosahexaenoic Acid Single Cell Oil (DHASCO). The optimum conditions were a mole ratio of 1:3 (DHASCO/CA) at a temperature of 45°C, and a reaction time of 24 h in the presence of 4% enzyme and 2% water content. Capric acid on the glycerol backbone of the modified structured lipid was present mainly in the sn-1,3 positions of the triacylglycerol molecules. Meanwhile, DHA was favorably present in the sn-2 position but also located in the sn-1 and sn-3 positions. DHASCO modified oil showed lower oxidative stability than the unmodified counterpart over the entire storage period.

Haman and Shahidi (2005) also studied the incorporation of capric acid into a commercial omega-3 oil derived from microalgae *Schizochytrium sp.* and determined the optimal reaction conditions, determined the positional distribution of fatty acids in the enzymatically modified oil, and evaluated the oxidative stability of the resulting structured lipid. The optimum conditions were a mole ratio of 1:3 oil to capric acid at 45°C, a reaction time of 24 h, 4% (w/w of substrates) PS-30 lipase from Pseudomonas sp. and 2% (w/w of substrates and enzyme) water content. Capric acid was present mainly in the sn-1,3 positions of the triacylglycerol molecules, while DHA and DPA were mainly esterified to the sn-2 position. The resulting structured lipid generally had higher conjugated diene (CD) and 2-thiobarbituric acid (TBA) values than its unmodified counterpart. Hamam and Shahidi (2006) studied the incorporation of EPA, DPA, and DHA into high-laurate canola oil (Laurical 35) by enzymatic modification using RSM. Effective incorporation of EPA, DPA, and DHA into Laurical 35 were carried out using *C. rugosa, M. miehei, Pseudomonas* sp. These authors pointed out that incorporation of n-3 fatty acids into Laurical 35 was in the order of EPA, DPA, then DHA, and this may be due to structural differences such as chain length and number of double bonds. Hamam and Budge (2010) studied the synthesis of structured lipids of fish oil and capric acid in packed bed bioreactors using one or two immobilized enzyme columns in a solvent-free environment and using RSM to optimize the process. These authors concluded that the one bed reactor was more efficient considering the reaction yield, amount of enzyme, shorter residence time, and lower amount of substrates.

The production of human milk fat substitutes has been reported through the synthesis of structured lipids with omega-3s. Sahin et al. (2006) used the enzyme Lypozyme RSM to carry out the incorporation of DHA, EPA, and oleic acid into tri-palmitin. The optimal conditions incorporated 5% of both DHA and EPA and 40% of oleic acid. Pina-Rodriguez and Akoh (2009) studied the incorporation of DHA in Amaranth Oil as a partial fat substitute in milk-based infant formula by increasing its palmitic acid content at the sn-2 position and incorporating DHA in order to deliver a lipid component more similar to that in breast milk. To accomplish this objective these authors used Novozym 435 lipase to incorporate palmitic acid in the sn-2 position and then they incorporated DHA, mainly at the sn-1,3 positions using Lipozyme RM IM. Furthermore, an optimization model was developed to determine the exact parameters to incorporate a specific amount of DHA (1.0%–2.5%). Chen et al. (2010) studied the enzymatic production of ABA-Type (n-3 PUFA in sn-1 and sn-3 positions and medium chain fatty acid in sn-2 position) structured lipids with EPA, DHA mixtures of fatty acids, and ethyl esters with tricaprin using Lipozyme RM IM and RSM to optimize the process. These authors found that the ethyl ester mixtures of EPA and DHA were more efficient than their respective fatty acids to produce ABA-type SL.

Some studies have reported on the modification of high-in- omega-6 borage oil (24% γ-linolenic acid) and evening primrose oil (7%–10% γ-linolenic acid) through the incorporation of omega-3s to produce a unique oil with specific nutritional health benefits of omega-3s and the omega-6 fatty acids (Akoh & Sista, 1995; Akoh et al., 1996; Ju et al., 1998; Senanayake & Shahidi, 1999; Senanayake & Shahidi, 2004). Akoh and Moussata (1998) studied the incorporation of capric acid and EPA in borage oil by enzymatic modification using IM60 and SP435 lipases. Both enzymes were able to incorporate both fatty acids, but just SP435 lipase was able to incorporate both fatty acids at the *sn*-2 position. Senanayake and Shahidi (2002c) studied the incorporation and optimization by acidolysis of EPA in borage and evening primrose oils. The highest EPA incorporations were 39.9 and 37.4% in borage oil and evening primrose oil, respectively, and occurred at the stoichiometric mole ratio of 1:3 for oil to EPA at 45–55°C and at 150–250 enzyme activity units using n-hexane as solvent. Also Senanayake and Shahidi (2002a) studied the incorporation of DHA into borage oil using Novozym 435 in hexane and used RSM to optimize the reaction process for the maximum incorporation of DHA while using the minimum amount of enzyme.

Optimum incorporation of DHA up to 34.1% was achieved at 165 units of enzyme concentration after 35 h at 50°C.

Other studies report on the synthesis of structured lipids with alpha linolenic acid (ALA, C18:3 n3) from perilla oil (Mitra et al., 2010). Perilla oil is traditionally used in Asian countries and contains up to 60% of α-linolenic acid (Shin & Kim, 1994). Kim et al. (2002) used perilla oil to synthesize structured lipids containing alpha linolenic acid with caprylic acid, using Lipozyme RM IM and Lipozyme TL IM in a solvent and solvent-free reaction. Lipozyme TL IM showed higher incorporation of caprylic acid into the perilla oil than its counterpart RM IL when solvent was present, but no difference was found in the incorporation in a solvent-free reaction. Jovica (2010) studied the incorporation of ALA in cocoa butter using an immobilized enzyme derived from *Rhizomucor miehei* and the highest incorporation of ALA was 77.3±1.3%.

Some authors have also focused on the oxidative stability of SL containing omega-3s. Lee and Akoh (1998) studied the oxidative stability of SL containing EPA, DHA, and caprylic acid. These authors found that caprylic acid tended to protect the SL against oxidation because of the saturated nature of this fatty acid. Akoh and Moussata (2001) also produced structured lipids of menhaden oil with caprylic acid and studied the effect of tocopherols and TBHQ and combinations on the oxidative stability of the modified lipids. The resulting modified lipids showed less oxidative stability than the unmodified counterparts. These authors pointed out the need to protect the SL containing highly unsaturated fatty acids during the synthesis process. Senanayake and Shahidi (2002c) evaluated the chemical and stability characteristics of structured lipids from borage and primrose oils under Schaal oven conditions at 60°C by measuring conjugated dienes, 2-thiobarbituric acid reactive substance (TBARS), and headspace volatiles. The modified lipids showed higher values of conjugated dienes, TBARS, and head space volatiles than their unmodified counterparts, which can be explain by the introduction of polyunsaturated fatty acids and removal of endogenous antioxidants during the enzymatic process. Pina-Rodriguez and Akoh (2010), as mentioned, synthesized a structured lipid from amaranth oil with DHA and palmitic acid esterified at the *sn*-2 position to match a breast milk fat analog. The modified lipid was incorporated in a prototype infant formula and its fatty acid composition and oxidative stability was compared with a control infant formula and a commercial formula. The modified lipid from the prototype infant formula showed the least oxidative stability compared to the fats of the control or commercial formulas. These authors point out the need to use antioxidants such as tocopherols to increase the oxidative stability of these modified fats. Mitra et al. (2010) studied the effects of antioxidants such as catechin, BHT, and rosemary on the oxidative stability of an SL of soybean oil and perilla oil. The highest oxidative stability was obtained when catechin was used. In summary, the enzymatic modification process for the production of structured lipids, as well as the purification processes (e.g., short path distillation), can lead to loss in natural antioxidants, suggesting that the addition of tocopherols and other antioxidants during the enzymatic process is necessary to prolong the shelf life of the modified lipids (Jennings & Akoh, 1999; Mitra et al., 2010).

Lipids are often used in oil-water systems, and some researches have focused on the oxidation of structured lipids when used in emulsions. Fomuso et al. (2002) studied the effect of emulsifier lecithin, Tween 20, whey protein isolate, mono-/diacylglycerols, and sucrose fatty acid ester on the oxidation stability of an oil-in-water emulsion prepared with a structured lipid of menhaden oil-caprylic acid. All structured lipid emulsions had similar droplet size distributions, except for the lecithin, and were stable to creaming over the 48-day period studied. Lipid oxidation was affected by emulsifier type and concentration; low concentration of emulsifier (0.25%) showed higher oxidation rate than high emulsifier concentration (1%).

Some authors have also studied the digestion, absorption, transport, and lipidemic effects of SL with omega-3s. Kenler et al. (1996) synthesized SL from fish oil and medium chain fatty acids and administered it to patients undergoing surgery for upper gastrointestinal malignancies and compared the effects with the ones of a control diet that differed only in its fat source. The SL diet was tolerated significantly better, led to improved hepatic and renal function, and reduced the number of infections per patient. These same authors studied the effect of SL with fish oil diet on prostaglandin release from mononuclear cells in cancer patients after surgery. The hepatic, renal, and immune functions improved, and there was a significant reduction in eicosanoid production from mononuclear cells with endotoxin stimulation in patients fed with the modified lipid. Lee and Akoh (1999) found that SL with EPA, DHA, and caprylic acid lowered triglycerides and low density lipoprotein (LDL) cholesterol levels more efficiently (when compared to soybean oil) when fed to mice after 21 days. Chopra and Sambaiah (2009) synthesized SL from rice bran oil with ALA from linseed oil and omega-3s from cod liver oil and studied their effect in liver and serum lipids when fed to rats. Total serum cholesterol and triglycerides of the rats fed with the resulting SL were lower than the ones fed with rice bran oil. Nagata et al. (2004) studied the effects of structured lipids containing eicosapentaenoic or docosahexaenoic acid and caprylic acid on serum and liver lipid profiles in rats. These authors observed significantly lower relative perirenal adipose tissue weights, serum cholesterol concentrations, and serum lipids concentrations of the rats fed with SL diets than those of the control group (soybean oil) over the experimental period. Kim et al. (2008) studied the effects in obese mice fed with structured lipids-diacylglycerol (SL-DG) enriched with DHA on hepatic lipid metabolism and the mRNA expression of genes involved in hepatic steatosis of mice and compared them to the ones fed with SL-triacylglycerol (SL-TG), soybean oil, and algae oil. Mice fed SL-DG showed a lower total white adipose tissue weight, plasma triglyceride concentration, and hepatic cholesterol levels compared to the soy bean oil diet. Furthermore, SL-DG lowered mRNA expressions of sterol regulatory element binding protein-1 and its target genes compared to the SL-TG, soy bean oil, and algae oil diets. Cho et al. (2009) studied the effects of a low-trans structured fat from flaxseed oil on plasma and hepatic lipid metabolism when fed to Apo E deficient mice and compared this to a shortening and commercial low trans fat diets. This study showed that this diet was highly effective for improving hyperlipidemia and hepatic lipid accumulation in Apo E deficient mice through significant suppression of hepatic lipid accumulation, decreasing liver weight, lowering plasma total cholesterol and free fatty acid, and increased HDL-C concentration when compared with the control groups.

Sengupta and Ghosh (2010) investigated the effects on platelet aggregation, haematological parameters, and the liver in male albino rats when fed capric acid-rich and EPA-/DHA-rich mustard oils and a control diet (mustard oil) for 28 days. These authors concluded that the medium chain and polyunsaturated rich structured lipids of mustard oil improved the haematological and histological conditions in rats in hypercholesterolaemic condition.

As shown by the numerous studies mentioned above, SLs containing MCFAs and n-3 PUFAs can have applications as therapeutic or medical lipid sources and may be useful in enteral and parenteral nutrition. Druschky and Pscheidl (2000) investigated the effects of SL esterified with omega-3s fatty acids in position sn-2, in comparison to SL containing omega-6s fatty acids in position sn-2, on protein and energy metabolism in a low dose continuous endotoxin rat model. Rats fed SL with omega-3s in the sn-2 position plus endotoxin showed a significant decrease in mean body weight and nitrogen balance compared with the control group fed just SL with omega-3s. No significant decrease was observed in the rats fed SL with omega-6s in the sn-2 position plus endotoxin and the control group. Also, rats fed SL with omega-6s plus endotoxin showed significantly higher liver weight means and rectus muscle protein contents compared with rats of group SL with omega-3s plus endotoxin.

In animal studies, Lepine et al. (1994) investigated the effect of dietary structured triacylglycerides synthesized by the random re-esterification of medium-chain triacylglycerides and menhaden oil on growth, performance, nitrogen retention, and apparent digestibility of nitrogen, lipid, and fatty acids. These authors found that apparent nitrogen digestibility and nitrogen retention were high but not affected by dietary treatment. Furthermore, the apparent fatty acid digestibilities were affected by a dietary lipid source and by the physical structure of the lipids in the pigs. Nielsen et al. (2005) studied the effect on the growth and fatty acid composition of rainbow trout fed six diets for 61 days: fish oil and rapeseed oil (FO diet), specific structured lipid (enzymatic process, based on fish oil and caprylic acid) and rapeseed oil (SL diet), randomized structured lipids (chemical process, based on tricaprylin and fish oil) and rapeseed oil (RL diet), medium chain triglyceride and fish oil (MCT diet), diacylglycerol and fish oil (DAG diet), and fish oil (FO max diet). FO and FO max fish were significantly bigger than the SL diet fish. Also, there was no higher incorporation of omega-3s in rainbow trout fillets, livers, carcasses, and visceras by feeding them specific structured lipids compared with randomized structured lipids or physical mixture lipids. Diacylglycerides did not reduce fish weight when added to the diets.

Synthesis of Structured Phospholipids

Phospholipids are the main component of lecithin which is mainly a byproduct from processing soybean oil. Lecithin is widely used in the food, cosmetic, and pharmaceutical industries because of its numerous surfactant and bioactive properties. Most lecithin is obtained from the process of degumming crude soybean and other vegetable oils. Lecithin is basically a mixture of several phospholipids and is commonly present in cells and plant membranes. Other minor commercial sources of lecithin are derived from egg yolk. Lecithin is usually defined as a mixture of acetone insoluble polar lipids and triglyceride oil together with other minor components produced by degumming of crude vegetable oils and animal fats (Szuhaj, 1988; Hernandez & Quezada, 2008). Phospholipids generally contain a glycerol or a sphingosyl backbone and are classified into four main classes: glycerophospholipids, sphingolipids, ether phospholipids, and phosphonopholipids, depending on their backbones and bonding types (Akoh & Min, 1998). More commonly derived from vegetable oils, glycerophospholipids are composed of glycerol, fatty acids, phosphoric acid, and a second hydroxyl compound which commonly contains nitrogen. The hydroxyl group can be choline, ethanolamine, serine, inositol, or glycerol as illustrated in Fig. 6.2. Table 6.B shows the ratio of the components in the hydroxyl group in commercial lecithin.

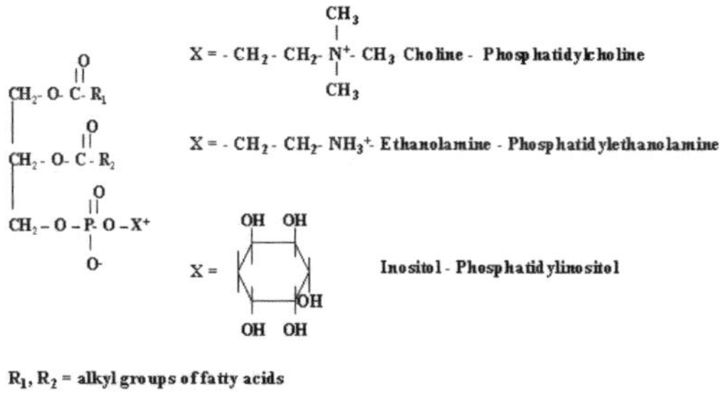

Fig. 6.2. Structure of common glycerophospholipids.

Table 6.B. Composition of Commercial Lecithin (Hernandez and Quezada, 2008).

	Regular	Deoiled
soybean oil	35%	3%
phosphtaidyl choline	16%	24%
phosphatidyl ethanolamine	14%	20%
phosphatidyl inositol	10%	14%
phosphatidic acid	5%	7%
phytoglycolipids	11%	15%
carbohydrates	5%	8%
moisture	1%	1%

Relatively little phospholipid variety exists in native lecithins. However, phospholipids can be modified as well as lipids to improve their properties or meet particular functional requirements. For this purpose, the fatty acids or head group of phospholipid structure can be changed or modified. These modifications can be performed by physical, chemical, and enzymatic modification (Doig & Dick, 2003). Physical processes such as solvent fractionation using acetone and alcohol, as well as preparatory chromatography have been used to obtain de-oiled and high purity phospholipids from commercial vegetable oil sources by (Doig & Dicks, 2003). The chemical modification of phospholipids usually allows little or no control over the positional distribution of fatty acids in the final product (Willis et al., 1998) and require high temperatures, solvents, and anhydrous conditions, making this process difficult to use in food or nutritional applications (Osborn & Akoh, 2002). Enzymatic modification offers a great advantage in catalyzed interesterification reactions because it can control the positional distribution of fatty acids in the final product, due to their selectivity and region specificity (Akoh & Min, 1998). Guo et al. (2005) published a review of enzymatic modification of phospholipids for functional applications and human nutrition. Fig. 6.3 shows the enzymes most used for phospholipid modification. Phospholipases and lipase1-specific selectively recognize each of the four individual ester bonds in a phospholipids molecule (Servi, 1999).

The enzymatic modification of the head group has been carried out using phospholipase D, which catalyzes the transphosphatidyl reaction and hydrolysis of the phosphoester bond (Frohman & Morris, 1999; Servi, 1999). Phospholipase D has been obtained from plants, such as cabbage and bacterial sources, such as *Streptomyces* sp. Low molecular weight primary alcohols to large secondary ones have been shown to be viable substrates for phospholipase D with different yields and selectivity (Takami et al., 1994; D'Arrigo et al., 1997; Servi, 1999; Iwasaki & Yamane, 2004; Piazza & Marmer, 2007). Ulbrich-Hoffman (2000) published a review of phospholipids used in lipid transformation and included a table with the head groups introduced by phospholipase D into glycerophospholipids. Hosokawa et al. (2000) investigated the transphosphatidylation of squid skin lecithin with L-serine using Phospholipase D to produce DHA-Phosphatidylserine. The modified phospholipid contained the same level of DHA as the original substrate. Hossen and Hernandez

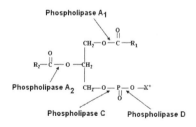

Fig. 6.3. Phospholipases and lipase in phospholipid modification.

(2004) synthesized a phosphatidyl-sterol by successfully incorporating -sitosterol into phosphatidylcholine using phospholipase D. Yamamoto et al. (2008) synthesized phosphatidylated terpenes and phospholipids with phenylalkanols and hydroquinone (Yamamoto et al., 2010) via phospholipase D-catalyzed transphosphatidylation reactions. See Table 6.C.

Enzymatic fatty acid modification of phospholipids has also been attained using lipases 1-specific and phospholipase A2 through hydrolysis and then direct interesterification or re-esterification in the sn-1 and sn-2 position (Aura et al., 1995). Structured phospholipids have been synthesized with diverse fatty acids such as caproic acid (Adlercreutz et al., 2002) using a lipase from *Rhizopus oryzae* in a water activity-controlled organic medium, caprylic acid using Lipozyme TL IM in

Table 6.C. Summary of factors that affect enzymatic modification of phospholipids (Guo et al., 2005).

Factors	Remarks
Enzyme	• Phospholipase A_1 and A_2 or lipases can be used. • Increasing enzyme dosage may result in higher incorporation. • Increasing of enzyme may result in increased hydrolysis. • Enzyme can have different effect on phospholipids with different head groups.
Water	• For some enzymes, low water activity is recommended for high yields. • Reaction time to reach equilibrium increases with decreasing water activity. • Water activity influences the molecular organization of phospholipids substrate. The packing density of phospholipids molecules increases with decreasing water activity.
Acyl donor	• By using a large excess of free fatty acids, hydrolysis reaction is inhibited • Very high concentration of acyl donors, slight decrease of in-reaction rate. • Generally free fatty acids are more efficient acyl donors than their esters. • Reactivity relates to chain length and degree of saturation.
Solvent	• Solvent reduces the viscosity of the system; as a consequence, the reaction rate is increased through mass transfer increase of substrates. • Reaction is solvent-type dependent, the rate being inversely proportional to solvent polarity. • Polar solvents should be avoided, as they compete with the enzyme on available water. • Solubility of the substrate depends on solvent type.
Reaction time	• The longer the reaction time the higher the incorporation of acyl donors into phospholipids can be expected. • Long reaction time, however, may also result in increased acyl migration.
Temperature	• Optimal temperature changes with enzyme source and type. • Increased temperature may result in higher acyl migration. • Higher temperature lowers viscosity of reaction medium. • Enzyme stability has reverse relationship with temperature.

continuous packed bed enzyme reactors in a solvent system and a solvent-free system (Vikbjerg et al. 2005) and using Lipozime RM IM (*Rhizomucor miehei*) by catalyzed transesterification with soybean phosphatidylcholine in solvent-free media (Vikbjerg et al., 2006), oleic acid (Yagi et al., 1990) using a lipase from Rhizopus delemar with a final content of 25% of oleic acid into phosphatidylcholine; heptadecanoic acid (Svensson et al., 1992) using Rhizopus arrhizus lipase with an incorporation nearly of 50% into phospholipid, conjugated linoleic acid using Thermomyces lanuginosa lipase (Hossen & Hernandez, 2005; Peng et al., 2002), and with conjugated linolenic acids (Quezada, 2005).

Structured Phospholipids with Omega-3s

Natural sources of phospholipids with omega-3s are found in krill oil. EPA and DHA in this oil are structurally attached to phospholipid molecules. By weight, krill oil is comprised of at least 20% of each EPA and DHA and 40% phospholipids, mostly in the form of phosphatidylcholine, and several health properties have been attributed to this form of omega-3s (Massrieh, 2008).

Numerous reports have been published on the synthesis of structured phospholipids with omega-3s and their beneficial effects. Hosokawa et al. (1995a) studied the incorporation of EPA and DHA into soy phosphatidylcholine using Lipozyme IM-60 in organic solvents with high dielectric constants (water mimics) in order to avoid hydrolysis. Increased transesterification of EPA into soy phosphatidylcholine was obtained when using appropriate amounts of water and organic solvents. The same authors investigated the preparation of phospholipids containing highly unsaturated fatty acids, such as EPA and DHA, using porcine pancreatic phospholipase A_2 and Lipozyme IM-60 in water mimic environments. Formamide and water-propylene glycol mixture were reported to enhance the incorporation of EPA and DHA into lysophosphatidylcholine and phosphatidylcholine, respectively (Hosokawa et al., 1995b).

Hosokawa et al. (1998) studied the synthesis of structured phospholipids with DHA using porcine pancreatic phospholipase A2 with soy phosphatidylcholine. DHA was successively incorporated into the sn-2 position of soy phosphatidylcholine with a rate of incorporation higher than 70% of total fatty acids at the sn-2. Haraldsson and Thorarensen (1999) used Lipozyme TM to catalyze the reaction between a concentrate of 55% EPA and 30% DHA and pure phosphatidylcholine from egg yolk in non-aqueous solvent-free system. Structured phospholipids with 32% EPA and 16% DHA content were reported. Peng et al. (2002) studied the incorporation of caprylic acid, conjugated linoleic acid (CLA), eicosapentaenoic acid (EPA), and docosahexaenoic acid (DHA) into phospholipids using Lipozyme TL IM in a solvent-free system and optimized with RSM. These authors found that different apparent incorporation rates depended on the individual fatty acids, their purity in the starting acyl donor mixtures, and the nature of the phospholipids.

Regarding applications of structured phospholipids with n-3 PUFAs, esterified PLs with fish was reported to have positive effects on children with impaired visual sustained attention performance. Blood lipid alterations in LC omega-3 fatty acids were correlated with these changes (Vaisman et al., 2008). Docosahexaenoic acid (DHA) containing PLs has shown to have potential medical applications such as improvement in brain function and prevention of cerebral apoplexy (Peng et al., 2002). Structured phospholipids have shown to be potentially useful for the treatment of cancer. Ishigamori et al. (2005) found that DHA-phosphatidylethanolamine might enhance differentiation of leukemia cell lines (HL-60) and growth inhibition by regulation of c-*jun* and c-*myc* expression. Hossain et al. (2006) converted squid meal phospatidylcholine into phosphatidylserine by phospholipase D-mediated transphosphatidylation and investigated the effect of DHA-phosphatidylcholine and DHA-phosphatidylserine on butyrate-induced growth inhibition, differentiation and apoptosis using Caco-2 cells. These authors found that DHA-phosphatidylcholine and DHA-phosphatidylserine can be promising compounds used for colon cancer chemoprevention.

References

Adlercreutz, D.; Budde, H.; Wehtje, E. Synthesis of phosphatidylcholine with defined fatty acid in the sn-1 position by lipase-catayzed esterification and transesterification reaction. Biotechnol. Bioeng. 2002, 78, 403–411.

Akoh, C. Structured lipids-enzymatic approach. Inform 1995, 6, 1055–1061.

Akoh, C.; Jennings, B.; Lillard, D. Enzymatic modification of trilinolein: Incorporation of n-3 polyunsaturated fatty acids. J. Am. Oil Chem. Soc. 1995, 72, 1317–1321.

Akoh, C.; Min, D. Food Lipids: Chemistry, Nutrition and Biotechnology. Marcel Dekker, Inc.; New York, 1998.

Akoh, C.; Moussata, C. Lipase-catalyzed modification of borage oil: Incorporation of capric and eicosapentaenoic acids to form structured lipids. J. Am. Oil Chem. 1998, 75, 697–701.

Akoh, C.; Moussata, C. Characterization and oxidative stability of enzymatically produced fish and canola oil-based structured lipids. J. Am. Oil Chem. Soc. 2001, 78, 25–30.

Akoh, C.; Nwosu, C. Fatty acid composition of melon seed oil lipids and phospholipids. J. Am. Oil Chem. 1992, 69, 314–316.

Akoh, C.; Sista, R. Enzymatic modification of borage oil: Incorporation of eicosapentanoic acid. J. Food Lipids 1995, 2, 231–238.

Aura, A.; Forsell, P.; Mustranta, A.; Poutanen, K. Transesterification of soy lecithin by lipase and phospholipase. J. Am. Oil Chem. Soc. 1995, 72, 1375–1379.

Bektaş, I.; Yucel, S.; Ustun, G.; Aksoy, A. Production of reduced calorie structured lipid by acidolysis of tripalmitin with capric acid: optimisation by response surface methodology. J. Sci. Food Agric. 2008, 88, 1927–1931.

Braden, L.; Carroll, K. Dietary polyunsaturated fat in relation to mammary carcinogenesis in rat. Lipids 1986, 21, 285–288.

Chen, B.; Zhang, H.; Cheong, L.Z.; Tan, T.; Xu, X. Enzymatic production of ABA-type structured lipids containing omega-3 and medium-chain fatty acids: Effects of different acyl donors on the acyl migration rate. Food Bioprocess. Technol. 2010. On line.

Cho, Y.Y.; Kwona, E.Y.; Kim, H.J.; Park, Y.B.; Lee, K.T.; Park, T.S.; Choi, M.S. Low trans structured fat from flaxseed oil improves plasma and hepatic lipid metabolism in apo E–/– mice. Food Chem. Toxicol. 2009, 47, 1550–1555.

Chopra, R.; Sambaiah, K. Effects of rice bran oil enriched with n-3 PUFA on liver and serum lipids in rats. Lipids 2009, 44, 37–46.

Conquer, J.A.; Holub, B.J. Dietary docosahexaenoic acid as a source of eicosapentaenoic acid in vegetarian and omnivores. Lipids 1997, 32, 341–345.

D'Arrigo, P.; Servi, S. Using phospholipases for phospholipid modification. Trends Biotechnol. 1997, 15, 90–96.

Ding, S.; Yan, Y.; Yangs, J. Optimization of lipase-catalyzed acidolysis of soybean oil to produce structured lipids. J. Food Biochem. 2009, 33, 442–452.

Doig, S.; Diks, R. Toolbox for exchanging constituent fatty acids in lecithins. Eur. J. Lipid Sci. Technol. 2003, 105, 359–367.

Doll, R. The lessons of life: keynote address to the nutrition and cancer conference. Cancer Res. 1992, 52, 2024S–2029S.

Druschky, K.; Pscheidl, E. Different effects of chemically defined structured lipids containing ω3 or ω6 fatty acids on nitrogen retention and protein metabolism in endotoxemic rats. Nutr. Res. 2000, 20, 1183–1192.

Fomuso, L. Lipase-catalyzed acidolysis of olive oil and caprylic acid in a bench-scale packed bed bioreactor. Food Res. Int. 2002, 35, 15–21.

Fomuso, L.; Akoh, C. Enzymatic modification of triolein: incorporation of caproic and butyric acids to produce reduced-calorie structured lipids. J. Am. Oil Chem. Soc. 1997, 74, 269–272.

Fomuso, L.; Corredig, M.; Akoh, C. Effect of emulsifier on oxidation properties of fish oil-based structured lipid emulsions. J. Agric. Food Chem. 2002, 50, 2957–2961.

Forssel, P.; Parovuori, P.; Linko, P.; Poutanen, K. Enzymatic transesterification of rapeseed oil and lauric acid in a continuos reactor. J. Am. Oil Chem. Soc. 1993, 75, 1573–1579.

Frohman, M.; Morris, A. Phospholipase D structure and regulation. Chem. Phys. Lipids 1999, 98, 127–140.

Ghazali, H.M.; Hamidah, S.; Che Man, J.B. Enzymatic transesterification of palm olein with nonspecific and 1,3-specific lipases. J. Am. Oil Chem. Soc. 1995, 72, 633–639.

Goli, H.; Kadivar, M.; Sahri, M. Enzymatic interesterification of structured lipids containing conjugated linoleic acid with palm stearin for possible margarine production. Eur. J. Lipid Sci. Technol. 2008, 110, 1102–1108.

González Moreno, P.; Robles Medina, A.; Camacho Rubio, F.; Camacho Páez, B.; Molina Grima, E. Production of structured lipids by acidolysis of an EPA-enriched fish oil and caprylic acid in a packed bed reactor: analysis of three different operation modes. Biotechnol. Prog. 2004, 20, 1044–1052.

Guo, Z.; Sun, Y. Solvent-free production of 1,3-diglyceride of CLA: Strategy consideration and protocol design. Food Chem. 2007, 100, 1076–1084.

Guo, Z.; Vikbjerg, A.; Xu, X. Enzymatic modification of phospholipids for functional applications and human nutrition. Biotechnol. Adv. 2005, 23, 203–259.

Hamam, F.; Budge, S. Structured and specialty lipids in continuous packed column reactors: comparison of production using one and two enzyme beds. J. Am. Oil Chem. Soc. 2010, 87, 385–394.

Hamam, F.; Shahidi, F. Lipase-catalyzed acidolysis of algal oils with capric acid: optimization of reaction conditions using response surface methodology. J. Food Lipids 2004, 11, 147–163.

Hamam, F.; Shahidi, F. Structured lipids from high-laurate canola oil and long-chain omega-3 fatty acids. J. Am. Oil Chem. 2005, 82, 731–736.

Hamam, F.; Shahidi, F. Synthesis of structured lipids containing medium-chain and omega-3 fatty acids. J. Agric. Food Chem. 2006, 54, 4390–4396.

Haraldsson, G.; Thorarensen, A. Preparation of phospholipids highly enriched with n-3 polyunsaturated fatty acids by lipase. J. Am. Oil Chem. Soc. 1999, 76, 1143–1146.

Harbige, L. Fatty acids, immune response, and autoimmunity: a question of n-6 essentiality and the balance between n-6 and n-3. Lipids 2003, 38, 323–341.

Hernandez, E. Structured lipids as delivery systems. Delivery and controlled release of bioactives in foods and nutraceuticals, N. Garti Eds., CRC Press: Boca Raton, FL, 2008.

Hernandez, E.; Quezada, N. Uses of phospholipids as functional ingredients. Phospholipid Technology and Applications, F.D. Gunstone, Eds., Oily Press:, Bridgewater, England, 2008.

Hernandez-Martin, E.; Hill, C.; Otero, C. Selective synthesis of diacylglycerols of conjugated linoleic acid. J. Am. Oil Chem. Soc. 2009, 86, 427–435.

Hosokawa, M.; Ito, M.; Takahashi, K. Preparation of highly unsaturated fatty acid-containing phosphatidylcholine by transesterification with phospholipase A_2. Biotechnol. Tech. 1998 12, 583–586.

Hosokawa, M.; Shimatani, T.; Kanada, T.; Inoue, Y.; Takahashi, K. Conversion to docosahexaenoic acid-containing phosphatidylserine from squid skin lecithin by phospholipase D-mediated transphosphatidylation. J. Agric. Food Chem. 2000, 48, 4550–4554.

Hosokawa, M.; Takahashi, K.; Miyazaki, N.; Okamura, K.; Hatano, M. Application of water mimics on preparation of eicosapentaenoic and docosahexaenoic acids containing glycerolipids. J. Am. Oil Chem. Soc. 1995a, 72, 421–425.

Hosokawa, M.; Takahash, K.; Kikuchi, Y.; Hatano, M. Preparation of therapeutic phospholipids through porcine pancreatic phospholipase A2-mediated esterification and lipozyme-mediated acidolysis. J. Am. Oil Chem. Soc. 1995b, 72, 1287–1291.

Hossain, Z.; Konishi, M.; Hosokawa, M.; Takahashi, K. Effect of polyunsaturated fatty acid-enriched phosphatidylcholine and phosphatidylserine on butyrate-induced growth inhibition, differentiation and apoptosis in Caco-2 cells. Cell Biochem. Funct. 2006 24, 159–165.

Hossein, S.; Sahri, M.; Kadivar, M. Enzymatic interesterification of structured lipids containing conjugated linoleic acid with palm stearin for possible margarine production. Eur. J. Lipid Sci. Technol. 2008, 110, 1102–1108.

Hossen, M.; Hernandez, E. Phospholipase D-catalyzed synthesis of novel phospholipid-phytosterol conjugates. Lipids 2004, 39, 777–82.

Hossen, M.; Hernandez, E. Enzyme-catalyzed synthesis of structured phospholipids with conjugated linoleic acid. Eur. J. Lipid Sci. Technol. 2005, 107, 730–736.

Huang, K.H.; Akoh, C. Lipase. Catalyzed incorporation of n-3 polyunsaturated fatty acids into vegetable oils. J. Am. Oil Chem. Soc. 1994, 71, 1277–1280.

Huang, K.H.; Akoh, C.; Erickson, M. Enzymatic modification of melon seed oil: Incorporation of eicosapentanoic acid. J. Agric. Food Chem. 1994, 42, 2646–2648.

Ishigamori, H.; Hosokawa, M.; Kohno, H.; Tanaka, T.; Miyashita, K.; Takahashi, K. Docosahexaenoic acid-containing phosphatidylethanolamine enhances HL-60 cell differentiation by regulation of c-*jun* and c-*myc* expression. Mol. Cell Biochem. 2005, 275, 127–133.

Iwasaki, Y.; Yamane. Enzymatic synthesis of structured lipids. Adv. Biochem. Engin/Biotechnol. 2004, 90, 151–171.

Jennings, B.; Akoh, C. Enzymatic modification of triacylglycerols of high eicosapentaenoic and docosahexaenoic acids content to produce structured lipids. J. Am. Oil Chem. Soc. 1999, 76, 1133–1137.

Jovica, F. Lipase-catalyzed synthesis of omega-3 vegetable oils. [MS Thesis]: Dalhousie University, 2010.

Ju, Y.H.; Huang, F.C.; Fang, C.H. The incorporation of n-3 polyunsaturated fatty acids into acylglycerols of borage oil via lipase-catalyyzed reactions. J. Am. Oil Chem. Soc. 1998, 75, 961–965.

Kenler, A.S.; Swails, W.S.; Driscoll, D.S.; DeMichele, S.J.; Daley, B.; Babinead, T.J.; Peterson, M.B.; Bistrian, B.R. Early enteral feeding in postsurgical cancer patients: fish oil structured lipid-based polymeric formula versus a standard polymeric formula. Ann. Surg. 1996, 223, 316–333.

Kim, I.H.; Kim, H.; Lee, K.T.; Chung, S.H.; Ko, S.N. Lipase-catalyzed acidolysis of perilla oil with caprylic acid to produce structured lipids. J. Am. Oil Chem. Soc. 2002, 79, 363–367.

Kim, H.J.; Lee, K.T.; Park, Y.B.; Jeon, S.M.; Choi, M.S. Dietary docosahexaenoic acid-rich diacylglycerols ameliorate hepatic steatosis and alter hepatic gene expressions in C57BL/6J-Lepob/obmice. Mol. Nutr. Food Res. 2008, 52, 965–973.

Konishi, H.; Neff, W.E.; Mounts, T.L. Chemical interesterification with regioselectivity for edible oils. J. Am. Oil Chem. Soc. 1993, 70(4), 411–415.

Lee, K.T.; Akoh, C. Immobilized lipase-catalyzed production of structured lipids with eicosapentaenoic acid at specific positions. J. Am. Oil Chem. Soc. 1996, 73, 611–615.

Lee, K.T.; Akoh, C. Characterization of enzymatically synthesized structured lipids containing eicosapentaenoic, docosahexaenoic and caprylic acids. J. Am. Oil Chem. Soc. 1998, 75, 495–499.

Lee, K.T.; Akoh, C.; Dawe, D. Effects of structured lipid containing omega-3 and medium chain fatty acids on serum lipids and immunological variables in mice. J. Food Biochem. 1999, 23, 197–208.

Lee, K.; Foglia T. Synthesis, purification, and characterization of structured lipids produced from chicken fat. J. Am. Oil Chem. Soc. 2000, 77, 1027–1034.

Lee, L.; Lee, K.; Akoh, C.; Chung, S.; Kim, M. Antioxidant evaluation and oxidative stability of structured lipids from extravirgin olive oil and conjugated linoleic acid. J. Agric. Food Chem. 2006, 54, 5416–5421.

Lee, J.H.; Shin, J.A.; Lee, J.; Lee, K.T. Production of lipase-catalyzed structured lipids from saf-flower oil with conjugated linoleic acid and oxidation studies with rosemary extracts. Food Res. Int. 2004, 37, 967–974.

Lepine, A.; Garlebt, K.; Reinhart, G.; Kresty, L. Dietary nitrogen and lipid utilization by growing pigs fed structured triacylglycerides synthesized from medium-chain triacylglycerides and men-haden oil. J. Anim. Sci. 1994, 72, 938–945.

Li, Z.Y.; Ward, O.P. Enzyme catalysed production of vegetable oils containing omega-3 polyun-saturated fatty acid. Biotech. Lett. 1993, 15, 185–188.

Lin, M.; Yeh, S.; Tsou, S.; Wang, M.; Chen, W. Effects of parenteral structured lipid emulsion on modulating the inflammatory response in rats undergoing a total gastrectomy. Nutrition 2009, 25, 115–121.

Liu, K.J., Chang, H.M., and Liu, K.M. Enzymatic synthesis of cocoa butter analog through inter-esterification of lard and tristearin in supercritical carbon dioxide by lipase. Food Chem 2007, 100, 1303–1311.

Liu, K.J.; Shaw, J.F. Synthesis of cocoa butter equivalent by lipase-catalyzed interesterification in supercritical carbon dioxide. J. Am. Oil Chem. Soc. 1997, 74, 1477–1482.

Lumor, S.; Akoh, C. Incorporation of γ-linolenic and linoleic acids into palm kernel oil/pal olein blend. Eur. J. Lipid Sci. Technol. 2005, 107, 447–454.

Malcata, X. Engineering of/ with lipases. Springer, 1996.

Massrieh, W. Health benefits of omega-3 fatty acids from neptune krill oil. Lipid Technol. 2008, 20, 108–111.

Mitra, K.; Lee, J.H.; Lee, K.T.; Kim, S.A. Production tactic and physicochemical properties of low $\omega 6/\omega 3$ ratio structured lipids synthesized from perilla oil and soy bean oil. Int. J. Food Sci. Tech. 2010, 45, 1321–1329.

Mori, T.A.; Beilin, L.J. Omega-3 fatty acids and inflammation. Curr. Atheroscler. Rep. 2004, 6, 461–467.

Mu, H.; Hoy, C. Effects of different medium-chain fatty acids on intestinal absorption of struc-tured triacylglycerols. Lipids 2000, 35, 83–89.

Nagao, K.; Yanagita, T. Conjugated fatty acids in food and their health benefits. J. Biosci. Bioeng. 2005, 100, 152–157.

Nagata, J.; Kasaib, M.; Negishib, S.; Saitoa, M. Effects of structured lipids containing eicosap-entaenoic or docosahexaenoic acid and caprylic acid on serum and liver lipid profiles in rats. BioFactors 2004, 22, 157–160.

Nielsen, N.; Göttsche, J.; Holmc, J.; Xub, X.; Mub, H.; Jacobsen, C. Effect of structured lipids based on fish oil on the growth and fatty acid composition in rainbow trout (Oncorhynchus mykiss). Aquaculture 2005, 250, 411–423.

Osborn, H.; Akoh, C. Structured lipids-novel fats with medical, nutraceutical, and food applica-tions. Compr. Rev. Food Sci. 2002, F 3, 93–103.

Osório, N.; Maeiro, I.; Luna, D.; Ferreira-Dias, S. Interesterification of fat blends rich in Omega-3 polyunsaturated fatty acids catalyzed by immobilized lipase on modified sepiolite. New Biotechnol. 2009, 25, S111–S112.

Peng, L.; Xu, X.; Mu, H.; Hoy, C.; Nissen, A. Production of structured phospholipids by lipase-catalyzed acidolysis: optimization using response surface methodology. Enzyme Microb. Technol. 2002, 31, 523–532.

Piazza, G.; Marmer, W. Conversion of phosphatidylcholine to phosphatidylglycerol with phospholipase D and glycerol. J. Am. Oil Chem. Soc. 2007, 84, 645–651.

Pina-Rodriguez, A.; Akoh, C. Synthesis and characterization of a structured lipid from amaranth oil as a partial fat substitute in milk-based infant formula. J. Agric. Food Chem. 2009, 57, 6748–6756.

Pina-Rodriguez, A.; Akoh, C. Composition and oxidative stability of a structured lipid from amaranth oil in a milk-based infant formula. J. Food Sci. 2010, 75(2), c140–c146.

Quezada, N. "Synthesis of structured phospholipids with conjugated linolenic acid, and evaluation of their physical properties". [MS Thesis]: Texas A&M University.

Rocha-Uribe, A.; Hernandez, E. Effect of conjugated linoleic acid and fatty acid positional distribution on physicochemical properties of structured lipids. J. Am. Oil Chem. Soc. 2008, 85, 997–1004.

Sahin, N.; Akoh, C.; Karaali, A. Enzymatic production of human milk fat substitutes containing γ-linolenic acid: optimization of reactions by response surface methodology. J. Am. Oil Chem. Soc. 2005, 82, 549–557.

Sahin, N.; Akoh, C.; Karaali, A. Human milk fat substitutes containing omega-3 fatty acids. J. Agric. Food Chem. 2006, 54, 3717–3722.

Sardesai, V.M. Nutritional role of polyunsaturated fatty acids. J. Nutr. Biochem. 1992, 3, 154–166.

Schorken; Kempers, P. Lipid biotechnology: Industrially relevant production processes. Eur. J. Lipid Sci. Technol. 2009, 111, 627–645.

Senanayake, N.; Shahidi, F. Enzymatic incorporation of docosahexaenoic acid into borage oil. J. Am. Oil Chem. Soc. 1999, 76, 1009–1015.

Senanayake, N.; Shahidi, F. Lipase-catalyzed incorporation of docosahexaenoic acid (DHA) into borage oil: optimization using response surface methodology. Food Chem. 2002a, 77, 115–123.

Senanayake, S.; Shahidi, F. Enzyme-catalyzed synthesis of structured lipids via acidolysis of seal (Phoca groenlandica) blubber oil with capric acid. Food Res. Int. 2002b, 35, 753–759.

Senanayake, S.; Shahidi, F. Chemical and stability characteristics of structured lipids from borage (Borago officinalis L.) and evening primrose (Oenothera Biennis L.) Oils. J. Food Sci. 2002c, 76, 2038–2045.

Senanayake, N.; Shahidi, F. Incorporation of docosahexaenoic acid (DHA) into evening primrose (Oenothera biennis L.) oil via lipase-catalyzed transesterification. Food Chem. 2004, 85, 489–496.

Sengupta, A.; Ghosh, M. Modulation of platelet aggregation, haematological and histological parameters by structured lipids on hypercholesterolaemic rats. Lipids 2010, 45, 393–400.

Servi, S. (1999). Phospholipases as synthetic catalysts. Top Curr. Chem. 1999, 200, 127–157.

Shimada, Y.; Suenaga, M.; Sugihara, A.; Nakai, S.; Tominaga, Y. Continuous production of structured lipid containing γ-linolenic and caprylic acids by immobilized rhizopus delemar lipase. J. Am. Oil Chem. Soc. 1999, 76, 189–193.

Shin, H.S.; Kim, S.W. Composition of perilla seed. J. Am. Oil Chem. Soc. 1994, 71, 619–622.

Shuang, D.; Jiang-Ke, Y.; Yun-jun, Y. Optimization of lipase-catalized acidolysis of soybean oil to produce structured lipids. J. Food Biochem. 2009, 33, 442–452.

Straarup, E.; Jakobsen, K.; Høy, C.; Danielsen, V. Dietary structured lipids for post-weaning piglets: fat digestibility, nitrogen retention and fatty acid profiles of tissues. J. Anim. Physiol. An. 2006, N 90, 124–135.

Svensson, I.; Adlercreutz, P.; Mattiasson, B. Lipase-catalyzed acidolysis of phosphatidylcholine at controlled water activity. J. Am. Oil Chem. Soc. 1992, 69, 986–991.

Szuhaj, B. Lecithins: Sources, manufacture and uses. AOCS: Champaign, IL, 1988.

Takami, M.; Hidaka, N.; Suzuki, Y. Phospholipase D-catalyzed synthesis of phosphatidyl aromatic compounds. Biosci. Biotech. Biochem. 1994, 58, 2140–2144.

Torres, C.F.; Lin, B.; Moeljadi, M.; Hill Jr., C.G. Lipase-catalyzed synthesis of designer acylglycerols rich in residues of eicosapentaenoic, docosahexaenoic, conjugated linoleic, and/or stearic acids. Eur. J. Lipid Sci. Technol. 2003, 105, 614–623.

Ulbrich-Hofmann, R. Phospholipases used in lipid transformations. In: Bornscheuer U. and Spenser F., editors. Enzymes in Lipid Modification. Wiley-VCH: Weinheim, Berlin. 2000. 219–262.

Vaisman, N.; Kaysar, N.; Zaruk-Adasha, Y.; Pelled,D.; Brichon, G.; Zwingelstein, G.; Bodennec, J. Correlation between changes in blood fatty acid composition and visual sustained attention performance in children with inattention: effect of dietary n-3 fatty acids containing phospholipids. Am. J. Clin. Nutr. 2008, 87, 1170–1180.

Vessby, B. Dietary fat, fatty acid composition in plasma and the metabolic syndrome. Curr. Opin. Lipidol. 2003, 14, 15–19.

Vikbjerga, A.; Mu, H.; Xu X. Elucidation of acyl migration during lipase-catalyzed production of structured phospholipids. J. Am. Oil Chem. Soc. 2006, 83, 609–614.

Vikbjerga, A.; Penga, L.; Mua, H.; Xua, X. Continuous production of structured phospholipids in a packed bed reactor with lipase from thermomyces lanuginose. J. Am. Oil Chem. Soc. 2005, 82, 237–242.

Villeneuve, P.; Barouh, N.; Barea, B.; Piombo, G.; Figueroa-Espinoza, M. Chemoenzymatic synthesis of structured triacylglycerols with conjugated linoleic acids (CLA) in central position. Food Chem. 2007, 100, 1443–1452.

Willis, W.; Lencki, R.; Marangoni, A. Lipid modification strategies in the production of nutritionally functional fats and oils. Crit. Rev. Food Sci. 1998, 38, 639–674.

Xu, X.; Balchen, S.; Høy, C.E.; Adler-Nissen, J. Production of specific-structured lipids by enzymatic interesterification in a pilot continuous enzyme bed reactor. J. Am. Oil Chem. Soc. 1998, 75, 1573–1579.

Xu, X.; Fomuso, L.; Akoh, C. Modification of menhaden oil by enzymatic acidolysis to produce structured lipids: optimization by response surface design in a packed bed reactor. J. Am. Oil Chem. Soc. 2000, 77, 171–176.

Yagi, T.; Nakanishi, T.; Yoshizawa, Y.; Fukui, F. The enzymatic acyl exchange of phospholipids with lipases. J. Ferment. Bioeng. 1990, 69, 23–25.

Yamamoto, Y.; Hosokawa, M.; Kurihara, H.; Miyashita, K. Preparation of phosphatidylated terpenes via phospholipase d-mediated transphosphatidylation. J. Am. Oil Chem. Soc. 2008, 85, 313–320.

Yamamoto, Y.; Kurihara, H.; Miyashita, K.; Hosokawa, M. Synthesis of novel phospholipids that bind phenylalkanols and hydroquinone via phospholipase D-catalyzed transphosphatidylation. New Biotechnology,2010, doi:10.1016/j.nbt.2010.06.014.

7

Applications of Omega-3 Fats in Foods

Ernesto M. Hernandez and Linda de Jong
Omega Protein Inc., Houston, Texas

Introduction

Lipids in general play important physiological roles in many biological processes of living organisms. Some of the most important functions include being a source of essential fatty acids, facilitating absorption of fat soluble nutrients, and being a source of energy via beta oxidation. Besides nutrition, fats and oils are essential components in food processing and food quality. Fats and oils define the organoleptic and texture properties of many food products, as well as their shelf life (Kinsella, 1988).

The increasing awareness of the health benefits of omega-3 fatty acids is reflected by the growth in consumption of omega-3 fats either as dietary supplements or in fortified foods. Omega-3 fatty acids, particularly the longer chain fatty acids eicosapentaenoic acid (EPA) and docosahexaenoic acid (DHA), have been studied in numerous health applications. The early studies on the benefits of essential fatty acids focused primarily on prevention of diseases such as neurological abnormalities (Burr & Burr, 1930; Holman et al., 1982; Holman, 1998). With regards to heart disease, more specific epidemiological studies conducted in the 1970s on omega-3 reported on the relatively low cardiovascular mortality in populations of Eskimos with high fish consumption (Dyerberg et al., 1975). These health benefits were associated with fish and fish oil with high content of EPA and DHA. Numerous clinical trials have been conducted since then to evaluate the effects of EPA and DHA, not just on cardiovascular disease but on other positive effects such as retina and brain development and cognitive function.

EPA is generally associated with cardiovascular protection and has been reported to have strong anti-inflammatory, anti-thrombotic, anti arrhythmic, and antiatherogenic effects (Kris-Etherton et al., 2002; Carlson et al., 1993; Trautein, 2001; Simopoulus, 1997; Dupont, 1996). More recently research of EPA has extended to study the effect on decrease of blood triglyceride levels, decrease growth rate of atherosclerotic plaque, lower blood pressure, and as modulators of gene expression. DHA, on

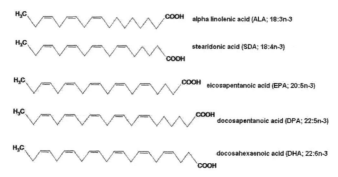

Fig. 7.1. Common types of Omega-3 fatty acids.

the other hand, is generally associated with cell structures and has been found to be particularly important in neurologically related metabolism, such as brain and retina development and function, and it has become an important component in prenatal and postnatal nutrition for mothers and children. (Wang et al., 2006 ; Chung, 2008; Carpentier et al., 2006, Kris-Etherton et al., 2002; Carlson et al., 1993; Trautein, 2001; Simopoulus, 1997; Dupont, 1996). EPA and DHA cannot be synthesized by the human body, so they have to be consumed directly thorough diet or derived from alpha linolenic acid (ALA) through metabolic reactions. Typical symptoms of deficiency of essential fatty acid include skin dryness, reduced growth rate, and susceptibility to infections. Fig. 7.1 shows the different types of omega-3 fatty acids according to their chain length and number of double bonds.

Sources of Omega-3

The main sources of omega-3 fatty acids for the general population are from vegetable oils and fish. Omega-3 from vegetable sources is generally in the form of shorter 18 carbon chain alpha-linolenic acid (ALA).

Vegetable oils, like soybean and canola, and other seeds, like flaxseed, walnut, and chia, are the common sources of short chain omega-3 fatty acids. The longer chain omega-3 fatty acids, EPA, and DHA are obtained primarily from cold-water fish.

Another source of long chain omega-3, mainly in the form of DHA, is obtained from micro algae oil and is widely used commercially as an infant formula ingredient (Carlson et al., 1994; Jensen et al., 2000). Oil from micro algae has over 40% DHA. A relatively new source of long chain omega-3s is krill oil. Krill are shrimp-like marine invertebrates and are important part of the zooplankton as a source of food for some large marine mammals. The main characteristic of this oil is that the EPA and DHA in it are structurally attached to phospholipid molecules. By weight, krill oil is comprised of at least 20% EPA and DHA and 40% phospholipids, mostly in the form of phosphatidylcholine (Massrieh, 2008).

Table 7.A. Sources of Oils with Omega 3 Fatty Acids

Fatty Acid	Menhaden (%)	Anchovy (%)	Cod Liver (%)	Micro algae (%)	Flax (%)
Lauric (C12:0)	—	—	—	6	—
Myristic (C14:0)	7	8	4	19	
Palmitic (C16:0)	19	16	9	17	6
Palmitoleic (C16:1)	12	9	10	2	
Stearic (C18:0)	3	3	4	1	4
Oleic (C18:1)	13	9	20	9	21
Linoleic (C18:2)	2	4	1	1	16
Linolenic (C18:3)	2	1	2	4	56
Eicosapentanoic (C20:5)	13	21	12	—	—
Docosahexanoic (22:6)	12	10	12	46	—

Vegetable-oil-derived omega-3, alpha-linolenic acid (ALA), is reported to be poorly converted by the human body to EPA and DHA, the actual fatty acids involved in many biological functions. On the other hand, studies show that there are important differences between men and women in their capacity for synthesis of EPA and DHA, and also between men of different ages in the conversion of ALA to EPA and DHA (Burdge & Calder, 2005).

With new advances in genetic manipulation, it is now possible to engineer common omega-6 rich oilseeds such as soybean and canola to produce longer chain omega-3 fatty acids such as stearidonic acid (SDA), EPA, and DHA (Damude & Kinney, 2007). SDA is now commercially available from genetically modified soybean oil (Harris et al., 2008; Ursin, 2003). Table 7.A shows common sources of long chain and short chain omega-3s.

Properties of Omega-3 Fatty Acids

Four main fatty acids can be considered the main precursors of the lipid metabolic pathways in mammals: palmitoleic, oleic, linoleic, and linolenic acids. Palmitoleic and palmitic can be synthesized endogenously. However, both linoleic (omega-6) and linolenic acids (omega-3) cannot be synthesized endogenously and must be obtained from the diet, hence the reason why they are considered essential fatty acids.

Metabolically, omega-3 fats, along with the other types of fatty acids, participate in several biological processes: 1) They can undergo β-oxidation to produce energy; 2) They can be recycled to make other fatty acids; 3) They can serve as a substrate for ketogenesis; 4) They can be stored in adipose tissue for later use by the body; 5) They can be incorporated into the phospholipids of cell membranes, where they participate in other membrane metabolic activities; and 6) In the case of ALA, can be converted into EPA, DPA, and DHA.

Another fatty acid that is considered essential is linoleic (LA) 18:2n-6 or omega-6 which has 2 double bonds and is 18 carbon chain length. As mentioned, this short chain omega-6 which, along with alpha linolenic (ALA), cannot be synthesized endogenously, must be obtained from diet. LA and ALA go through a series of metabolic steps such as elongation and desaturation reactions to be converted to the more bioactive longer chain fatty acids, LA to arachidonic acid (AA) and ALA to EPA and DHA.

The omega-3 and omega-6 metabolic pathways involve the same enzymatic reactions, thus competing for the same elongases and desaturases enzymes, as shown in Fig. 7.2. EPA and DHA are the precursors of many bioactive compounds known as eicosanoids, which are very specialized group of cell signaling molecules.

Omega-3 and Omega-6 Metabolism

The metabolism of the shorter chain, omega-6 LA for the production of arachidonic acid, begins with the catalysis by the enzyme 6-desaturase, which acts on carbon 6; then the enzymes 5-desaturase and elongase continue the reaction cascade to synthesize arachidonic acid, 20:4n-6, ending with four double bonds and 20 carbon chain length (i.e., arachidonic acid). One important factor to consider is that the desaturation steps in the metabolism of essential fatty acids are rate limiting. This is one of the reasons why in some cases it is recommended that diet be supplemented with oils rich in gamma-linolenic acid for the more direct production of arachidonic acid

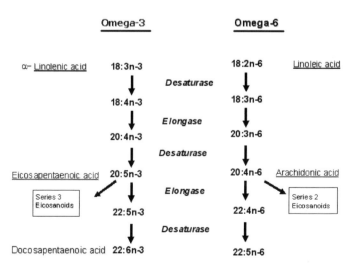

Fig. 7.2. Metabolic pathways of Omega-3 and Omega-6 fatty acids.

(Johnson, et al., 1997; Carter, 1988). This competition for enzymatic activity in the omega-3 and omega-6 pathways is one of the main arguments for recommending a balanced ratio of omega-3 to omega-6. As a result of these enzymatic reactions, it is also recommended that the long chain omega-3 fatty acids such as DHA and EPA found in marine oils be consumed directly (Wang et al., 2006; Jump, 2002), especially when it is considered that the rate of these enzymatic processes decrease even more with age.

As mentioned above, when the long chain omega-6 (arachidonic acid) and omega-3 (eicosapentanoic, docosahexanoic acids) PUFAs (polyunsaturated fatty acids), enter the metabolic pathways, they become part of the synthesis cascade for the formation of eicosanoids. This reaction cascade results in a family of oxygenated compounds with very potent bioactive properties (See Fig. 7.3) (Dupont, 1987). Eicosanoids are signaling molecules that basically result from the enzymatic oxidation of the long chain essential fatty acids. These signaling molecules play a crucial role in the control over many bodily systems such as inflammation, immune responses, and as messengers in the central nervous system. The main enzymes that catalyze the formation of eicosanoids are cyclooxigenases and lipoxygenases. The relationship between essential fatty acids and eicosenoids was first reported in the mid-1960s and were described as bioactive compounds derived originally from arachidonic acid (AA) (Van Dorp et al., 1964). The eicosanoids from omega-3 and omega-6 can act

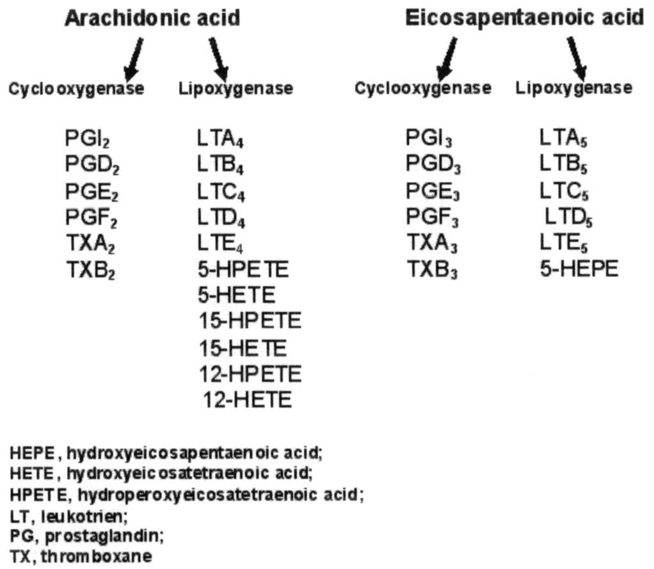

Arachidonic acid		Eicosapentaenoic acid	
Cyclooxygenase	Lipoxygenase	Cyclooxygenase	Lipoxygenase
PGI_2	LTA_4	PGI_3	LTA_5
PGD_2	LTB_4	PGD_3	LTB_5
PGE_2	LTC_4	PGE_3	LTC_5
PGF_2	LTD_4	PGF_3	LTD_5
TXA_2	LTE_4	TXA_3	LTE_5
TXB_2	5-HPETE	TXB_3	5-HEPE
	5-HETE		
	15-HPETE		
	15-HETE		
	12-HPETE		
	12-HETE		

HEPE, hydroxyeicosapentaenoic acid;
HETE, hydroxyeicosatetraenoic acid;
HPETE, hydroperoxyeicosatetraenoic acid;
LT, leukotrien;
PG, prostaglandin;
TX, thromboxane

Fig. 7.3. Eicosanoids from arachidonic and eicosapentaenoic acids.

antagonistically, and, as result, an excessive and imbalanced production or deficiency can result in the onset of several pathological situations like thrombosis, arthritis inflammation, or immune suppression (Jump, 2002).

Examples of common eicosanoids include prostaglandins (PGs), thromboxanes (TXs), leukotriens (LTs), lipoxins (LXs), hydroperoxyeicosatetraenoic (HPETEs), hydroxyeicosatetraenoic (HETEs), and hydroxyeicosapentaenoic acids (HEPEs). The basic mechanism of the synthesis of these eicosanoids from their precursors, either EPA or AA, involves their release from phosphatidylcholine in the cell membranes by the action of phospholipase A_2 or from membrane phosphatidylinositol-4,5-bisphosphate by the actions of phospholipase C, and a diacylglycerol lipase.

The action of cyclooxygenase (COX) on long chain EFAs yields prostaglandins (PGs) and thromboxanes (TXs). When EPA or AA is exposed to lipooxygenases, the reactions yield, HEPE, HETE, HPETE and leukotrienes LTs (Fig. 7.3). The more powerful series 2 PGs and series 4 LT derivatives are generated from arachidonic acid. The series 3 PGs and series 5 LT derivatives are generated from eicosapentaenoic. These compounds usually have a very short half-life and act in the immediate proximity of the cells where they are produced (Tapiero et al., 2002).

Besides mediation of allergic and inflammatory reactions, the bioactivity of eicosanoids can be very specialized; for example, they play an essential role in the modulation of renal function, induction and inhibition of thrombotic processes, and regulation of smooth muscle cell tone leading, for example, to vaso- and bronchoconstriction reactions (Sellmayer & Koletzko, 1999). Eicosanoids also act as important intracellular mediators in relation to several metabolic factions and responses. They have been identified as novel intracellular second messengers in inflammatory and mitogenic signaling (Kahn, 1995).

Typically, inflammation is characterized by the production of inflammatory agents such as cytokines and arachidonic acid-derived eicosanoids (prostaglandins, thromboxanes, leukotrienes). Studies show that, at sufficiently high intakes, long-chain omega-3 fatty acids decrease the production of these inflammatory eicosanoids and other reactive oxygen species. Long-chain omega-3 PUFAs can act both directly by replacing arachidonic acid as an eicosanoid substrate and inhibiting arachidonic acid metabolism and indirectly by altering the expression of inflammatory genes through effects on transcription factor activation (Calder, 2006). The role eicosanoids play in inflammation has been extensively researched for applications in the prevention of some diseases. Increasing the (EPA +DHA)/AA ratio in membrane phospholipids has been confirmed as a form of protection in inflammatory states (Gewirtz et al., 2002).

More recently, novel anti-inflammatory eicosanoids and docosanoids derived from EPA and DHA resolvins and protectins have been identified. These agents are also highly potent inhibitors of inflammation. These new signaling agents are made by the human body from both EPA and DHA. They are produced by the COX-2 pathway, especially in the presence of aspirin. Resolvins have been reported to reduce cellular inflammation by inhibiting the production and transportation of inflammatory cells and chemicals to the sites of inflammation. These eicosanoids have also been studied in the protection of the kidneys and could be used as a tool against acute renal failure (Levy, 2010; Singer et al., 2008; Ariel, 2007; Serhan, 2005). This further illustrates the importance of sustaining a balance of omega-3 and omega-6 fatty acids, not just because they compete for the same enzymes (desaturaeses and elongases) but because they can produce antagonistic types of eicosanoids who play an important role in the onset of inflammation-related diseases. This can be achieved either through delivery systems in foods or through supplements. The different roles that omega-3 fatty acids play in health and disease is further discussed in Chapters 1 and 2.

Omega-3 and Nutrition

A balanced diet in general includes a wide array of well know essential nutrients such a macronutrients (Carbohydrates, fats, proteins), vitamins, minerals, electrolytes, essential amino acids, and essential fatty acids. With the exception of essential fatty acids, the Institutes of Medicine has established minimum daily requirements of these nutrients. Likewise, the US Departments of Agriculture and Health and Human Services jointly developed the Food Guide Pyramid, which recommends servings from various food classes. Generally these recommendations call for 45% to 65% of daily calories from carbohydrates, 10% to 20% from protein, and 20 to 35% from fat (IOM, 2010).

In order to assist the consumer with their food choices and to improve nutrition, the US Food and Drug Administration (FDA) has required since 1994 that all food packages carry a Nutrition Facts Label. Food labels are required to show the DV (Daily Value) for specific nutrients based on Recommended Dietary Allowances (RDAs) set by the Food and Nutrition Board (FNB) of the National Academy of Sciences. In the case of nutrients without set standards, such as carbohydrates, protein, fat, and fiber, the FDA uses Daily Reference Values (DRVs), which are calculated as the proportion of calories that these nutrients should represent in a 2000-calorie reference diet, based on the Food Guide Pyramid recommendations.

The original objective of RDAs was to prevent classic nutritional deficiency diseases, such as anemia and rickets. However, due to the current problems of obesity and cardiovascular disease, the nutritional recommendations now include dietary guidelines to prevent chronic diseases such as osteoporosis, cancer, and diabetes.

With regards to omega-3 fatty acids, even though no RDAs for essential fatty acids have been established, their consumption as a means of improving heart health as well as other areas of health and wellness has increased appreciably. However, there are several issues with regards to proper processing and handling of omega-3 rich oils and consumer awareness of the different types of omega-3 fatty acids that can present obstacles for the wider consumption of omega-3 products.

Recommended Intakes of Omega-3 Fatty Acids

For the first time, in the 2002 Dietary Reference Intakes Report for Energy and Macronutrients, The National Academies released recommendations for adequate intakes (AI) of omega-3 (as ALA) and Omega-6 (as LA) (IPM, 2010). So far, however, they have not issued recommendations for the longer chain omega-3, EPA and DHA. AI is defined as nutrient intake estimate observed in healthy individuals where sufficient scientific data is not available to suggest a RDA (recommended daily allowance). The AIs set for LA were 17 and 12 g/day for men and women aged 19–50 years, respectively. The AI for ALA was 1.6 and 1.1 g/day for men and women aged 19 to 70 years, respectively (IOM, 2010) (See Table 7.B). Also in 1999, the National Institutes of Health (NIH) published a report from a sponsored international workshop on the essentiality and recommended dietary intakes for n-6 and n-3 fatty acids. This NIH Working Group proposed adequate intakes of 2%–3% of total calories for LA, 1% of total calories for ALA, and 0.3% of total calories for EPA and DHA. The NIH-Working Group further recommended intakes of EPA and DHA of 650 mg/day and a minimum of 300 mg DHA/day during pregnancy and lactation. Other countries such as Health Canada suggests a minimum of 3% of energy from n-6 fatty acids and 0.5% from n-3 fatty acids or 1% for infants who do not receive a preformed source of EPA and DHA. The United Kingdom recommends intakes of 1% of energy be from ALA and 0.5% from EPA and DHA combined. (Gebauer et al., 2006; Kris-Etherton et al., 2002).

The World Health Organization has also issued recommendations on the basis of the ratio of n-6 to n-3 fatty acids of 5:1–10:1. Other countries such as Sweden recommend a ratio of 5:1, Canada recommends 4:1–10:1, and Japan recently changed

Table 7.B. Adequate Intakes for ALA.

Adequate Intake (AI) for Omega-3 Fatty Acids				
Life Stage	Age	Source	Males (g/day)	Females (g/day)
Infants	0–6 months	ALA, EPA, DHA*	0.5	0.5
Infants	7–12 months	ALA, EPA, DHA	0.5	0.5
Children	1–3 years	ALA	0.7	0.7
Children	4–8 years	ALA	0.9	0.9
Children	9–13 years	ALA	1.2	1.0
Adolescents	14–18 years	ALA	1.6	1.1
Adults	19 years and older	ALA	1.6	1.1
Pregnancy	All ages	ALA	—	1.4
Breastfeeding	All ages	ALA	—	1.3

(IOM, 2002)

its recommendation from 4:1 to 2:1. On the basis of ratio of Omega-6 to Omega-3 intake, the western diet has been estimated at 20:1.

For patients with documented heart disease the American Heart Association recommends intake of about 1 g of EPA+DHA per day. For people who need to lower their blood triglycerides it is recommend an intake of 2 g to 4 g of EPA+DHA per day under a physician's care. They also recommend that patients without documented coronary heart disease get omega-3s through the consumption of a variety of fish (preferably fatty fish) at least twice a week and include in their diet oils rich in alpha-linolenic acid (flaxseed, canola, and soybean oils; flaxseed and walnuts). England's Scientific Advisory Committee on Nutrition (SACN) recommends intake of 450 mg of DHA+EPA per day.

More recently the Scientific Panel from EFSA (European Food Safety Authority) issued an opinion recognizing that a cause and effect relationship has been established between the consumption of EPA and DHA and maintenance of normal cardiac function and that it considers that intakes of EPA and DHA of about 250 mg per day are required to obtain the claimed effect (EFSA, 2010).

As a food source, there have been concerns regarding the risks involved in the consumption of fish and fish oil versus plant origin PUFAS. Some types of fish have been reported to contain high levels of mercury, PCBs (polychlorinated biphenyls), dioxins, and other environmental contaminants (Melanson et al., 2005). Of particular concern are fish that may contain significant levels of methylmercury, considered one of the most dangerous food contaminants. In general, older, larger, predatory fish and marine mammals, such as swordfish and seals, tend to contain the highest levels of these contaminants (Foran et al., 2006). PCBs and methylmercury are reported to have long half-lives in the body and can accumulate in people who consume contaminated fish on a frequent basis. PCBs are highly oil soluble and accumulate preferentially in fatty tissue of fish. Methylmercury, on the other hand, represents a more serious problem since it is distributed throughout skin, muscle, and organs of the fish. Both of these contaminants, mercury and PCBs can be efficiently removed with current adsorptive and molecular distillation techniques. Contaminants such as dioxins, furans, dibenzo brominated biphenyls and pesticides can also be efficiently removed from fish oil by molecular distillation (See Chapter 5).

Production and processing of fish products is globally regulated by several national and international agencies. In the US the Federal Drug Administration regulates the safety of all commercial fish, including ocean-caught, farm-raised, and imported fish. One example is the recommendation for pregnant women and nursing mothers to limit their consumption of sport-caught fish to one 6-ounce meal per week. The Environmental Protection Agency recommends that young children consume less than 2 ounces of sport-caught fish per week. The FDA also recommends that women who are pregnant or nursing and young children eliminate shark, swordfish, king mackerel in the mackerel family, and tilefish from their diets completely. Also consumption of other fish should be limited to 12 ounces per week to minimize exposure to methylmercury. The FDA has concluded that the safe level of methylmercury is less than 1 ppm in 7 ounces of fish. Sources of omega-3 grown under controlled conditions, such as vegetable oils or from microlagae, are reported to be free of, or very low in contaminants.

Omega-3s in Foods

Lipids in general are a required ingredient of almost every diet. As a result, fats and oils are prevalent in many foods throughout the world and play significant roles in nutrition and preparation methods of food products. As mentioned above, fats and oils are a major source of storage energy and have important roles in the body's metabolic processes, especially the essential fatty acids omega-3s, as outlined in Chapters 1 and 2. Fats and oils in general can have a positive effect as common ingredients in food formulations for processing, quality, organoleptic, and texture properties of food products. They can also have a negative effect when oxidized, particularly in the case of omega-3s. Fats and oils in general have to be extracted and processed through a series of pre-set steps in order to ensure that they meet the minimum specifications for human consumption (Hernandez, 2005a). More specific processing steps for omega-3 oils are described in Chapter 5.

Cooking with oils is one of the more common methods of preparing foods. The major application of cooking oils is frying, sautéing, and baking. In this case fats and oils function as a heat transfer medium and contribute flavor and texture to foods. Cooking oils are required to stand heating temperatures of up to 180°C during frying. More saturated oils are preferred as frying and cooking oils because they are less prone to oxidative, thermal or hydrolytic breakdown. However, this means that most cooking oils in general will be lower in the more unstable, more unsaturated essential fatty acids.

The current trend in the industry is to introduce healthier fats in food products, especially in foods that are traditionally high in fat. Common foods with a higher content of fats and oils include margarines, shortenings, butterfat, fried foods, mayonnaises, salad dressings, baked products, snacks, and confectionary products.

In the past, the food industry used fats that were more stable against oxidation such as partially hydrogenated oil and saturated fats. This was done in order to ensure longer shelf life of one or more years. However, since it was reported that *trans* fats can increase the risk of cardiovascular disease, several changes have taken place in the way companies formulate foods. These changes were also prompted when governments started requiring food companies, since 2006, to include the amounts of *trans* fats in the food facts labels. Oils high in saturated fat, such as palm oil, are being commonly used to replace hydrogenated oil. Also, oils high in linoleic acid, omega-6, such as corn oil and sunflower oil, and high in oleic acid, such as modified sunflower and canola oil, are being used to replace some high in *trans* frying oils. There are also now available in the market high oleic versions of soybean and safflower oils that have been produced either by genetic manipulation or natural breeding.

Following the consumer demand for healthier oils in food products, more recently manufacturers have introduced into the market many new foods fortified with omega-3 fatty acids. Fish oil from menhaden was the first oil approved as GRAS (generally recognized as safe) food ingredient in 1997 (CFR 184.1472) by the US government. Later on several other fish oils from salmon, tuna, anchovy, mackerel, andoil from micro algae were granted self affirmed GRAS status by the FDA. This has allowed fortification of foods with fish oil omega-3s from several sources and approved for use in a wide variety of foods. However limits were set by the FDA that the combined intake of EPADHA from these sources must not exceed 3 g/person/day.

Omega-3s in Beverages, Dairy, and Baby Formula

Dairy products and beverages fortified with omega-3s are the two categories that have seen the highest sales increase in the omega-3 market. Products that have shown the highest percentage increase in usage over these 2 years are also omega-3 fortified dairy products, followed by beverages and baked goods. Clinical/ enteral nutrition products come next, followed by prescription omega-3 products and infant formulas (Frost & Sullivan, 2010).

Within omega-3-fortified product categories that have multiple sources, there are consumer preferences for each. For dairy products yogurt is the most preferred source. Within beverages, natural juices followed by nutritional drinks are the most popular (Frost & Sullivan, 2010).

Globally, the market for omega-3-enriched beverages is growing steadily, especially in North America; the omega-3-enriched beverage market generated global sales in the region of over $7 billion during 2006. North America accounted for nearly 33% of global volume in 2006, followed by Western Europe at 24% and Asia/Australasia at 23%. In North America, the current 2.2 liters annual average is expected to rise to at least 2.9 liters per person by 2011 (Cosgrove, 2008).

Although getting the omega-3s into a palatable beverage can be challenging, in recent years ingredient formulators and beverage manufacturers have been working together to discover new ways to mask the fishy taste and odor of fortifying with EPA and DHA without compromising its health benefits (Cosgrove, 2008).

Consumers are becoming more aware of food products with added nutritional benefits such as vitamins, minerals, fibers and omega-3s. Nutritional beverages are an example of food products that usually contain multiple beneficial ingredients. A trend among these beverages is to be high in protein and fiber to induce satiety. Many also have added vitamins and minerals. Manufacturers of these beverages are currently developing and marketing beverages fortified with omega-3s, as well. With the existing supportive data indicating omega-3 benefits for the cognitive development in children, especially nutritional beverages targeted at children are well established in the market (Drover et al., 2009; Cheatham et al., 2006).

As mentioned, DHA is an important nutrient in prenatal and postnatal nutrition for mothers and children. It is generally associated with cell structures and has been found to be particularly important in neurologically related metabolism such as brain and retina development and function. DHA is naturally occurring in human breast milk. It is therefore one of the nutrients that is being added to most infant formulas on the market today.

The concentration of DHA in human breast milk is an important indicator for infant formula DHA formulations. Studies suggest that neural maturation of breast-fed infants is linked to breast-milk LCPUFA concentrations. The mean (+/–SD) concentration of DHA in breast milk (by wt) is 0.32 +/– 0.22% (range: 0.06–1.4). The highest DHA concentrations are primarily seen in coastal populations and were associated with marine food consumption (Brenna et al., 2007). Supplementing infant formula with DHA at the above mentioned mean concentration of 0.32 percent of total fatty acids has been reported to improve visual acuity in infants (Birch et al., 2010).

When fortifying infant formula for pre-term infants with DHA, growth of the infants was significantly enhanced compared to pre-term infants fed with non-fortified formula. Weight of the algal-DHA and fungal AA supplemented infants was also greater than fish-DHA and fungal AA supplemented infants. The observed weight increase did not differ from breast-fed term infants, which were included in the study as a reference group (Clandinin, 2005). These studies support the choice of many infant formula manufacturers to include omega-3s in their products.

Dairy is another group of food products on which scientists are focused. For many years, fluid milk has been one of the most commonly fortified foods. Lower-fat versions are fortified with enough fat-soluble vitamin A to make the per serving level of this nutrient equivalent to whole milk, and milk varieties typically have vitamin D added to prevent rickets. Dairy producers are also supplementing fluid milk as well as other dairy foods, such as cheese and yogurt, with extra calcium. More recently, omega-3s have been added to that list of beneficial ingredients to fortify dairy products.

Currently there are several products on the market fortified with EPA, DHA, and/or ALA, such as milk, buttery spreads, ice cream, and yogurt. Dairy foods have the reputation of being good for you. They deliver essential nutrients to consumers. Because dairy foods are typically refrigerated or frozen, these lower temperatures help protect the polyunsaturated fat against oxidation during storage over time. Furthermore, milk proteins have been shown to provide added protection against oxidation in food products that have been fortified with omega-3s (Livney, 2010). Dairy foods can also be easily flavored to help mask any off-odors or flavors that may develop over time. Examples are ice cream, flavored milks, and yogurts. Furthermore, dairy products provide a fat basis that makes incorporation of omega-3 fatty acids easier compared to non-fat products.

Dairy products fortified with omega-3 fatty acids has been studied as a good delivery system that benefits heart health and may reduce the risk of cardiovascular disease. Three grams of omega-3 per day in a dairy drink was associated with improvements in a range of cardiovascular risk factors, including cholesterol levels, triacylglyceride levels, and the ratio of arachidonic acid (AA) to eicosapentaenoic acid (EPA) (Dawczynski et al., 2010).

Research that assessed the feasibility of omega-3 enrichment in dairy products found that yogurt with added omega-3 fatty acids are less susceptible to oxidation than milk containing the same amount of omega-3. Differences in oxidative stability are suggested to be due to the different protein composition of milk and yogurt. Smaller fraction peptides (30, 10, and 3 kDa) were more effective antioxidants than larger fractions, especially in terms of iron-chelating activity and reducing power, of which Fe has previously been suggested to be an important catalyst in fish-oil oxidation. These findings suggest that the higher oxidative stability of yogurt might be also due to antioxidant peptides released during the fermentation of milk by lactic acid bacteria, and these peptides could be used to offer oxidative stability to food products enriched with fish oil. All of the peptides found to be anti-oxidative contained residues of the amino acid proline, a finding that supports previous research suggesting antioxidant activity could be linked to proline content. Also, a large proportion of antioxidant peptides were found to contain the hydrophobic amino acid residues valine or leucine at the N-terminus, which is consistent with the findings of previous research that suggests they have a strong link to antioxidants. In addition, the yogurt contained a considerable amount of histidine, tyrosine, methionine, and cysteine. These free amino acids have also been reported to have antioxidant properties (Farvin et al., 2010a; 2010b).

These findings suggest that the combined presence of antioxidant peptides and free amino acids in yogurt enriched with fish oil significantly contribute to its high oxidative stability. These naturally occurring antioxidant peptides as an ingredient in foods in order to increase their oxidative stability could make the enrichment with omega-3s more likely.

These antioxidant properties were further supported from studies that tested the smaller proteins with high antioxidant properties in omega-3-enriched milk which found that they had appreciable protection against oxidation of fish oil. (Farvin et al., 2010a; 2010b).

Omega-3s in Baked Goods, Cereals, and Nutritional Bars

Omega-3 can be delivered through baked goods in a variety of systems for snacks, baked goods and cereals. The fact that these foods are low in water activity with a long shelf life makes them better candidates for omega-3 fortification. Current status and future projections of marine functional ingredients in bakery and pasta products are reviewed by Kadam and Prabhasanka (2010). Straight molecularly distilled and deodorized fish oil paired with effective antioxidant blends is now effectively being used in several commercial baked products. Also, blends of fish oil with other meals such as flax meal can be blended for easier incorporation and simplify handling.

The main challenge for incorporation of omega-3 into baked goods and snacks is the extended shelf life required, which is sometimes over a year. Another challenge in fortifying baked goods and foods in general is to have the ability to handle and store refrigerated or frozen ingredients such as is required for most fish oils. So far flax has been the most widely used omega-3 source for a wide variety of baked goods, cereals, and snack bars (Chen et al., 1992).

Flax meal has the advantage that it can be used to substitute for flour in several baked goods. This is due to the fact that the meal contains soluble fiber and other carbohydrates. The soluble fiber is reported to have gums and lignans that can act as stabilizers and has been shown to improve loaf volume, oven spring, and it significantly improves bread characteristics. Recommended usage is up to 15%. Milled flax is also reported to have good water-binding properties, which enhances its ability to extend shelf life in products. Milled flax is reported to be able to replace 10 to 15% of the flour used in yeast-bread formulations. Yeast added had to be increased by 25% to maintain the same proof time, texture, and consistency (Shearer, 2005).

Either liquid fish oil in liquid or powder form functions well in baked goods. These can include breads, muffins, buns, pastries, bagels, cookies, and pizza dough. These baked goods can be refrigerated or frozen but have a short shelf life and should be well packaged. In terms of handling, fish oil is recommended to be refrigerated for short-term storage, although freezing is recommended for longer periods. Situations that may induce oxidation such as extended mix times or long-term exposure to the atmosphere should be avoided to protect product integrity. Also chelators, such as EDTA and citric acid, may be added to a mix to retard oxidation. If vitamins and minerals have to be blended in the baking mix it is recommended that they be encapsulated to prevent triggering an oxidation reaction.

Omega-3s in Pharmaceuticals and Enteral Foods

There is a large variety of enteral foods available targeted at patients suffering different types of diseases such as diabetes, cancer, and renal disease. Malnutrition is a common phenomenon for hospitalized patients in general, and there is a wide variety of products available with high levels of energy, fat, protein, and/or fiber. Next to these more general nutrients, products may have disease-targeted nutrients added, as well. An example is an immune health shake by Abbott Nutrition containing 320 mg of the plant-based omega-3 fatty acid ALA, which claims to support heart health.

Norwegian omega-3 specialist Smartfish is set to launch an omega-3 drink containing as much as 1100 mg of DHA and EPA, 10 mcg of vitamin D3, 8 g of proteins, and 200 calories. This juice-based beverage will be sold in pharmacies, hospitals, and rest homes as a medical food targeting the malnourished elderly (Starling, 2010). A recent study evaluated the effect of another Smartfish drink enriched with 500 mg EPA and DHA on the omega-3 index, a biomarker that measures EPA and DHA status in an individual. Daily intake of the convenience drink supplemented with omega-3 fatty acids leads to a significant increase of the omega-3 index. However, this study also shows a high variability response of omega-3 index among the individuals participating (Köhler et al., 2010).

In 2009, Nestlé launched a 200mL nutrient drink called Resource SeniorActiv in Switzerland targeting the malnourished elderly. Besides containing EPA and DHA omega-3 fatty acids, it is fortified with vitamin D, calcium, and protein (Starling, 2010).

They report that administration of a mixed fish oil preparation (providing around 2.2 g EPA and 1.4 g DHA daily) and a pure EPA preparation (6 g EPA daily) stabilizes weight in patients with unresectable pancreatic cancer. However, in order to lay down new tissue and thereby increase body weight additional macronutrients needed to be consumed. It has been also reported that after administration of a fish oil-enriched supplement, patients had significant weight-gain, dietary intake, performance status and appetite. These findings suggest that an EPA-enriched nutritional supplement may reverse cachexia (loss of weight in someone not actively trying to lose weight) in advanced pancreatic cancer (Barber et al., 1999).

Since the 1990s, researchers have been trying to determine whether omega-3 oils, through fish or fish oil consumption could retard the decline in cognitive function that might otherwise occur in an elderly population, which is subject to Alzheimer's disease. Several studies suggest that the omega-3 fatty acids DHA and EPA delayed the decline in cognition over time (Connor & Connor, 2007). Nutricia's (Danone) Souvenaid is one of the products high in EPA and DHA marketed for Alzheimer's disease. Other active ingredients in this beverage are uridine monophosphate, choline, phospholipids, B-vitamins, and antioxidants. The product is not yet commercially available and clinical studies are ongoing. One of the studies showed that supplementation with Souvenaid for 12 weeks improved memory (delayed verbal recall) in mild Alzheimer's disease patients. This proof-of-concept study justified further clinical trials (Scheltens et al., 2010).

Omega-3s in the form EPA/DHA ethyl esters concentrates have been approved by the United States Food and Drug Administration for the treatment of very high triglyceride levels. This prescription product is reported to contain a total of 84% (weight basis) of EPA+DHA in every 1 g capsule of omega-3 fatty acids. The dose recommended of EPA/DHA for triglyceride lowering is approximately 2-4 g/day. An alternate source of EPA and DHA from fish in the triglyceride form at lower concentrations (approx. 30%) are available without a prescription, but, as mentioned before, the American Heart Association advises that therapy with EPA and DHA to lower very high triglyceride levels should be used only under a physician's care. It is reported that in patients with triglyceride levels above 500 mg/dl, approximately 4 g/day of EPA and DHA reduces triglyceride levels 45% and very low-density lipoprotein cholesterol levels by more than 50%. Low-density lipoprotein cholesterol levels may increase depending on the baseline triglyceride level, but the net effect of EPA and DHA therapy is a reduction in non-high-density lipoprotein cholesterol level. It is also reported that in controlled trials, prescription omega-3 fatty acids were well tolerated, with a low rate of both adverse events and treatment-associated discontinuations (McKenney & Sicca, 2007). These omega-3 ethyl ester concentrates also have been reported as an effective adjuvant therapy in adult patients for secondary prevention post-myocardial infarction. The addition of 1000 mg/day of these ethyl ester concentrates to standard medical therapy provided secondary prevention benefits in post-myocardial infarction in adult patients. The benefits were attributable to reductions in death and cardiovascular death (including sudden death). Omacor®, the commercial name of ethyl ester concentrates, was reported to also reduce the incidence of severe arrhythmic events and mortality besides the improvements in hypertriglyceridemia (Hoy & Keating, 2009; Rupp, 2009).

One advantage to availability of omega-3 fatty acids in the form of prescription is that it ensures consistent quality and purity of the product. A disadvantage, on the other hand, it is higher in price and a prescription by a physician be needed.

Omega-3s as Supplements

The metabolic role of lipids and the manner in which they operate in cellular structures is increasingly better understood. This understanding has allowed for the developments of a diverse number of applications both in the pharmaceutical and cosmetic fields. The major types of lipids that are present in the human body and/or play major roles in metabolic processes are triacylglycerols (TAG), free fatty acids, phospholipids, sphingolipoids, bile salts, steroids and sterols, cholesterol, eicosanoids, and fat soluble vitamins (Hernandez, 2005b). As suggested above, essential fatty acids such as omega-3s and omega-6 play important metabolic roles as TAG, free fatty acids, and also as phospholipids.

Even though the number of foods fortified in the market is steadily increasing, the general consumer still prefers supplements as their main source of omega-3. Supplements are the most commonly used non-vitamin/non-mineral natural product taken by adults according to surveys by the National Center for Health Statistics and NCCAM on Americans' use of complementary and alternative medicine (CAM), and the second most commonly taken by children.

Though many leading health authorities and nutrition and health organizations have developed specific dietary recommendations for omega-3 fatty acids, few government health institutions have issued daily recommended allowances for EPA/DHA. As mentioned above, the US Institutes of Medicine have published recommendations for the adequate intake (AI) of short chain omega-3, linolenic acid, but none for the longer chain EPA/DHA.

Consumer awareness on the association of omega-3s with heart health is mainly responsible for the consumption increase of this supplement. One common way the general consumer prefers to take omega-3 in the form ofgel capsules. In 2009, 21% of the general population of the US self-reports to taking some form of omega-3 supplements, up from 14% in 2004 (NMI, 2009). Specific cardiovascular issues are the primary reason consumers report taking omega-3; however, consumers also report taking omega-3 supplements as help for other variety of ailments including arthritis, joint pain, memory, and mental concentration.

There are several types of omega-3 supplements available to the consumer. With regards to chain length, the short chain omega-3 in the form of alpha-linolenic acid (18 carbon long) is generally obtained from flax seed oil. Also, chia oil is now available in health stores. The longer chain omega-3, in the form of EPA and DHA, is generally supplied from fish oil in the form of gel capsules or as straight oil. The micro algae based DHA supplement source is also supplied in gel capsules.

Omega-3 supplements are also classified with regards to concentration and ratio of EPA/DHA. The most common concentration is 30% (area) or about 27% weight. The typical ratio is 18:12, EPA:DHA. Other concentrations also available include 50% EPA/DHA in different ratios of EPA/DHA.

Supplements of EPA/DHA in the form of ethyl esters have recently being introduced into the market. The concentration of EPA/DHA will vary between 50 and 84%. The 84% EPA/DHA concentration is aimed more at the pharmaceutical market and available through a doctor's prescription, but a supplement version of this product is also available through the internet. Studies on the bioavailability of different sources of marine omega-3s has been reported recently (Dyerberg et al., 2010; Neubronner et al., 2011). They compared three concentrated preparations of omega-3s (i.e., ethyl esters, free fatty acids and re-esterified triglycerides), with placebo oil in a double-blinded design, and with fish body oil and cod liver oil in a single-blinded test. They found that bioavailability of EPA+DHA from re-esterified triglycerides was superior (124%) compared with natural fish oil, whereas the bioavailability from ethyl esters was inferior (73%). No major difference was observed between free fatty acids and triglycerides.

Other studies have reported on the in vitro bioaccessibility of n-3 oils (salmon oil, tuna oil), enriched-n-3 oil and enriched-n-3 oil as ethyl esters into micellar structures after digestion. They found that after treatment with pancreatin, the highest degree of hydrolysis and inclusion of lipid products in the micellar structure was found for enriched n-3 oil, but compared to regular fish oils, long times of digestion were required. They also found the presence of n-3 concentrates in the form of EEs in the micellar structures (Martin et al., 2010).

More recently, krill oil has entered the supplements market as a new source of EPA/DHA. This oil is being marketed as having superior absorption and metabolic properties because it is in the form of phospholipids. Krill oil has been reported to increase plasma concentrations of EPA and DHA without adversely affecting indicators of safety, tolerability, or selected metabolic parameters. Reported results showed that plasma EPA and DHA concentrations increased significantly when consuming krill oil (P<0.001) when compared to other fish oils. Systolic blood pressure declined significantly more (P<0.05) (Maki et al., 2009). Krill has also been touted as being an abundant source of nutrients not just for the omega-3 phospholipid content but as a source of a variety of components, such as high-quality protein, with the advantage over other animal proteins of being low in fat and a rich source of omega-3 fatty acids. Also, antioxidant levels, such as astaxanthins, are reported to be high

Table 7.C. Types of Omega-3 Supplements.

	% EPA+DHA	Comments
Straight oil Anchovy Cod liver Tuna Salmon Menhaden	26–30%	Typical EPA:DHA ratio in oil supplements is: anchovy 18:12; cod liver 8:12; tuna 16:6; tuna head 24:5; salmon (wild) 16:11; menhaden 12:13
Oil Blends	5–30%	Blends of fish with other oil such as canola and olive oil are available commercially
Concentrate	50%	Ratio of EPA:DHA will reflect the oil source
Ethyl ester concentrates	50–84%	Used mostly in pharmaceutical applications
Krill oil	16–19%	Mostly in phospholipid form

providing additional protection to the omega-3s in the oil (Tou et al., 2007). Table 7.C shows the different types of omega-3 supplements currently available and their concentrations.

Application Techniques

As mentioned above, over the last few years scientists and manufacturers have overcome many of the challenges that previously limited the application of omega-3s in food and beverage products. Technological advances, such as microencapsulation, and the development of new antioxidants to protect polyunsaturated oils have been major contributors to the increase in omega-3 fortified beverages because they help reduce the fishy odors and taste that tend to arise as omega-3 fatty acids oxidize (Cosgrove, 2008). However, before an omega-3 oil can be stabilized and protected against oxidation it has to be properly treated to remove impurities and potentially harmful contaminants using processing techniques described in Chapter 5.

Regardless of the type of food to fortify, omega-3 fats should be added as close to the end of the process as possible. By doing so, the oil comes into minimum contact with pro-oxidants such as heat, light, and oxygen. It is also recommended that the omega-3 oil be blended with any other existing fat in the product before adding to the process. Blending omega-3 oils with other sources of fat increases thermal stability of the oil, providing added protection, or a buffer, during processing. This is especially true in the case when omega-3 oils are blended with types of fats and oils that have a high stability themselves.

Regarding processing environment and conditions, it is suggested that all equipment be free of metal ions such as copper and iron to reduce the chance of oxidation. Another important processing factor is to avoid any incorporation of air into the product. Furthermore, particularly in case of dairy and beverage manufacturing, it is important to de-aerate the product in the sterile tank before pasteurization and homogenization. Filling all tanks from the bottom is also advised, as is flushing tanks with nitrogen, where possible.

In case of homogenized beverages, enteral nutrition and most dairy products, the oil should be added prior to the homogenization step. For other products, like fruit yogurts and smoothies, omega-3s can be blended into the flavoring system, such as the fruit preparation in yogurt, and then added to the dairy food.

In the case of cheese production, it is advised to add the oil to a small portion of the milk and disperse it with high shear and then add it to the rest of the milk as it is pasteurized. This helps with the dispersion of the oil into the whole batch. Alternatively, the oil can be introduced using in-line mixers, if available. The final option is to mix the oil with the curd, but care must be taken since salt can be a pro-oxidant. The oil can be pre-mixed with a small portion of the curd and then added back to the whole. This approach is used in companies that have closed systems and re-use the whey or whey cream, since some (2% to 5%) oil will be lost to the whey (Berry, 2007).

Ice cream is another excellent fortification vehicle, since it is a frozen product. This creates the best storage option for marine EPA and DHA fortified foods, as increasing temperatures—even room temperature—may act as pro-oxidants. It is advised to add the omega-3 oil shortly before homogenization, which is relatively mild in this manufacturing process. When the ice cream base is stored for aging, it is essential to minimize exposure to oxygen.

There are also fully prepared sterile emulsions available. These can be dosed in with special equipment, even after the heating stage, which makes it ideal for aseptic products (Berry, 2007). Available emulsions are also ideal for application in beverages and dairy products that do not include a homogenization step in their current processing.

Powdered fish oil is another advisable delivery system, next to oil and emulsions. A dry product is especially convenient for manufacturers with limited frozen or refrigerated storage capacity, such as in the bakery industry. Next to powdered fish oil (with EPA and DHA) and flax seed oil (with ALA), as mentioned above, there are products available that offer the three most popular omega-3s: EPA, DHA, and ALA. These powders are also suitable for application in beverages and dairy. Another convenient application is the addition of the powder to instant drinks and soups. Regardless of the type of product to be fortified with omega-3 powders, it is advised to add the powder as late in the process as possible.

Analysis of Omega-3s

Due to the small amounts of EPA and DHA usually found in fortified foods, it is sometimes a big difference between the calculated levels of EPA and DHA in the products formulated and the experimental results from the lab. There are two different testing methods used to analyze for EPA and DHA content of oils: area percent and weight percent. The area percent method calculates EPA and DHA as a fraction of only the rest of the acids in the oil. This method does not take into account the non-fatty acid components of the oil such as glycerol, cholesterol, alcohols, hydrocarbons, and squalene, also naturally found in fish oil. These other components may account for as much as 5–15% of the total.

The weight percent method for EPA and DHA analysis takes into account the weight of the fatty acids plus glycerol and the rest of unsaponifiable matter when calculating the EPA/DHA content. This method basically reports on the weight percent of each fatty acid as compared to the overall weight of the oil (mg/g). The type of analytical method used becomes important when formulating for fortified foods, for quality control, and for product validation. This is particularly important when the levels of fortification in foods are extremely low and the fat content of the product is also very low, as in the case of fortified low fat milk. Fortified milk can have as little as 32 mg of EPA and DHA in a 250 g serving.

The fact that the fatty acids are imbedded in the matrix in the food components (i.e., complexes of protein, carbohydrates and other fats) makes the analysis of EPA and DHA ever more difficult. A method commonly used for analysis of EPA and DHA is based on the American Oil Chemists Society Method, Ce 1b-89; here the fatty acids in the oil are converted first to methyl esters. This method provides with percent area distribution of all the fatty acids in the oil including EPA and DHA, and when an internal standard is used this analysis provides with percent weight content.

A common method for recovery of the oil fraction from a food for analysis is by extraction with a solvent such a petroleum ether or hexane. However, this method is not effective to extract the oil samples from some food matrices and acid, or alkaline hydrolysis has to be performed before solvent extraction and derivatization for GC analysis. AOCS Official Method Ce 1k-09 offers the option for the preparation of fatty acid methyl esters directly from food matrices where the oil may be released from the matrix by in situ acid digestion followed by alkali hydrolysis and methylation or by simultaneous alkali hydrolysis and methylation without prior digestion. Then the fatty acid methyl esters (FAMEs) are quantitatively determined by capillary gas chromatography. AOCS method Ce 1i-07 can be used for marine and PUFA-containing oils.

Table 7.D shows the methods used for EPA/DHA and for other quality control analysis.

Table 7.D. Analytical Methods for Quality Parameters of Fish Oil.

Parameter	Method
EPA/DHA	AOCS Ce 1i-07
Acid value	AOCS Cd 3d-63
Peroxide value (PV)	AOCS Cd 8-53
Anisidine value (AV)	AOCS Cd 18-90
TOTOX	AOCS AV + 2PV
Color	AOCS Cc 13e-92

Labeling with Omega-3s

As mentioned above, omega-3 fatty acids have become one of the most popular functional ingredients incorporated into food products. It has also become a marketing opportunity to include a healthy message on the food facts label. This is evident looking at the growing number of new omega-3 products that have been introduced into the food market. However, there have been some limitations in the US regarding omega-3 health claims that can be included in a food label because of the fact that no recommended daily allowances (RDA) have yet been assigned for EPA and DHA.

Recently, Europe and some countries in Asia have issued recommended intake levels of EPA and DHA. As mentioned before, the European Food Safety Authority (EFSA) has released a series of scientific opinions on Article 13 regarding health claims for omega-3s, including three that addressed EPA and/or DHA claims for a variety of health conditions. It also established daily recommendations of 250 mg/day of EPA/DHA for cardiovascular health and 100 mg/day for children and infant development (EFSA, 2010).

In the US only qualified health claims are allowed for omega-3s. This basically means the health claim made on the label is expressed by implication as opposed to "authorized health claims" that include "significant scientific agreement" (SSA). Qualified health claims do not require highly scientific support. However they do not have the advantage of the use of factual scientific evidence to be able to recommend RDAs. The qualified health claim for omega-3s states: "Supportive but not conclusive research shows that consumption of EPA and DHA omega-3 fatty acids may reduce the risk of coronary heart disease. One serving of [Name of the food] provides [x] gram of EPA and DHA omega-3 fatty acids" (FDA/CFSAN, 2004).

There are also other labeling preconditions such as for dietary supplements that should not recommend/suggest in their labeling a daily intake exceeding 2 grams of EPA and DHA. Foods need to be low in saturated fat and low in sodium. Conventional foods also must be low in fat, saturated fat, cholesterol, and sodium, and meet 10% minimum nutrient requirements such as 10% of the Daily Value per RACC of vitamin A, vitamin C, iron, calcium, protein, or dietary fiber. Also, as already mentioned, in general, the FDA recommends that consumers not exceed more than a total of 3 grams per day of EPA and DHA omega-3 fatty acids.

Labels can also include nutrient content claims, such as "Excellent Source of EPA & DHA Omega-3s" or "Rich in EPA & DHA Omega-3s" for omega-3s EPA/DHA as long as content in the food exceeds 32 mg of EPA and DHA per serving. This "excellent source" claim is based on the AI (adequate intake) level of 160mg/day established for linolenic acid by the US Institute of Medicine (IOM). This follows the same principle for all other nutrient content claims in the US, namely that a product can claim to be an "excellent source" of a nutrient if it contains 20% of the average daily requirement levels of said nutrient.

Structure function claims can also be included in some labels; however, these are considered a more tentative claim. It basically describes the role of a nutrient or functional component in affecting or maintaining normal body structure or function or general well-being. However, it cannot describe or imply that a nutrient or functional component affects a disease or health-related condition via diagnoses, cure, mitigation, treatment, or prevention. For omega-3s such as EPA and DHA, structure function claims may include "may contribute to maintenance of heart health" and "may contribute to maintenance of mental and visual function."

References

Ariel, A.; Serhan, C.N. Resolvins and protectins in the termination program of acute inflammation. Trends Immunol. 2007, 28, 176–183.

Arterburn, L.M.; Hall, E.B.; Oken, H. Distribution, interconversion, and dose response of n-3 fatty acids in humans. Am. J. Clin. Nutr. 2006, 83(6 Suppl), 1467S–1476S.

Barber, M.D.; Ross, J.A.; Voss, A.C.; Tisdale, M.J.; Fearon, K.C.H. The effect of an oral nutritional supplement enriched with fish oil on weight–loss in patients with pancreatic cancer. Br. J. Cancer 1999, 81(1), 80–86.

Berry, D. DHA and EPA in dairy foods: Omega-3s are the trend in 2007. Dairy Foods, February 2007.

Birch, E.E.; Carlson, S.E.; Hoffman, D.R.; Fitzgerald-Gustafson, K.M.; Fu, L.N.V; Drover, J.R.; Castañeda, Y.S.; Minns, L.; Wheaton, D.K.H.; Mundy, D.; Marunycz, J.; and Diersen-Schade, D.A. The DIAMOND (DHA Intake And Measurement Of Neural Development) Study: a double-masked, randomized controlled clinical trial of the maturation of infant visual acuity as a function of the dietary level of docosahexaenoic acid. Am. J. Clin. Nutr. 2010, 91(4). 848–859.

Breivik, H.; Boslashed, B.; Dahl; Helk, K.; Krokan, H. E.; Kaare, H. Treatment and prevention of risk factors for cardiovascular diseases US Patent 5,698,594, 1997.

Brenna, JT.; Varamini, B.; Jensen, R.G.; Diersen-Schade, D.A.; Boettcher, J.A.; Arterburn, L.M. Docosahexaenoic and arachidonic acid concentrations in human breast milk worldwide. Am. J. Clin. Nutr. 2007 June, 85(6), 1457–1464.

Burdge, G.C.; Calder, P.C. a-Linolenic acid metabolism in adult humans: the effects of gender and age on conversion to longer-chain polyunsaturated fatty acids. Eur. J. Lipid Sci. Technol. 2005, 107, 426–439.

Burdge, G.C.; Wootton, S.A. Conversion of alpha-linolenic acid to eicosapentaenoic, docosapentaenoic and docosahexaenoic acids in young women. Br. J. Nutr. 2002, 88, 411–420.

Burr, R.R.; Burr. M.M. On the nature and role of the fatty acids essential in nutrition. J. Biol. Chem. 1930, 86, 587–621.

Calder, P.C. n-3 polyunsaturated fatty acids, inflammation, and inflammatory diseases. Am. J. Clin. Nutr. 2006, 83(6 Suppl), 1505S–1519S.

Carlson, S.E.; Werkman, S.H.; Rhodes, P.G.; Tolley EA. Visual-acuity development in healthy preterm infants: effect of marine-oil supplementation. Am. J. Clin. Nutr. 1993, 58, 35–42.

Carpentier, Y.A; Portois, L.; Malaisse, W.J. n–3 Fatty acids and the metabolic syndrome. Am. J. Clin. Nutr. 2006, 83(6), S1499–1504S.

Carter, J.P. Gamma linolenic acid as a nutrient. Food Tech. 1988, 42(6), 72.

Chang, S.S.; Bao, Y.; Pelura, T.J. Fish oil Antioxidants. US Patent 5,023,100, 1991.

Cheatham, C.L.; Colombo, J.; Carlson, S.E. n-3 Fatty acids and cognitive and visual acuity development: methodologic and conceptual considerations. Am. J. Clin. Nutr. 2006, 83(suppl), 1458S–1466S.

Chen, Z.Y.; Ratnayake, W.M.N.; Cunnane, S.C. Stability of flaxseed during baking. J. Am. Oil Chem. Soc. 1992, 71, 629–632.

Chung, M.K. Omega-3 fatty acid supplementation for the prevention of arrhythmias. Curr. Treat. Options Cardiovasc. Med. 2008, 10, 398–407.

Clandinin, M.T.; van Aerde, J.E.; Merkel, K.L.; Harris, C.L.; Spreinger, M.A.; Hansen, J.W.; Diersen-Schade, D.A. Growth and development of preterm infants fed infant formulas containing docosahexaenoic acid and arachidonic acid. J. Pediatr. 2005, 146(4), 461–468.

Connor, W.E.; Connor, S.L. The importance of fish and docosahexaenoic acid in Alzheimer Disease. Am. J. Clin. Nutr. 2007, 85, 929–930.

Cosgrove, J. Omega 3 beverages: New technologies to make this venerable health ingredient 'less fishy' are leading to increased beverage applications. Nutraceuticals World Iss. June 2008.

Damude, H.G.; Kinney, A.J. Engineering oilseed plants for a sustainable, land-based source of long chain polyunsaturated fatty acids. Lipids, 2007, 42, 179–185.

Dawczynski, C.; Martin, L.; Wagner, A.; Jahreis, G. n-3 LC-PUFA-enriched dairy products are able to reduce cardiovascular risk factors: A double-blind, cross-over study. Clin. Nutr. 2010 Oct., 29(5), 592–599.

Drover, J.; Hoffman, D.R.; Castaneda, Y.S.; Morale, S.E.; Birch, E.E. Three randomized controlled trials of early long-chain polyunsaturated fatty acid supplementation on means-end problem solving in 9-Month-Olds. Child Dev. 2009, 80(5), 1376–1384.

Dupont, J.; et al. Fatty acid-related functions. Am. J. Clin. Nutr. 1996, 63(6), 991S–993S.

Dupont, J. Essential fatty acids and prostaglandins. Prev. Med. 1987, 16(4), 485–492.

Dyerberg, J.; Bang, H.O.; Hjorne, N. Fatty acid composition of the plasma lipids in Greenland Eskimos. Am. J. Clin. Nutr. 1975, 28, 958–966.

Dyerberg, J.; Madsen, P.; Møller, J.M.; Aardestrup, I.; Schmidt, E.B. Bioavailability of marine n-3 fatty acid formulationsProstaglandins, Leukotrienes Essent. Fat. Acid. 2010, 83(3), 137–141.

EFSA. Scientific Opinion on the substantiation of health claims related to eicosapentaenoic acid (EPA), docosahexaenoic acid (DHA), docosapentaenoic acid (DPA) and maintenance of normal cardiac function.... EFSA J. 2010, 8(10), 1796.

Farvin, K.H.S.; Baron, C.P.; Nielsen, N.S.; Jacobsen, C. Antioxidant activity of yoghurt peptides: Part 1-in vitro assays and evaluation in n-3 enriched milk. Food Chem. 2010a, 123(4), 1081–1089.

Farvin, K.H.S.; Baron, C.P.; Nielsen, N.S.; Jacobsen, C. Antioxidant activity of yoghurt peptides: Part 1-in vitro assays and evaluation in n-3 enriched milk. Food Chem. 2010b, 123(4), 1090–1097.

FDA/CFSAN. Letter Responding to Health Claim Petition dated June 23,2003 (Wellness petition): Omega-3 Fatty Acids and Reduced Risk of Coronary Heart Disease (Docket No. 2003Q-0401), 2004.

Frost & Sullivan. US Consumers' Choice: Omega-3 Nutrient Products, August 2010.

Gebauer, S.K.; Psota, T.L.; Harris, W.S.; Kris-Etherton, P.M. n-3 Fatty acid dietary recommendations and food sources to achieve essentiality and cardiovascular benefits. Am. J. Clin. Nutr.
2006, 83, 1526S–1535S.

Gewirtz, A.T.; Collier-Hyams, L.S.; Young, A.N. Lipoxin A4 analogs attenuate induction of
intestinal epithelial proinflammatory gene expression and reduce the severity of dextran sodium
sulfate-induced colitis. J. Immunol. 2002, 168, 5260–5267.

Hardman, W.E. (n-3) Fatty Acids and Cancer Therapy. J. Nutr. 2004, 134, 3427S–3430S.

Harris, W.S.; Lemke, S.L.; Hansen, S.N.; Goldstein, D.A.; DiRienzo, M.A.; Su, H.; Nemeth,
M.A.; Taylor, L.; Ahmed, G.; George, C. Stearidonic acid-enriched soybean oil increased the
omega-3 index, an emerging cardiovascular risk marker. Lipids 2008, 43, 805–811.

Hernandez, E. Production, Processing and Refining of Oils. Healthful Lipids, C. Akoh, O.M. Lai,
Eds.;AOCS: Champaign IL, 48–64.

Hernandez, E. Pharmaceutical and Cosmetic Use of Lipids. Bailey's Industrial Oils and Fat Products, F. Shahidi, Ed. John Wiley & Sons: New York.

Holman, R.T.; Johnson, S.B.; Hatch, T.F. A case of human linolenic acid deficiency involving
neurological abnormalities. Am. J. Clin. Nutr. 1982, 35, 617–623.

Holman, R.T. The slow discovery of the importance of omega 3 essential fatty acids in human
health. J. Nutr. 1998, 128(2S), 427S–433S.

Hoy, S.M.; Keating, G.M. Omega-3 ethylester concentrate: a review of its use in secondary
prevention post-myocardial infarction and the treatment of hypertriglyceridaemia. Drugs. 2009,
May 29, 69(8), 1077–105.

IOM (Institutes of Medicine). Dietary Reference Intakes for Energy, Carbohydrate, Fiber, Fat,
Fatty Acids, Cholesterol, Protein, and Amino Acids. National Academies Press: Washington
D.C., 2010.

Jeun-Horng, L.; Yuan-Hui, L.; Chun-Chin, K. Effect of dietary fish oil on fatty acid composition,
lipid oxidation and sensory property of chicken frankfurters during storage, Meat Sci. 2002,
60(2), 161–167.

Johnson, M.; Swan, D.D.; Surette, M.E.; Stegner, J.; Chilton, T.; Fonteh, A.N.; Chilton, F.H.
Dietary supplementation with gamma-linolenic acid alters fatty acid content and eicosanoid
production in healthy humans. J. Nutr. 1997, 127(8), 1435–1444.

Kadam, S.U.; Prabhasanka, P. Marine foods as functional ingredients in bakery and pasta products. Food Res. Inter. 2010, 43(8), 1975–1980.

Kahn, W.A.; Blobe, G.C.; Hannun, Y.A. Arachidonic acid and free fatty acids as second messengers and the role of protein kinase C. Cell. Signal. 1995, 7, 171–184.

Kinsella, J.E. Food lipids and fatty acids: importance in food quality, nutrition, and health. Food
Technol. 1988, 42(10), 124–142.

Köhler, A.; Bittner, D.; Löw, A.; von Schacky, C. Effects of a convenience drink fortified with n-3
fatty acids on the n-3 index. Br. J. Nutr. 2010 Sep, 104(5), 729–736.

Kolanowski, W.; Swiderski, F.; Berger, S. Possibilities of fish oil application for food products
enrichment with omega-3 PUFA. Intern. J. Food Sci. Nutr. 2002, 50(1), 39–49.

Kris-Etherton, P.M.; Harris, W.S.; Appel, L.J. Fish consumption, fish oil, omega-3 fatty acids, and
cardiovascular disease. Circ. 2002, 106(21), 2747–2757.

Levy, B.D. Resolvins and protectins: Natural pharmacophores for resolution biology. Prostagland. Leukotrienes Essent. Fat. Acids. 2010, 82, 327–332.

Livney, Y.D. Milk proteins as vehicles for bioactives. Curr. Opin. Colloid Inter. Sci. 2010, 15(1–2), 73–83.

McKenney, J.M.; Sicca, D. Role of prescription omega-3 fatty acids in the treatment of hypertri- glyceridemia. Pharmacotherapy 2007, 27(5), 715–728.

Martin, D.; Nieto-Fuentes, J.A.; Señoráns, F.J.; Reglero, G.; Soler-Rivas, C. Intestinal digestion of fish oils and ω-3 concentrates under in vitro conditions. Eur. J. Lipid Sci. Technol. 2010, 112, 1315–1322.

Melanson, S.F.; Lewandrowski, E.L.; Flood, J.G.; Lewandrowski, K.B. Measurement of organo- chlorines in commercial over-the-counter fish oil preparations: implications for dietary and therapeutic recommendations for omega-3 fatty acids and a review of the literature. Arch. Pathol. Lab Med. 2005, 129, 74–77.

Maki, K.C.; Reeves, M.S.; Farmer, M.; Griinari, M.; Berge, K.; Vik, H.; Hubacher, R.; Rains, T.M. Krill oil supplementation increases plasma concentrations of eicosapentaenoic and docosahexaenoic acids in overweight and obese men and women. Nutr. Res. 2009 Sep, 29(9), 609–615.

Massrieh, W. Health benefits of omega-3 fatty acids from Neptune krill oil. Lipid Technol. 2008, 20, 108–111.

Muller, R.H.; Keck, C.M. Challenges and solutions for the delivery of biotech drugs–a review of drug nanocrystal technology and lipid nanoparticles. J. Biotechnol. 2004, 113, 151–170.

Natural Marketing Institute, NMI's 2009 Health and Wellness Trends Database(tm) (HWTD) and NMI's 2009 Supplement/OTC/Rx Database(tm) (SORD)Natural Marketing Institute (NMI), April 2010.

Neubronner, J.; Schuchardt, J.P.; Kressel, G.; Merkel, M.; von Schacky, C.; Hahn, A. Enhanced increase of omega-3 index in response to long-term n-3 fatty acid supplementation from triacyl- glycerides versus ethyl esters. Eur. J. Clin. Nutr. Published on line ahead of print doi: 10.1038/ ejcn.2010.239.

Rupp, H. Omacor(r) (prescription omega-3-acid ethyl esters 90): From severe rhythm disorders to hypertriglyceridemia. Adv. Ther. 2009, 26(7), 675–690.

Scheltens, P.; Kamphuis, P.J.G.H.; Verhey, F.R.J.; Olde Rikkert, M.G.M.; Wurtman, R.J.; Wilkinson, D.; Twisk, J.W.R.; Kurz, A. Efficacy of a medical food in mild Alzheimer's disease: A randomized controlled trial. Alzheimer's & Dementia: J. Alzheimer's Assoc. 2010, 6(1), 1–10.

Sellmayer, A.; Koletzko, B. Long-chain polyunsaturated fatty acidy and eicosanoids in infant- physiological and pahtophysiological aspects and open questions. Lipids, 1999, 34, 199–205.

Serhan, C.N.; Hong, S.; Gronert, K. Resolvins: a family of bioactive products of omega-3 fatty acid transformation circuits initiated by aspirin treatment that counter proinflammation signals. J. Exp. Med. 2002, 196(8), 1025–1037.

Serhan, C.N. Novel omega-3– derived local mediators in antiinflammation and resolution. Phar- macol. Ther. 2005, 105, 7–21.

Shearer, A.E.H.; Davies, C.G.A. Physicochemical properties of freshly baked and stored whole- wheat muffins with and without flaxseed meal. J. Food Qual. 2005, 28, 137–153.

Simopoulos, A.P. Essential fatty acids in health and chronic disease. Food Rev. Inter. 1997, 13(4), 623–631.

Singer, P.; Shapiro, H.; Theilla, M.; Anbar, R.; Singer, J.; Cohen, J. Anti-inflammatory properties of omega-3 fatty acids in critical illness: novel mechanisms and an integrative perspective. Intensive Care Med. 2008, 34, 1580–1592.

Starling, S. Scandinavians target elderly with 1000mg+ omega-3 drink. NutraIngredients, 26 August 2010.

Sundram, K.; Perlman, D.; Hayes, K. Blends of palm fat and corn oil provide oxidation-resistant shortenings for baking and frying. US Patent 5,874,117.

Tapiero, H.; Nguyen, B.G.; Couvreur, P.; Tew, K.D. Polyunsaturated fatty acids (PUFA) and eicosanoids in human health and pathologies. Biomed. Pharmacother. 2002, 56, 215–222.

Tou, J.C.; Jaczynski, J.; Chen, Y.C. Krill for human consumption: nutritional value and potential health benefits. Nutr. Rev. 2007, 65(2), 63–77.

Tecelãoa, C.; Silva, J.; Dubreucqc, E.; Ribeirod, M.H; Ferreira-Dias, S. Production of human milk fat substitutes enriched in omega-3 polyunsaturated fatty acids using immobilized commercial lipases and Candida parapsilosis lipase/acyltransferase. J.Mol. Catalys. B: Enzymatic, 2010 (in press).

Trautwein, E.A. N-3 Fatty acids – physiological and technical aspects for their use in food. Eur. J. Lipide. Sci. 2001, 103, 45–55.

Ursin, V.M. Modification of plant lipids for human health: development of functional land-based omega-3 fatty acids. J. Nutr., 2003, 133, 4271–4274.

Vaisman, N.; Kaysar, N.; Zaruk-Adasha, Y.; Pelled, D.; Brichon, G.; Zwingelstein, G.; Bodennec, J. Correlation between changes in blood fatty acid composition and visual sustained attention performance in children with inattention: effect of dietary n-3 fatty acids containing phospholipids. Am. J. Clin. Nutr. 2008, 87, 1170–1180.

Venkateshwarlu, G.; Let, M.B.; Meyer, A.S.; Jacobsen, C. Modeling the sensory impact of defined combinations of volatile lipid oxidation products on fishy and metallic off-flavors. J. Agric. Food Chem. 2004, 52(2), 311–317.

Wang, C.; Harris, W.S.; Chung, M.; Lichtenstein, A.H.; Balk, E.M.; Kupelnick, B.; Jordan, H.S.; Lau, J. n-3 Fatty acids from fish or fish-oil supplements, but not α-linolenic acid, benefit cardiovascular disease outcomes in primary- and secondary-prevention studies: a systematic review. Am. J. Clin. Nutr. 2006, 84, 5–17.

Warnants, N.; Van Oeckel, M.J.; Boucque. C.V. Effect of incorporation of dietary polyunsaturated fatty acids in pork backfat on the quality of salami. Meat Sci. 1999, 49(4), 435–445.

8

Synergistic/Additive Health Effects of Fish Oil and Bio-Active Compounds

Tomoko Okada, Nana Mikami, Tomoyuki Okumura, Masashi Hosokawa, and Kazuo Miyashita

Graduate School of Fisheries Sciences, Hokkaido University, Hakodate, Hokkaido, Japan

Introduction

Fish oil with high levels of long chain n-3 polyunsaturated fatty acids (LC n-3 PUFA) has been shown in numerous research studies to be of nutritional importance in the diets of both animal and human subjects. The nutritional significance of fish oil is principally associated with the n-3 PUFA eicosapentaenoic acid (EPA) and docosahexaenoic acid (DHA). Consumption of fish oil has been linked to several beneficial physiological effects including improvement of lipid metabolism (Eslick et al., 2008; Nieuwehuys et al., 2001), prevention of coronary heart diseases (Erkiläa et al., 2008; Harris, 2007; Holub & Holub, 2004; Jacobson, 2007; Schacky, 2007), reduction of inflammatory response (Brunborg et al., 2008; Innis et al., 2006), remediation of diabetes (Montori et al., 2000; Nettleton & Katz, 2005), delayed onset of Alzheimer's disease (Boudrault et al., 2008; Cole et al., 2005), and inhibition of various types of cancers (Berquin et al., 2008; Roynette et al., 2004).

Given that the favorable bioactivity of LC n-3 PUFAs has been well established, research interests have moved towards achieving the maximum physiological benefits of fish oil. Thus far, a limited amount of studies have shown synergistic/additive physiological effects of fish oil and other bio-active agents such as vitamins, carotenoids, and commercially available drugs (e.g., statins), as well as bio-matrix such as fiber. Although this area of study is still in its early stages, the growing body of research on synergistic/additive health effects of fish oil and bio-active agents shows a new promising trend of study related to human nutrition.

This chapter will summarize the current body of knowledge regarding the additive effects of fish oil and bio-active agents on the prevention and remediation of clinical disorders. In each section, the singular effect of fish oil consumption in relation to a specific illness will be briefly discussed, followed by the reported synergistic/additive effects of fish oil combined with bio-active agents. We will discuss only the diseases in which studies have reported combinative effects of fish oil and bio-active agents, such as dyslipidemia, diabetes, obesity and others. As the majority of research in this subject area relates to lipid metabolism disorders, the opening section briefly introduces some common risk factors and health problems associated with the metabolic syndrome in relation with dyslipidemia. This will provide a basis for discussing the attenuation of specific diseases by fish oil combined with bio-active agents. In the discussion of the combinative therapy of fish oil and bio-active agents, it should be noted that some of the studies were not carried out specifically with fish oil, but rather with the isolated fatty acid ethyl esters of the predominant LC n-3 PUFAs in fish oil.

Fish Oil and Metabolic Syndrome in Relation with Dyslipidemia

The metabolic syndrome is a worldwide problem characterized by impaired dyslipidaemia, insulin sensitivity, abdominal obesity and hypertension. These symptoms are common precursors to type 2 diabetes mellitus, coronary heart diseases and premature death (Roche, 2005). They are associated with increased levels of circulating metabolites, such as free fatty acids and triacylglycerols (TG), and unfavorable cytokines, such as tumor necrosis factor-α (TNF-α) and interleukin-6 (IL-6) (Esposito & Giugliano, 2006). One of the main factors of metabolic syndrome disorder is atherogenic dyslipidemia, which is characterized by elevated TG levels (hypertriglyceridemia), reduced high-density lipoprotein cholesterol (HDL-C) levels, and increased concentrations of low-density lipoprotein cholesterol (LDL-C) (Krauss, 2005). Hypercholesterolemia is one of the most important coronary risk factors for cardiac morbidity and mortality, and it is directly related to serum cholesterol levels (Chen et al., 1999). Hypertriglyceridemia, the most common type of dyslipidemia, is a risk factor for progression of atherosclerosis and a strong independent predictor of future myocardial infarction (Qi et al., 2008). Furthermore, the high cardiovascular mortality rate in diabetes has been partly attributed to such unfavorable lipoprotein profiles.

Consumption of dietary fats has a profound effect on plasma lipids and lipoproteins and may explain many of the effects of lipids on risk factors for several major diseases. Fish oil consumption has a substantial effect on improvement of dyslipidaemia. It has been shown to lower levels of plasma TG (Eslick et al., 2008; Harris & Bulchandani, 2007; Moore et al., 2006), HDL-C, and LDL-C levels in experimental animals and in normal and hypertriglyceridemic men (Harris et al., 2008; Kris-Etherton et al., 2003; Mozaffarian & Rimm, 2006).

Effect of Fish Oil and Vitamins on Dyslipidemia

Vitamin E

Vitamins possess numerous beneficial biofunctionalities and have important roles in lipid metabolism. At the foremost, vitamin E and its isomers are well known as the most abundant lipid-soluble antioxidants in plasma and LDL-C. As a highly potent antioxidant, vitamin E inhibits lipid peroxidation and attenuates atherosclerosis (Pryor, 2000). Several human and animal studies have suggested that vitamin E have beneficial effects on atherosclerosis and coronary heart disease (Bron & Asmis, 2001; Kritharides & Stocker, 2002; Kushi et al., 1996; Mune et al., 1999; Steinberg & Workshop-participants, 1992; Stephens et al., 1996). However, some studies have reported that vitamin E act as a pro-oxidant (Chen et al., 1999; Schneider, 2005), and in controlled trials vitamin E was found to have no effects on cardio vascular diseases (Eidelman et al., 2004; Lee et al., 2005; Morris & Carson, 2003; Yusuf et al., 2000). In addition to its biofunctionality, vitamin E has been widely used as a food additive to prevent lipid oxidation of oil-rich products. Vitamin E is especially useful in fish oil, which has high potential for oxidation due to its high levels of LC n-3 PUFAs.

Taking into account that vitamin E has been suggested to reduce the peroxidative damage that can occur in fish oil, a few investigations have been carried out on the concomitant intake of fish oil and vitamin E in terms of lipid metabolism. The aim of these investigations was to improve or enhance the physiological benefits of fish oil by preventing lipid peroxidation with vitamin E. Hsu and others (2001) conducted a study to determine if both fish oil and vitamin E intake can boost the antioxidant defense system in hypercholesterolemic-induced rabbits. Male New Zealand white rabbits were randomly divided into 5 groups, each of which was fed a different diet. The reference rabbits were fed regular laboratory rabbit chow, and the experimental groups were fed one of the following diets: (1) high fat and cholesterol, (2) high fat cholesterol + vitamin E, (3) high cholesterol + fish oil, (4) high cholesterol + fish oil + vitamin E. The authors found a synergistic effect of fish oil and vitamin E intake on glutathione reductase activity, but not on glutathione peroxidase or catalase activity. They also determined plasma lipid peroxide levels by measuring plasma thiobarbituric acid-reactive substances (TBARS). Based on the results of this test, the authors found that feeding rabbits high-cholesterol diets containing fish oil resulted in a significant increase in TBARS compared to rabbits fed high-cholesterol diets without fish oil and high-cholesterol diets with vitamin E. However, dietary intake of high cholesterol, fish oil and vitamin E significantly decreased TBARS in relation to the high-cholesterol and fish oil diet. Rabbits fed diets high in cholesterol showed marked atheroma formation in the ascending aorta; however, addition of fish oil into the diet attenuated this effect. Furthermore, supplementation of vitamin E to the diet containing cholesterol and fish oil showed a favorable synergistic effect, with decreases in atheroma formation. The synergistic effects observed for vitamin E and fish oil were suggested to be due to enhanced glutathione reductase activity along with potent antioxidant activity exhibited by vitamin E (Hsu et al., 2001).

Mune and others (1999) focused on the effects of concomitant application of fish oil and vitamin E, as well as probucol on renal injury from hypercholesterolemia. The study utilized nephrectomized cholesterol-fed Sprague Dawley rats as a model for progressive kidney disease. Rats were fed either a control diet or a cholesterol-supplemented diet (2%) that contained either fish oil, fish oil with vitamin E, or fish oil with probucol. The focus of the study was on the attenuation of renal injury, which is associated with an increased serum cholesterol level. Plasma lipid profile was determined after an experimental feeding period of 4 weeks. Rats fed fish oil with probucol showed the lowest serum total cholesterol (1.73 ± 0.44 mmol/L) and LDL-C + very low-density lipoprotein cholesterol (VLDL-C) (1.04 ± 0.34 mmol/L) among all groups, with ranges of 2.59–3.21 mmol/L and 1.24–2.15 mmol/L, respectively. Diets containing fish oil with probucol also resulted in the lowest TG levels among all the groups. Overall, the study demonstrated that dietary fish oil could slow the progression of glomerulosclerosis in rats with remnant kidney nephrons. There were no supplementary effects of vitamin E addition (500 IU vitamin/kg body) to fish oil. However, the study demonstrated that the synergistical antioxidative effects of fish oil and 1% probucol further reduced glomerulosclerosis in the remnant kidney model of progressive kidney disease in the rat.

Vitamin B₃

Vitamin B_3 (Niacin) has been used clinically for 4 decades and is the most effective treatment currently available to increase levels of HDL-C. It has also been reported to reduce coronary heart disease morbidity and mortality when taken either alone or in combination with statins (Taylor et al., 2004). Niacin has recently attracted renewed interest because, in addition to being the most potent drug available for the treatment of low HDL-C, it can also lower levels of LDL-C and TG. To this regard, an arterial disease multiple intervention trial evaluated the effect of niacin on lipid and lipoprotein levels in patients with diabetes and peripheral arterial disease (Elam et al., 2000). The study found that treatment with niacin significantly decreased TG and LDL-C and significantly increased HDL-C in subjects with and without diabetes. Recent studies demonstrated that treatment with niacin raises adiponectin secretion by 52–95% in patients with metabolic syndrome and high-molecular weight adiponectin (the fraction most affected by niacin treatment) (Westphal et al., 2006; Westphal & Luley, 2008). Another study revealed that niacin could stimulate the transcription of several genes involved in lipid metabolism, including peroxisome proliferators activated receptor (PPAR) γ, cluster of differentiation 36, and ATP binding cassette transporter A1 in monocytoid cells (Yang et al., 2008).

Based on the beneficial physiological functionalities of niacin, Isley and others (2007) studied the effects of combined intake of niacin and LC n-3 PUFA on dyslipidemia by measuring TG, HDL-C and LDL-C concentrations, as well as serum phospholipids fatty acid composition. In this study, 29 patients with atherogenic dyslipidemia were given either a dual placebo, LC n-3 PUFA (Omacor(tm), Pronova Biocare, Norway), crystalline niacin, or a combination of LC n-3 PUFA and crystalline niacin daily for 12 weeks. Total cholesterol increased significantly (by 13%) in the LC n-3 PUFA group compared to the niacin and combination group, while a 52% decrease in serum TG was observed in the combination group. The significant findings of this study were that combined daily treatment with 3.4 g LC n-3 PUFA plus 3 g crystalline niacin lowered serum triglyceride levels by :50% and raised HDL-C by: 30% (Isley et al., 2007).

Niacin-associated flushing, induced by drug-mediated release of vasodilatory prostaglandins synthesized from phospholipids, is frequently reported by patients (Taylor & Stanek, 2008). In relation to this matter, Qi and others (2008) found that diets containing pure fish oil or 15% fish oil in terms of total fat significantly reduced plasma TG concentrations by 40–60% in C57 BL/6J mice in either the fasting or postprandial state. The authors suggested that consumption of LC n-3 PUFA, which is present in high amounts in fish oil, reduces both fasting and postprandial TG concentrations (Qi et al., 2008). As prostaglandin $(PGD)_2$ are linked to niacin-associated flushing, as well as the fact that PGD_3 is produced by LC n-3 PUFAs, Isley and others (2007) focused on whether the niacin flush could be diminished by increasing tissue levels of LC n-3 PUFA. Based on their results, the authors did not find such an effect of LC n-3 PUFA, although they suggested that the applied dose of LC n-3 PUFA may have been too small to sufficiently shift membrane ratios of arachidonate to EPA.

Effect of Fish Oil and Commercially Available Drugs in Dyslipidemia

Statins

Statins are a class of hypolipidemic drugs that inhibit 3-hydroxy-3methylglutaryl-CoA (HMG-CoA) reductase. They are able to decrease LDL levels primarily by raising the number of LDL receptors, and consequently enhancing the removal of LDL from plasma. Research reported by Heart Protection Study Collaborative Group (2002) has shown that statins are beneficial to patients for primary prevention of coronary artery disease. Statins have a beneficial effect on lipid levels but also decrease cardiovascular events and mortality because of their pleiotropic effects (improved endothelial function, antithrombotic and antioxidant effects, anti-inflammatory properties, and stabilization of atherosclerotic plaque) (Corsini et al., 2007; Davignon, 2004). Several studies have evaluated whether co-administration of LC n-3 PUFA with statins improves the lipid profile in patients with hypertriglyceridemia (Becker et al., 2008; Chan et al., 2002a; 2002b; Durrington et al., 2001; Grekas et al., 2000; Nordoy et al., 1998; Saify et al., 2003; 2005).

A comparative study with white male rabbits was conducted to examine the potential for fish oil, lovastatin and/or gemfibrozil to lower cholesterol with high-potency and minor side effects. The study found synergistic effects of fish oil and lovastatin: their combined administration resulted in a significant reduction in HDL-C level (by 50.67%), while fish oil treatment alone showed only 34.20% reduction (Saify et al., 2003). The authors conducted a subsequent study with the same three compounds in which they measured total cholesterol and total lipid levels. They found that the most significant cholesterol-lowering effect occurred with the concomitant administration of fish oil and lovastatin. The total cholesterol lowering mechanism was suggested to be due to overall selective inhibition of HMG-CoA reductase by lovastatin (Saify et al., 2005).

Another study using human subjects evaluated the efficacy of combined treatment with low-dose pravastatin and fish oil in post-renal transplantation dyslipidemia subjects (Grekas et al., 2000). The study found that the concomitant intake of fish oil and pravastatin for 8 weeks after a 4-week washout period showed significant reductions in total cholesterol, TG, LDL-C, as well as apolipoprotein (apo) B concentration. The authors concluded that the combined treatment with fish oil and pravastatin is superior to the pravastatin treatment alone in correcting hyperlipidemia in renal transplant patients, especially in those who are resistant to lipid-lowering diets (Grekas et al., 2000). In a different study, 48 male subjects with insulin-resistant obese dyslipidemia were used to test combinative effects of fish oils (4g/day) and atorvastatin (40 mg/day) on apoB kinetics (Chan et al., 2002a). Compared with the placebo group, combined treatment decreased VLDL-apoB secretion, as well as percent conversion of VLDL-C to LDL-C. A possible mechanism was suggested to be due to differential effects of fish oil and atorvastatin, where the latter decreases cholesterol synthesis and the former reduces TG synthesis. Durrington and others (2001) reported that two grams of LC n-3 PUFAs (Omacor(tm): 44% EPA and 36% DHA) administered twice a day for 48 weeks decreased serum TG by 20–30% and VLDL-C by 30–40% in patients with combined hyperlipidemia and coronary heart disease who had already been receiving simvastatin 10–40 mg daily. Nordoy and others (1998) carried out a double-blind placebo controlled randomized study evaluating the effect of simvastatin separately and in combination with n-3 PUFAs (4 g/day, Omacor(tm): 45% EPA and 39% DHA) in 41 healthy patients with defined hyperlipidemia. Simvastatin (20 mg/day) reduced serum TC, TG, apoB and apoE and increased HDL-C and apoA$_1$. The study found a further reduction of serum TG, TC, and apoE through the addition of LC n-3 PUFAs, illustrating the additive effects of fish oil and simvastatin. It was suggested that redundant effects of LC n-3 PUFAs were suppressed when the cholesterol synthesis was inhibited by simvastatin. This was based on the result that supplementation with LC n-3 PUFA slightly decreased TC without increasing LDL-C levels. Furthermore, 65% of the patients given the combinative treatment of simvastatin and LC n-3 PUFAs showed a significant reduction of LDL-C in addition to the reduction in total cholesterol. A further study showed that LC n-3 PUFA can enhance the effectiveness of atorvastatin on LDL-subfractions and postprandial hyperlipemia in patients with combined hyperlipemia (Nordoy et al., 2001). In this study, LC n-3 PUFAs (2 g/day, Omacor(tm)) with either atorvastatin (10 mg/day) or a placebo (corn oil, 2 g/day) were administered to 42 patients with combined dyslipidemia. Concomitant intake of fish oil with atorvastatin significantly increased HDL-C levels and reduced small dense LDL particles and postprandial hypertriglyceridemia compared to the atorvastatin therapy alone. Similarly, Chan and others (2002b) tested the effect of atorvastatin and LC n-3 PUFA co-administration with 52 obese men (body mass index : 29 kgm^{-2}) with dyslipidemia. They were subjected to atorvastatin (40 mg/day), LC n-3 PUFAs (4g/day, Omacor(tm)), a combination of these or placebo for 6 weeks. The combination treatment of both atorvastatin and LC n-3 PUFAs showed additive effects on plasma TG and HDL-C concentration, as well as RLP-C and apoC-III levels, as compared with monotherapy of atorvastatin or fish oil. For a general overview on the combined use of LC n-3 PUFAs and statins, see Nambi and others (2006).

Acetylsalicylic Acid

Antiplatelet therapy with acetylsalicylic acid (aspirin) is commonly used to reduce the risk of cardio-and cerebrovascular events. There are only a few reports discussing the combination of acetylsalicylic acid with fish oil in terms of lipid composition and platelet function, and the results have not been consistent (Engstrom et al., 2001; Harris et al., 1990; Svaneborg et al., 2002). One study was conducted with 18 healthy men to examine the acute and short-term effects of supplementation with LC n-3 PUFA and intravenous acetylsalicylic acid on platelet function, platelet fatty acid composition and plasma lipids (Svaneborg et al., 2002). The LC n-3 PUFA group received 10 g LC n-3 PUFAs (a fish oil TG concentrate, Epax 5500 TG, produced by Pronova Biocare, Norway) while the placebo group received 10 g of n-6 PUFAs. The supplementation periods were 14 h for the acute effects and 14 days for short-term effects. An intravenous injection of 100 mg acetylsalicylic acid was administered. Blood samples were collected before and after intravenous injections. No significant effects were found for the acute state, but after 14 days of LC n-3 PUFA administration, LDL-C decreased significantly as compared to the control group. The authors concluded that the study did not show any additive effects of LC n-3 PUFA and acetylsalicylic acid (Svaneborg et al., 2002).

Effect of Fish Oil, Capsaicin, and Caffeine in Diabetes

LC n-3 PUFAs are known to lower TG levels and possibly increase HDL-C levels in subjects with type 2 diabetes, which is typically associated with dyslipidemia and obesity (Nettleton et al., 2005). Capsaicin and caffeine are known to improve lipid metabolism similarly to fish oil. Therefore, as part of our ongoing research, the combined effect of fish oil, capsaicin and caffeine on visceral fat weight and blood glucose levels in KK-A^y mice has been studied. Capsaicin, which is the primary compound responsible for the pungency of chilies, has been found to be effective for improving glucose tolerance and glucose-stimulated insulin response and can also reduce diabetic hyperglycemia in animal models (Gram et al., 2005; Tolan et al., 2001). Tolan and others (2001) reported that capsaicin intake decreased blood glucose levels to 4.91 ± 0.52 (n = 6) mmol/dL versus 6.40 ± 0.13 mmol/dL (n = 6) for the control group (p < 0.05) in male mongrel dogs. Results were obtained after a 2.5 h time interval with an oral glucose tolerance test. Relative to the control group, plasma insulin levels increased to 5.78 ± 0.76 µIU/mL (n = 6) for the capsaicin-treated dogs compared to 3.70 ± 0.43 µIU/mL (n = 10) for the control (p<0.05).

As a widely used psychoactive stimulant drug, the effect of caffeine as well as coffee consumption has been the focus of many studies, although the outcomes are debatable. Some studies have reported deleterious effects of caffeine including impairment of glucose metabolism in both animal model and human subjects (Biaggioni & Davis, 2002; Lane et al., 2004). On the other hand, several studies have demonstrated that caffeine actually improves glucose homeostasis in rats, as well as in both healthy and obese man (Graham et al., 2001; Keijzers et al., 2002; Park et al., 2007; Petrie et al., 2004). Studies have also demonstrated that caffeine intake is associated with a substantially lower risk of clinical type 2 diabetes (Dam & Hu, 2005). In an animal model using obese Zucker rats, capsaicin intake induced sensory nerve desensitization, which consequently improved glucose tolerance. This nerve desensitization occurred due to the binding of capsaicin to small unmyelinated sensory nerves. This binding leads to a reduction in the levels of calcitonin gene-related peptide, which is released from sensory nerves upon stimulation and has been shown in vitro to induce insulin resistance and inhibit insulin secretion (Gram et al., 2005).

We have been testing the combinative effect of fish oil with capsaicin and caffeine in terms of white adipose tissue (WAT) weight and blood glucose level in KK-A^y mice. The mice were divided into 8 groups, with the control group diet consisting of 14% soybean oil (SO). Four experimental diets were designed to contain either 7% SO + 7% fish oil (FO), or 7% SO + 7% FO with (1) capsaicin, (2) capsaicin + caffeine, or (3) caffeine. The other three groups were fed 14% SO with either (4) capsaicin, (5) capsaicin + caffeine, or (6) caffeine (with 0.07% capsaicin and 0.054% caffeine). Significant reductions were found in body and WAT weights for mice fed fish oil with capsaicin and caffeine. Furthermore, blood glucose levels in all the experimental groups except the fish oil monotheraphy and the soybean oil + capsaicin diet were significantly lower relative to the control. Interestingly, there was an additive effect of fish oil + capsaicin as well as fish oil + capsaicin + caffeine on blood glucose levels in relation to the fish oil administration alone (43.0% reduction and 59.7% reduction, respectively) (Fig. 8.1). These results suggest that concomitant intake of

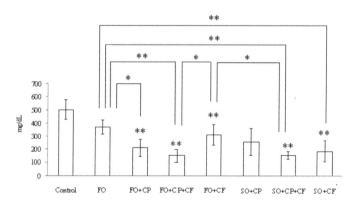

Fig. 8.1. Blood glucose levels of KK-A^y mice fed the control or experimental diets: (1) control [AIN-93G diet containing 14% soy bean oil (SO)], (2) 7% soybean oil + 7% fish oil (FO), (3) 7% SO + 7% FO + 0.007% capsaicin (CP), (4) 7% SO + 7% FO + 0.007% CP + 0.054% caffeine (CF), (5) 7% SO + 7% FO + 0.054% CF, (6) 14% SO + 0.007% CP, (7) 14% SO + 0.007% CP + 0.054% CF, (8) 14% SO + 0.054% CF for 30 days. Each data point represents the mean ± SE of seven mice. *, P < 0.05; **, P < 0.01.

fish oil and capsaicin + caffeine may possibly improve lipid metabolism disorders including diabetes. Plausible mechanisms of this combinative therapy of fish oil with capsaicin and caffeine has been under investigation.

Effect of Fish Oil and Marine-derived Agents in Obesity

Obesity increases the risk of developing type 2 diabetes (Nettleton et al., 2005) and has been identified as the key causative agent in the development of the metabolic syndrome (Nettleton et al., 2005; Roche, 2005). Changes in adipose tissue mass are associated with changes in the metabolic functions of adipose tissue, whereby increased adipocyte volume is positively correlated with leptin production, a key regulator of metabolic rate. It is also known that adipose tissue from obese animals/humans expresses increased amounts of proinflammatory proteins (e.g., TNF-α, IL-6, inducible nitric oxide synthase, transforming growth factor β1), and procoagulant proteins (e.g., plasminogen activator inhibitor type-1) (Weisberg et al., 2003).

There are a few reports concerning the antiobesity effects of LC n-3 PUFAs (Garaulet et al., 2001; Lombardo & Chicco, 2006; Soria et al., 2002; Wang et al., 2002). Soria and others (2002) demonstrated that fish oil administration resulted in reduction of fat pad adiposity, cell hypertrophy and lipolysis, as well as the normalization of insulin action in epididymal and whole body peripheral tissue. Wang and others (2002) reported that fish oil (EPA-28; from Yamanouchi Pharmaceutical, Tokyo, Japan) intake resulted in significant reduction of adipose tissue (epididymal, perirenal, and inguinal) accumulation in C57BL/6J male mice over the course of 7 weeks, as compared with a low-fat diet and a highly saturated diet with the same levels of energy intake. Furthermore, a four-week dietary reversal showed that replacing the fish oil diet with the saturated fat diet resulted in significant body fat accumulation. There is now evidence that fish oil may work via PPARs, which control the expression of genes involved in lipid and glucose metabolism and adipogenesis (Soria et al., 2002). Neschen and others (2006) suggest that fish oil stimulates adiponectin secretion in epididymal fat in a PPARγ-dependent and PPARα-independent manner, and that part of the anti-inflammatory, antiatherogenic, and antidiabetic effects of fish oil may be mediated by this mechanism.

It is now clear that LC n-3 PUFAs possess an anti-inflammatory effect and that they act directly on transcription factors to influence gene expression, which is mediated through eicosanoids. Studies show that dietary fish oil intake results in altered lymphocyte function and in suppressed production of proinflammatory cytokines by macrophages (Calder, 2001). Fish oil with more than 2.4 g of EPA+DHA administered daily decreased the production of TNF-α, IL-1β, and IL-6 by LPS-stimulated mononuclear cells; decreased lymphocyte proliferation and decreased the production of IL-2; and interferon-γ. Most often, the finding that LC n-3 PUFAs diminish inflammatory and immune cell functions is interpreted in a favorable way, with the conclusion that these fatty acids are anti-inflammatory and so will be beneficial to health (Wallace et al., 2003). As obesity is associated with chronic inflammatory mediators, such as TNFα and monocyte chemoattractant protein 1 (Nagao &

Yanagita, 2008), there are some studies showing that LC n-3 PUFA may offer a useful anti-inflammatory dietary strategy to decrease obesity-related disease (Browning, 2003).

Fucoxanthin

Obesity causes adipocyte hypertrophy, which is associated with decreased adiponectin secretion accompanied by increased secretion of unfavorable cytokines. The result is induction of insulin resistance, as well as atherosclerosis. Adipocyte hypertrophy increases with fat accumulation and, therefore, excess accumulation of adipose tissue in obesity can lead to numerous problematic consequences for human health. The human body has two kinds of adipose tissue: brown adipose tissue (BAT) and white adipose tissue (WAT). BAT accumulates lipids in lower amounts than WAT and instead contains more mitochondria. BAT also contains the unique uncoupling protein 1 (UCP1) and, because of this protein, lipid oxidation is poorly coupled to ATP synthesis and energy is dissipated. Overall, the main function of BAT is fat consumption while the main function of WAT is fat accumulation.

In recent years, our focus has been on increasing UCP1 expression in both BAT and WAT. The target compound has been a carotenoid from seaweed called fucoxanthin. Similar to LC n-3 PUFA-rich fish oil, fucoxanthin has exhibited very unique biological activities, such as suppressive effects on adipocyte differentiation, antimutagenicity, anti-ocular inflammation, and cancer growth (Das et al., 2008). Fucoxanthin (Fig. 8.2) is a naturally abundant pigment found in different types of edible brown seaweed such as *Undaria pinnatifida, Laminaria japonica,* and *Hijikia fusiformis.* In our previous studies we reported the ability of fucoxanthin to induce apoptosis in human leukemia cells (Hosokawa et al., 1999), as well as in colon cancer cells (Hosokawa et al., 2004). Fucoxanthin has also exhibited a suppressive effect on lipopolysaccharide-induced inflammation both in vitro and in vivo (Shiratori et al., 2005). Furthermore, fucoxanthin has been shown to possess antiobesity properties both in vivo (Maeda et al., 2005; Maeda et al., 2007) and in vitro (Maeda et al., 2006).

We have proposed a partial mechanism of antiobesity action demonstrated by fucoxanthin through our previous reports (Maeda et al., 2005; 2006). The abdominal fat weight reduction was caused by fucoxanthin or fucoxanthin-containing Wakame (*Undaria pinnatifida*) lipid feeding in KK-A^y mice, and weight reduction might be linked to energy dissipation by means of the generation of heat by UCP1 expression in WAT. Furthermore, we also conducted a study to elucidate whether fucoxanthinol (Fig. 8.2), a metabolite of fucoxanthin, significantly suppressed lipid accumulation in 3T3-L1 cells. The results showed that glycerol 3-phosphate dehydrogenase (GPDH) activity, which plays a key role in TG synthesis, was decreased due to fucoxanthinol treatment (Fig. 8.3). According to these results, we propose that the fucoxanthinol may be the active compound for the antiobesity effect shown by fucoxanthin, as demonstrated in the animal study. In the same study, PPARγ

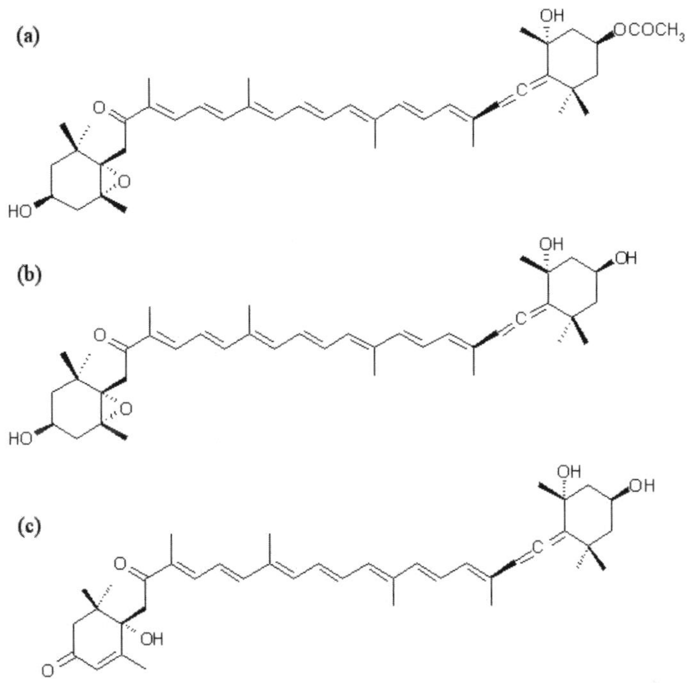

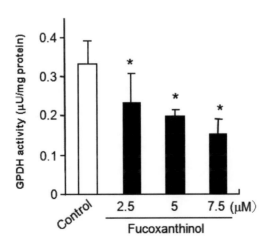

Fig. 8.2. Structure of (a) fucoxanthin and its metabolites; (b) fucoxanthinol and (c) amarouciaxanthin A.

Fig. 8.3. Effect of fucoxanthinol on GPDH activity in 3T3-L1 cells induced during adipocyte differentiation. 3T3-L1 cells were treated with fucoxanthinol for 120 h and measured intercellular GPDH activity. The values (n=3) are expressed as micro units per milligram total cell protein. Asterisks indicate significantly different values: *$P<0.05$.

levels were down-regulated in 3T3-L1 cells treated with fucoxanthinol. Thereby, it is assumed that fucoxanthinol and other metabolites of fucoxanthin are able to influence levels of PPARγ, a known nuclear transcription factor that regulates adipogenic gene expression (Paul, 2001).

To further examine the mechanism of action, fucoxanthin metabolites were identified using NMR analysis in our laboratory and quantified with reversed high performance liquid chromatography. The results of NMR showed that fucoxanthin was metabolized into three different compounds: fucoxanthinol, amarouciaxanthin A, and an unknown compound in agreement with a previous study (Asai et al., 2004). Quantification of these compounds revealed that fucoxanthinol was the predominant fraction in all organs, while amarouciaxanthin A was primarily accumulated in WAT (Fig. 8.4). We have proposed that carotenoids possessing certain structural characteristics (i.e., allene bond and hydroxyl substituent on the adjacent end group) may exhibit antiobesity properties, such as the ability to suppress adipocyte differentiation (Okada et al., 2008). Interestingly, amarouciaxanthin A, which showed the highest concentration of fucoxanthin metabolites in WAT, possesses these structural characteristics. For that reason, amarouciaxanthin A may be somewhat associated with the antiobesity properties linked to WAT weight reduction; however, this issue is still under investigation in our laboratory. According to recent findings, we hypothesize that the beneficial anti-obesity effect of orally-administered fucoxanthin may be associated with fucoxanthin metabolites rather than fucoxanthin itself.

Given that both fish oil and fucoxanthin show antiobesity effects with similar proposed mechanisms of action involving induced expression of UCP1, we conducted a study to elucidate whether or not there exists an additive antiobesity effect when dietary intakes of fish oil and fucoxanthin are combined. After administration

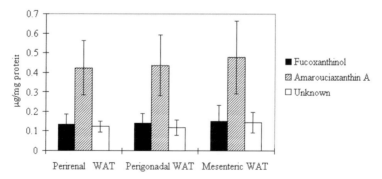

Fig. 8.4. Fucoxanthin metabolites (expressed as μg/mg protein) accumulated in white adipose tissue (perirenal WAT, perigonadal WAT, and mesenteric WAT) of KK-A^y fed experimental diet containing 13.4% soybean oil supplemented with 0.1% fucoxanthin for 3 weeks.

of either 0.1% fucoxanthin, fish oil (half of the amount of 14% soybean oil from the control diet was replaced by fish oil), or 0.1% fucoxanthin plus fish oil, WAT weights (uterine, mesentery, perirenal, and retroperitoneal) were determined (Fig. 8.5). In terms of uterine WAT, 0.1% fucoxanthin plus fish oil diet remarkably showed a suppressive effect on weight gain of WAT compared to 0.1% fucoxanthin or 7% fish oil, showing possible additive effect of fish oil and fucoxanthin. Studies have reported that LC n-3 PUFAs induce the expression of lipolytic genes, which is under PPAR regulation, thereby enhancing β-oxidation of fatty acids (Nagao et al., 2008; Tsub-oyama-Kawaoka et al., 1999). It also has been reported that LC n-3 PUFAs including EPA can bind to PPARα with reasonable affinity, and their metabolites have greater affinity for PPARα than the parent fatty acids. Along these lines, we can assume that fish oil intake could also induce UCP1 expression in a similar manner as fucoxanthin. However, the results show that fish oil alone does not increase UCP1 expression compared to the control at the doses used in this study. This is in contrast to fucoxanthin, which showed a tendency to increase UCP1 expression, as demonstrated by the 0.1% fucoxanthin plus fish oil diet relative to fish oil intake alone. With this in mind, we suggest that fucoxanthin and fish oil may possess different mechanisms of action in

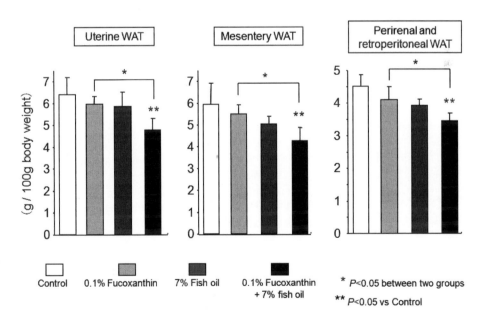

Fig. 8.5. Adipose tissue weight of KK-A^y mice fed fucoxanthin and/or fish oil. KK-A^y mice were fed the experimental diets for 4 weeks.

terms of gene regulation, in spite of the fact that the outcomes observed in WAT weight reduction were identical for fucoxanthin and fish oil administration.

Taurine

Taurine is another unique bio-active compound that has been of interest in our laboratory in recent years. It is a small amino sulphonic acid (2-amino ethane sulphonic acid) that is ubiquitous in mammalian tissues. Taurine is the most abundant free amino acid in the heart, retina, skeletal muscle, brain, and leukocytes, and it participates in the transformation of cholesterol to water-soluble bile salts (Schuller-Levis & Park, 2003). Taurine has been found to be a key compound in various biological systems, such as the formation of bile acids, cholesterol excretion, osmoregulation, cell volume regulation, ion transport (Na+ Clx, Na^+-Ca^{2+}-exchange), inhibition of protein phosphorylation, formation of N-chlorotaurine in leukocytes as part of the host defense, scavenging of reactive carbonyl compounds, and pre- and postnatal development (Hansen, 2001).

Studies have demonstrated that taurine has various physiological properties. It is anti-hypertensive and anti-hypercholesterolemic, it improves insulin sensitivity in animal models, and it attenuates lipid metabolism in both animal and human subjects (Harada et al., 2004; Matsushima et al., 2003; Nandhini et al., 2002; Zhang et al., 2004). A recent study showed that taurine possesses an antiobesity effect by reducing high fat diet-induced increases in body weight, parametrial WAT weight, estimated percentage of body fat, and adipocyte size (Tsuboyama-Kasaoka et al., 2006). In comparing these results to a few previous studies that reported no anti-obesity effect of taurine, the authors suggested that a large amount of taurine (~3 mg/g BW per day) is perhaps required to observe an anti-obesity effect, especially in the taurine-deficient state accompanied with extreme obesity status.

The results of our study supported the anti-obesity effect of marine-derived taurine, and also suggested a combinative effect of taurine and fish oil in an animal model. KK-A^y mice were divided into 4 groups: (1) control (AIN-93G diet containing 14% soy bean oil), (2) taurine (4% taurine + 14% soy bean oil), (3) fish oil (7% fish oil + 7% soy bean oil) and (4) taurine + fish oil (7% fish oil + 7% soy bean oil + 4% taurine). The phenotypic changes and WAT weight of each group were determined after a 4-week feeding period. The data showed that intake of fish oil and taurine + fish oil resulted in a significant reduction in the total WAT weight (by 14.8% and 20.0%, respectively) relative to the control group. In addition, intake of both fish oil alone and taurine + fish oil resulted in a 16% reduction in the total WAT compared to intake of taurine alone. Further analysis revealed that intake of taurine + fish oil resulted in a significant reduction in the leptin mRNA expression in epididymal WAT relative to taurine alone, as well as intake of fish oil alone (Fig. 8.6). As leptin mRNA and plasma leptin levels correlate with increased bodyweight and decreased weight loss, the results of our study suggest an encouraging combinative anti-obesity effect of taurine and fish oil.

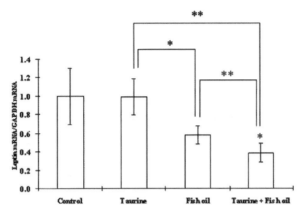

Fig. 8.6. Changes in level of leptin mRNA expression in epididymal WAT of KK-A^y mice fed the control or experimental diets: (1) control (AIN-93G diet containing 14% soy bean oil), (2) taurine (4% taurine + 14% soy bean oil), (3) fish oil (7% fish oil + 7% soy bean oil), and (4) taurine + fish oil (7% fish oil + 7% soy bean oil + 4% taurine) for 4 weeks. Each data point represents the mean ± SE of seven mice. * P < 0.05; ** P < 0.01.

Additional Effects of Fish Oil with Bioactive Compounds

Cancer Prevention

Although emphasis has been placed on the effect of the total amount of fat in the diet, it also appears that fat quality is an important factor in terms of cancer predisposition (Roynette et al., 2004). One study reported that frequent fish intake appears to be a favorable indicator for reduced risk of several common cancers, mostly those of the digestive tract, but also of the endometrium, ovary, and prostate (Vecchia, 2004). Epidemiological studies have suggested that higher intake of saturated fatty acid is associated with increased risk of breast cancer (Holnes & Willett, 2004; Roynette et al., 2004), and several studies report protective properties of LC n-3 PUFA in the early stages of colon cancer development (Roynette et al., 2004). Also, it is well established that DHA and EPA intake (from 5% to 20% wt:wt as fish oil) reduces the growth of tumors in rodents, including those of the mammary gland colon, prostate, liver, and pancreas. Research with human cancer cell lines has convincingly demonstrated that DHA and/or EPA can reduce the growth of many different human tumor types, including breast, colon, pancreatic, chronic myleogenous leukemic, and melanoma cell lines (Field & Schley, 2004). Many factors may influence tumor induction and growth in the case of colon cancer, including a range of cytokines and growth factors, along with genotoxic and oxidative stress. LC n-3 PUFAs have been proposed to influence colon carcinogenesis by one or both of the following mechanisms: (1) altering enzyme expression and/or activity and, consequently, the concentrations of end products and/or (2) by modulating the levels of available precursors for biosynthetic pathways (Roynette et al., 2004).

A few studies have considered the potential additive effects of fish oil with bio-active agents in targeting cancer prevention. One study was conducted using an animal model to determine whether the combinative treatment of fish oil with the antioxidant vitamin E and concomitant use of conventional cytotoxic agent cisplatin (CP) would have a suppressive effect on tumor growth. Yam and others (2001) used C57 BL/6J female mice that were fed one of three isocaloric diets: (1) 5% soybean oil supplemented with 40 mg/kg α-tocopherol acetate (SO diet), (2) 4% fish oil plus 1% corn oil, and a basal amount of vitamin E (FO diet), or (3) FO diet supplemented with vitamin E and C (FO + E + C diet). These diets were tested in combination with CP in a series of regimens. The authors found that the FO diet followed by CP treatment and the FO + E + C diets showed the best anticancer activity. This finding suggests that the anticancer activity of CP can be more effective when supported by an FO diet with low levels of antioxidants.

Another antioxidative vitamin, vitamin D, has been tested for a synergistic effect with fish oil in prostate cancer prevention. Instead of focusing on the antioxidant properties of vitamin D, Istfan and others (2007) were interested in the fact that human prostate cells contain receptors for the active form of vitamin D, 1α, 25-dihydroxyvitamin D [1α, 25(OH)$_2$D], which decreases invasiveness and metastasis of prostate cancer cells. They investigated the influence of fish oil and vitamin D on cell cycle kinetics in cultured androgen-dependent LNCaP c-38 and androgen-independent LNCaPc-115 prostate cancer cells. The study showed that fish oil specifically inhibited the G1/S-phase transition and moreover, 1α, 25(OH)$_2$D$_3$ synergistically enhanced this effect in androgen-independent prostate cancer cells (Istfan et al., 2007).

Similar to LC n-3 PUFAs in fish oil, melatonin has been reported to show an anti-inflammatory effect (Mayoa et al., 2005) and suppresses cancer progression in the case of breast cancer (Cos et al., 2004; Sánchez-Barceló et al., 2005) and prostate cancer (Shiu, 2007). Melatonin acts by inhibiting growth, slowing tumor progression, and enhancing the effects of chemotherapy at decreased toxicity and antiangiogenic activity (Lissoni et al., 2001). Melatonin is a naturally synthesized neurohormone that participates in the control of cell differentiation and proliferation. In addition to being an oncostatic agent, it is also a potent antioxidant and a free radical scavenger (Karbownik et al., 2001). Persson and others (2005) conducted a study using human subjects to examine the combination effect of fish oil and melatonin on cachexia and biochemistry variables, including TNFα, IL-1β, soluble IL-2 receptor, IL-6, and IL-8, in patients with advanced gastrointestinal cancer. Twenty-four patients were randomly placed in two groups: one group received 30 mL/day of fish oil (EPA: 4.9 g, DHA: 3.2 g) and another group received 18 mg/day of melatonin. The intervention period was carried out for 4 weeks. Approximately 63% of patients showed weight stabilization or gain after fish oil and melatonin combined intervention, but there were no changes in C-reactive protein, fibrinogen, and cytokines that indicate inflammation suppression (Persson et al., 2005).

Lycopene, which is found in tomatoes, has demonstrated anticarcinogenesis activity in prostate cancer (Giovannucci et al., 2002; Kanagaraj et al., 2007), colon cancer (Shlomo et al., 2007) and breast cancer (Chalabi et al., 2006). Tang and others (2009) demonstrated that lycopene (10 μM) could effectively suppress the proliferation of colon cancer cells up to 47% during a 24-h period. It was suggested that lycopene could suppress proliferation of human colon cancer cells via modulation of cell signaling pathways (Tang et al., 2009). The combined effects of lycopene and EPA on the blockade of colorectal cancer were further evaluated by a cell line study. The study demonstrated that lycopene (2 μM) and EPA (25 μM) synergistically inhibited cell proliferation of human colon cancer cell HT-29 by approximately 70%, while treatment with lycopene or EPA alone blocked proliferation of human colon cancer cells by only 18% and 20%, respectively. The authors utilized a western blotting assay to further investigate if this action is associated with the PI-3K/Akt/mTOR signaling pathway, which plays an important role in tumor progression. The results showed that concomitant intake of lycopene and EPA significantly suppressed the activation of Akt compared with untreated colon cancer Ht-29 cells. Immunocytochemical staining results revealed that lycopene and EPA could also up-regulate the expression of apoptotic proteins such as Bax and Fas ligands to suppress cell survival. They concluded that decreased activity of Akt enhanced the Bax and Fas ligand proteins, thus preventing cell survival (Tang et al., 2008).

A recent study targeting colon cancer prevention demonstrated a combinative effect of dietary intake of fish oil and pectin in an animal model. Vanamala and others (2008) demonstrated that fish oil and pectin administration synergistically protected radiation-enhanced colon cancer by up-regulating apoptosis in colonic mucosa. They used forty male Sprague-Dawley rats that were fed either fish oil (18.2% EPA and 11.3% DHA in a 15% fat diet) + pectin or corn oil (55.4% LA in a 15% fat diet) + cellulose prior to irradiation treatment (1 Gy, 1 GeV/nucleon Fe-ion). Rats were injected with 15 mg/kg of a colon-specific carcinogen, azoxymethane (AOM), 10 and 17 days after irradiation. The results showed that the fish oil with pectin diet induced a higher apoptotic index in both AOM injection alone and irradiation treatment as compared to the corn oil + cellulose diet. This was associated with suppression of anti-apoptotic mediators in the cyclooxygenase pathway and the Wnt/β-catenin pathway. This study revealed that the concomitant intake of fish oil and pectin may be used as a countermeasure against radiation-enhanced colon carcinogenesis by maintaining protective levels of apoptosis (Vanamala et al., 2008).

Anti-inflammatory Effects

Likely due to its anti-inflammatory properties, fish oil has been shown to be beneficial for reducing proteinuria in some models of nephritis, treating human primary glomerulonephritis and preventing the evolution of glomerulosclerosis in a rat nephritis model (Mune et al., 1999). Over the last 10 years there has been increasing interest in the effects of LC n-3 PUFAs on human immune function. Eicosanoids are hormone-like compounds, made from C20 fatty acids (mainly arachidonic acid in current western diets). The eicosanoids induced by LC n-3 PUFA [PGE_3 and pro-inflammatory $(LTB)_5$] have a lower potency as pro-inflammatory mediators than PGE_2 and LTB_4 derived from n-6 (Camuesco et al., 2005). Previous studies demonstrated a beneficial effect of LC n-3 PUFAs on treatment of inflammatory bowel disease (IBD) patients owing to their ability to impair the colonic antioxidant system and thereby promote oxidative injury at the site of inflammation. To overcome this problem, Camuesco and others (2005) found that a diet rich in olive oil resulted in a lower colonic inflammatory response in rats with DSS-induced colitis compared to the control group, which was fed a soybean oil-based diet. This beneficial effect was increased by the dietary incorporation of fish oil rich in EPA and DHA. The levels of TNFα and LTB_4 were further reduced by the addition of fish oil, but not soybean oil, in colitis rats. The researchers further studied this matter by administering LC n-3 PUFAs with quercitrin, which is a potent antioxidative flavonoid. This study by Camuesco and others (2006) demonstrated an additive effect of fish oil and quercitrin in preventing colonic damage induced by DSS. This was thought to be due to the combination of the different mechanisms of action of quercitrin (inhibition of colonic TNFα, IL-1β production, and its antioxidant activity) and fish oil (inhibition of colonic TNFα and LTB_4 production and its antioxidant activity). The study implied that these parameters may synergistically act to ameliorate the colonic damage induced by DSS. The main advantage of the combinative treatment was suggested to be reduced IL-1β levels, which may have explained the faster recovery obtained in the combined treatment.

Concomitant intake of fish oil and atorvastatin was also tested for additive effects on inflammatory markers in the case of 48 obese individuals (body mass index > 29 kg/m^2 with dyslipidemia). Participants were randomly assigned to a 6-week treatment with either (1) atorvastatin (40 mg/day), (2) LC n-3 PUFA (4g/day, Omacor(tm): 45% EPA and 39% DHA), (3) concomitant intake of atorvastatin and LC n-3 PUFA, or (4) a placebo. Astorvastatin treatment revealed significant effects in terms of decreasing total cholesterol, TG, LDL-C, apoB and lathosterol and increasing HDL-C, while fish oil also showed a significant effect in lowering plasma TG and increasing HDL-C. Atorvastatin, but not fish oil, significantly lowered plasma high-sensitivity C-reactive protein (hs-CRP) concentration with a concurrent reduction of plasma IL-6 concentrations. Although fish oil intake alone had no effect on plasma hs-CRP, the authors concluded that addition of fish oil to statins may supplementary optimize the lipid-regulating effects by enhancing a decrease in plasma TG and increase in HDL-C (Chan et al., 2002).

In addition, a relatively small-scaled study found that intake of acetylsalicylic acid in combination with fish oil resulted in improvements in the whole blood production of eicosanoids in 4 voluntary, healthy, non-smoking subjects with a mean age of 42.8+6.1 years. Intake of acetylsalicylic acid in combination with fish oil resulted in a significant reduction in thromboxane B_2 (by 62%), whereas acetylsalicylic acid alone reduced thromboxane B_2 levels by only 40%. LTB_4 levels were significantly increased (by 19%) with acetylsalicylic acid alone, while combined acetylsalicylic acid and fish oil intake resulted in a significant reduction (69%) compared to the control (before acetylsalicylic acid, no fish oil). Low-doses of acetylsalicylic acid are associated with inflammatory side-effects, including asthma; however, fish oil exhibits an anti-inflammatory effect due to its inhibitory action on the production of LTB_4. Based on the results of this study, it was suggested that fish oil intake may prevent acetyl-salicylic acid-induced injury of the gastric mucosa and possibly acetylsalicylic acid-induced asthma (Engstrom et al., 2001).

Implications for Infant Nutrition

Intake of fish oil that is especially rich in DHA has shown beneficial effects on infant development. Brain and retina development occurs *in utero* and in infancy, and most DHA accumulates in the fetus between the 26th and 40th week of gestation. Infants born before week 32 of gestation have low brain concentrations of DHA (Sanders, 1993). Several studies show a positive relation between LC n-3 PUFA intake and infant development (Jensen, 2006; Lotte et al., 2005). Krauss-Etschmann and others (2007) conducted a multicenter (Germany, Hungary, and Spain), randomized, double-blind, 2×2 factorial, placebo-controlled study using 315 pregnant women to elucidate the combined effect of fish oil and methyltetrahydrofolic acid (MTHF). Although no significant difference was found in terms of maternal or fetal compli-cations and birth outcomes, fish oil supplementation significantly increased both maternal DHA and EPA during the supplementation period (at gestation week 20, gestation week 30, and delivery). In all cases, FO plus MHTF resulted in even higher levels of both DHA and EPA. The authors concluded that daily supplementation of 0.5 g DHA and 0.15 g EPA from gestation week 22 until delivery increases maternal and fetal plasma DHA and maternal plasma EPA levels. It was also suggested that the addition of MTHF may enhance maternal LC n-3 PUFA levels.

Conclusions

There is now substantial evidence that fish oil possess various physiological activity benefits in human health. This chapter has discussed concomitant applications of fish oil with bio-active compounds for the remediation of a variety of clinical disorders. By combinative therapy, well-documented beneficial health effect of fish oil can pos-sibly be maximized in various approaches. Although this area of research is still in its early stages, it has been receiving increasing attention with the realization that com-binative application of fish oil and bio-active compounds may provide vital outcomes and facilitate progress towards the design of beneficial clinical therapies for human health. Studies aimed at finding superior combinative effects of fish oil and bio-active compounds along with their mechanisms of action may eventually lead to reductions in the mortality and morbidity rates associated with human health disorders and complications that are of growing concern in modern society.

References

Asai, A.; Sugawa, T.; Ono, H.; Nagao, A. Biotransformation of fucoxanthin into amarouciaxanthin A in mice and HEPG2 cells: formation and cytotoxicity of fucoxanthin metabolites. Drug. Metab. Dipos. 2004, 32, 205–211.

Becker, D.J.; Gordon, R.Y.; Morris, P.B.; Yorko, J.; Gordon, Y.J. ;Li, M.; Iqbal, N. Simvastatin vs therapeutic lifestyle changes and supplements: randomized primary prevention trial. Mayo. Clin. Proc. 2008, 83, 758–764.

Berquin, I.M.; Edwards, I.J.; Chen, Y.Q. Multi-targeted therapy of cancer by omega-3 fatty acids. Cancer Letters 2008, 269, 363–377.

Biaggioni, I.; Davis, S.N. Caffeine: A cause of insulin resistance? Diabetes Care 2002, 25, 399–400.

Boudrault, C.; Bazinet, R.P.; Ma, D.W.L. Experimental models and mechanisms underlying the protective effects of n-3 polyunsaturated fatty acids in Alzheimer's disease. J. Nutr. Biochem. 2008, Article in press.

Bron, D.; Asmis, R. Vitamin E and the prevention of atherosclerosis. Int. J. Vitam. Nutr. Res. 2001, 71, 18–24.

Browning, L.M. n-3 polyunsaturated fatty acids, inflammation and obesity-related disease. Proc. Nutr. Soc. 2003, 62, 447–453.

Brunborg, L.A.; Madland, T.M.; Lind, R.A.; Arslan, G.; Berstad, A.; Froyland, L. Effects of short-term oral administration of dietary marine oils in patients with inflammatory bowel disease and join pain: A pilot study comparing seal oil and cod liver oil. Clin. Nutr. 2008, 27, 614–622.

Calder, P.C. Polyunsaturated fatty acids, inflammation, and immunity. Lipids 2001, 36, 1007–1024.

Camuesco, D.; Comalada, M.; Concha, A.; Nieto, A.; Sierra, S.; Xaus, J.; Zarzuelo, A.; Galvez, J. Intestinal anti-inflammatory activity of combined quercitrin and dietary olive oil supplemented with fish oil, rich in EPA and DHA (n-3) polyunsaturated fatty acids, in rats with DSS-induced colitis. Clin. Nutr. 2006, 25, 466–476.

Camuesco, D.; Galvez, J.; Nieto, A.; Comalada, M.; Rodriguez-Cabezas, M.E.; Concha, A.; Xaus, J.; Zarzuelo, A. Dietary olive oil supplemented with fish oil, rich in EPA and DHA (n-3) polyunsaturated fatty acids, attenuates colonic inflammation in rats with DSS-induced colitis. J. Nutr. 2005, 135, 687–694.

Chalabi, N.; Delort, L.; Corre, L.L.; Satih, S.; Bignon, Y.; Bernard-Gallon, D. Gene signature of breast cancer cell lines treated with lycopene. Pharmacogenomics 2006, 7, 663–672.

Chan, D.C.; Watts, G.F.; Barret, P.H.; Beilin, L.J.; Redgrave, T.G.; Mori, T.A. Regulatory effects of HMG CoA reductase inhibitor and fish oils on apolipoprotein B-100 kinetics in insulin-resistant obese male subjects with dyslipidemia. Diabetes 2002, 51, 2377–2386.

Chan, D.C.; Watts, G.F.; Barrett, P.H.R.; Beilin, L.J.; Mori, T.A. Effect of atorvastatin and fish oil on plasma high-sensitivity C-reactive protein concentrations in individuals with visceral obesity. Clin. Chem. 2002a, 48, 877–883.

Chan, D.C.; Watts, G.F.; Mori, T.A.; Barrett, P.H.R.; Beilin, L.J.; Redgrave, T.G. Factorial study of the effects of atorvastatin and fish oil on dyslipidaemia in visceral obesity. Eur. J. Clin. Inv. 2002b, 32, 429–436.

Chen, M.; Hsu, H.; Liau, C.; Lee, Y. The role of vitamine E on the anti-atherosclerotic effect of fish oil in diet-induced hypercholesterolemic rabbits. Pros.Other Lipid Mediat. 1999, 57, 99–111.

Cole, G.M.; Lim, G.P.; Yang, F.; Teter, B.; Begum, A.; Ma, Q.; Harris-White, M.E.; Frautschy, S.A. Prevention of Alzheimer's disease: Omega-3 fatty acid and phenolic anti-oxidant interventions. Neurobiol. Aging 2005, 26, 133–136.

Corsini, A.; Ferri, N.; Cortellaro, M. Are pleiotropic effects of statins real? Vasc. Health Risk Manag. 2007, 3, 6 11–613.

Cos, S.; Martínez-Campa, C.; Mediavilla, M.d.; Sánchez-Barceló, E.J. Melatonin modulates aromatase activity in MCF-7 human breast cancer cells. J. Pineal Res. 2004, 38, 136–142.

Dam, R.M.v.; Hu, F.B. Coffee Consumption and Risk of Type 2 Diabetes. JAMA 2005, 294, 97–104.

Das, S.K.; Hashimoto, T.; Kanazawa, K. Growth inhibition of human heaptic carcinoma HepG2 cells by fucoxanthin is associated with down-regulation of cyclin D. Biochim. Biophys. Acta. 2008, 1780, 743–749.

Davignon, J. Cardioprotective and other emerging effects of statins. Int. J. Clin. Pract. Suppl. 2004, 143, 49–57.

Durrington, P.N.; Bhatnagar, D.; Mackness, M.I.; Morgan, J.; Julier, K.; Khan, M.A.; France, M. An omega-3 polyunsaturated fatty acid concentrate administered for one year decreased triglycerides in simvastatin treated patients with coronary heart disease and persisting hypertriglyceridaemia. Heart 2001, 85, 544–548.

Eidelman, R.S.; Hollar, D.; Hebert, P.R.; Lamas, G.A.; Hennekens, C.H. Randomized trials of vitamin E in the treatment and prevention of cardiovascular disease. Arch. Intern. Med. 2004, 164, 1552–1556.

Elam, B.M.; Hunninghake, D.B.; Davis, K.B.; Garg, R.; Johnson, C.; Egan, D.; Kostis, J.B.; Sheps, D.S.; Brinton, E.A. Effect of niacin on lipid and lipoprotein levels and glycemic control in patients with diabetes and peripheral arterial disease. JAMA 2000, 284, 1263–1270.

Engstrom, K.; Wallin, R.; Saldeen, T. Effect of low-dose aspirin in combination with stable fish oil on whole blood production of eicosanoids. Prosta. Leuk. Essent. Fatty Acids 2001, 64, 291–297.

Erkiläa, A.; Mello, V.D.F.d.; Risérusd, U.; Laaksonen, D.E. Dietary fatty acids and cardiovascular disease: An epidemiological approach. Prog. Lipid Res. 2008, 47, 172–187.

Eslick, G.D.; Howe, P.R.C.; Smith, C.; Priest, R.; Bensoussan, A. Benefits of fish oil supplementation in hyperlipidemia: systematic review and meta-analysis. Int. J. Card. 2008, Article In Press.

Esposito, K.; Giugliano, D. Diet and inflammation: a link to metabolic and cardiovascular diseases. Eur. Heart. J. 2006, 27, 15–20.

Field, C.J.; Schley, P.D. Evidence for potential mechanisms for the effect of conjugated linoleic acid on tumor metabolism and immune function: lessons from n-3 fatty acids. Am. J. Clin. Nutr. 2004, 79, 1190S–1198S.

Garaulet, M.; Prez-Llamas, F.; Perez-Ayala, M. Site-specific differenes in the fatty acid composition of abdomical adipose tissue in an obese population from Mediterranean area: relation with dietary fatty acids, plasma lipid profile, serum insulin, and central obesity. Am. J. Clin. Nutr. 2001, 74, 585–591.

Giovannucci, E.; Rimm, E.B.; Liu, Y.; Stampfer, M.J.; Willett, W.C. A Prospective Study of To-mato Products, Lycopene, and Prostate Cancer Risk. J. Natl. Cancer Inst. 2002, 94, 391–398.

Graham, T.E.; Sathasivam, P.; Rowland, M.; Marko, N.; Greer, F.; Battram, D. Caffeine ingestion elevates plasma insulin response in humans during an oral glucose tolerance test. Can. J. Physiol. Pharmacol. 2001, 79, 559–565.

Gram, D.X.; Hansen, A.J.; Wilken, M.; Elm, T.; Svendsen, O.; Carr, R.D. Plasma calcitonin gene-related peptide is increased prior to obesity, and sensory nerve desensitization by capsaicin improves oral glucose tolerance in obese Zucker rats. Eur. J. Endocrinol. 2005, 153, 963–969.

Grekas, D.; Kassimatis, E.; Makedou, A.; Bacharaki, D.; Bamichas, G.; Tourkantonis, A. Com-bined treatment with low-dose pravastatin and fish oil in post-renal transplantation dilipidemia. Nephron. 2000, 88, 329–333.

Heart Protection Study Collaborative Group. MRC/BHF heart protection study of cholesterol lowering with simvastatin in 20,536 high-risk individuals: a randomized placebo-controlled trial. Lancet 2002, 360, 7–22.

Hansen, S.H. The role of taurine in diabetes and the development of diabetic complications. Diabetes Metab. Res. Rev. 2001, 17, 330–346.

Harada, H.; Tsujino, T.; Watari, Y.; Nonaka, H.; Emoto, N.; Yokoyama, M. Oral taurine supple-mentation prevents fructose-induced hypertension in rats. Heart Vessels 2004, 19, 132–136.

Harris, W.S. Omega-3 fatty acids and cardiovascular disease: A case for omega-3 index as a new risk factor. Pharmacol. Res. 2007, 55, 217–223.

Harris, W.S.; Bulchandani, D. Why do omega-3 fatty acids lower serum triglycerides? Curr. Opin. Lipidol. 2007, 17, 387–393.

Harris, W.S.; Miller, M.; Tighe, A.P.; Davidson, M.H.; Schaefer, E.J. Omega-3 fatty acids and coronary heart disease risk: Clinical and mechanistic perspectives. Atherosclerosis 2008, 197, 12–24.

Harris, W.S.; Silveira, S.; Dujovne, C.A. The combined effects of n-3 fatty acids and aspirin on hemostatic parameters in man. Thromb. Res. 1990, 57, 517–526.

Holnes, M.D.; Willett, W.C. Does diet affect breast cancer risk? Breast Cancer Res. 2004, 6, 170–178.

Holub, D.J.; Holub, B.J. Omega-3 fatty acids from fish oils and cardiovascular disease. Mol. Cell. Biochem. 2004, 263, 217–225.

Hosokawa, M.; Kudo, M.; Maeda, H.; Koboyashi, H.; Kohono, H.; Tanaka, T.; Miyashita, K. Fucoxanthin induces apoptosis and enhances the antiproliferative effect of PPARgamma ligand, troglitazone, on colon cancer cells. Biochim. Biophys. Acta. 2004, 1675, 113–119.

Hosokawa, M.; Wanezaki, S.; Miyauchi, K.; Kurihara, H.; Koho, H.; Kawabata, J.; Odashima, S.; Takahashi, K. Apoptosis-inducing effect of fucoxanthin on human leukemia cell HL-60. Food Sci. Techonol. Res. 1999, 5, 243–246.

Hsu, H.C.; Lee, Y.T.; Chen, M.F. Effects of fish oil and vitamin E on the antioxidant defense system in diet-induced hypercholesterolemic rabbits. Prostaglan.Other Lipid Med. 2001, 66, 99–108.

Innis, S.M.; Pinsk, V.; Jacobson, K. Dietary lipids and intestinal inflammatory disease. J. Ped. 2006, 149, S89–S96.

Isley, W.L.; Miles, J.M.; Harris, W.S. Pilot study of combined therapy with omega-3 fatty acids and niacin in atherogenic dyslipidemia. J. Clin. Lipid 2007, 1, 211–217.

Istfan, N.W.; Person, K.S.; Holick, M.F.; Chen, T.C. 1a, 25-dihydroxyvitamin D and fish oil synergistically inhibit G_1/S-phase transition in prostate cancer cells. J. Steroid Biochem. Mol. Biol. 2007, 103, 726–730.

Jacobson, T.A. Beyond lipids: The role of omega-3 fatty acids from fish oil in the prevention of coronary heart disease. Curr. Athero. Rep. 2007, 9, 145–153.

Jensen, C.L. Effects of n–3 fatty acids during pregnancy and lactation. Clin. Nutr. 2006, 83, 1452–1457.

Kanagaraj, P.; Vijayababu, M.R.; Ravisankar, B.; Anbalagan, J.; Aruldhas, M.M.; Arunakaran1, J. Effect of lycopene on insulin-like growth factor-I, IGF binding protein-3 and IGF type-I receptor in prostate cancer cells. J. Cancer Res. Clin. Oncol. 2007, 133, 351–359.

Karbownik, M.; Reiter, R.; Burkhardt, S.; Gitto, E.; Tan, D.; Lewiński, A. Melatonin Attenuates Estradiol-Induced Oxidative Damage to DNA: Relevance for Cancer Prevention. Exp. Biol. Med. 2001, 226, 707–712.

Keijzers, G.B.; Galan, B.E.D.; Tack, C.J.; Smits, P. Caffeine can decrease insulin sensitivity in humans. Diabetes Care 2002, 25, 364–369.

Krauss, R.M. Dietary and genetic probes of atherogenic dyslipidemia. Arterial. Thrombi. Vascular Biol. 2005, 25, 2265–2272.

Krauss-Etschmann, S.; Shadid, R.; Campoy, C.; Hoster, E.; Demmelmair, H.; Jimenez, M.; Gil, A.; Rivero, M.; Veszpremi, B.; Decsi, T.; et al. Effect of fish-oil and folate supplementation of pregnant woman on maternal and fetal plasma concentrations of docosahexaenoic acid and eicosapentaenoic acid: a European randomized multicenter trial. Am. J. Clin. Nutr. 2007, 85, 1392–1400.

Kris-Etherton, P.M.; Harris, W.S.; Lawrence, J. Fish Consumption, Fish Oil, Omega-3 Fatty Acids, and Cardiovascular Disease. Arterial. Thrombi. Vascular Biol. 2003, 23, e20–e30.

Kritharides, L.; Stocker, R. The use of antioxidant supplements in coronary heart disease. Atherosclerosis 2002, 164, 211–219.

Kushi, L.H.; Folsom, A.R.; Prineas, R.J. Dietary antioxidant vitamins and death from coronary heart disease in postmenopausal women. N. Engl. J. Med. 1996, 334, 1156–1162.

Lane, J.D.; Barkauskas, C.E.; Surwit, R.S.; Feinglos, M.N. Caffeine impairs glucose metabolism in type 2 diabetes. Diabetes Care 2004, 27, 2047–2048.

Lee, I.-M.; Cook, N.R.; Gaziano, J.M. Vitamin E in the primary prevention of cardiovascular disease and cancer. The Women's Health Study: A randomized controlled trial. ACC Curr. J. Rev. 2005, 14, 10–11.

Lissoni, P.; Rovelli, F.; Malugani, F. Anti-angiogenic activity of melatonin in advanced cancer patients. Neuroendocrinol. Lett. 2001, 22, 45–47.

Lombardo, Y.B.; Chicco, A.G. Effects of dietary polyunsaturated n-3 fatty acids on dyslipidemia and insulin resistance in rodents and humans. J. Nutr. Biochem. 2006, 17, 1–13.

Lotte, L.; Camila, H.; Traarup, S.; Marie, E.; Kim, M. Maternal fish oil supplementation in lactation and growth during the First 2.5 Years of Life. Pediatr. Res. 2005, 58, 235–242.

Maeda, H.; Hosokawa, M.; Sashima, T.; Funayama, K.; Miyashita, K. Fucoxanthin from edible seaweed, Undaria pinnatifida, shows antiobesity effect through UCP1 expression in white adipose tissues. Biochem. Biophys. Res. Commun. 2005, 332, 392–397.

Maeda, H.; Hosokawa, M.; Sashima, T.; Funayama, K.; Miyashita, K. Effect of medium-chain triacylglycerols on anti-obesity effect of fucoxanthin. J. Oleo Sci. 2007, 56, 615–621.

Maeda, H.; Hosokawa, M.; Sashima, T.; Takahashi, N.; Kawada, T.; Miyashita, K. Fucoxanthin and its metabolite, fucoxanthinol, suppress adipocyte differentiation in 3T3-L1 cells. Int. J. Mol. Med. 2006, 18, 147–152.

Matsushima, Y.; Sekine, T.; Kondo, Y.; Sakurai, T.; Kameo, K.; Tachibana, M.; Murakami, S. Effects of taurine on serum cholesterol levels and development of atherosclerosis in spontaneously hyperlipidaemic mice. Clin. Exp. Pharmacol. Physiol. 2003, 30, 295–299.

Mayoa, J.C.; Sainza, R.M.; Tanc, D.X.; Hardelandd, R.; Leonc, J.; Rodrigueza, C.; Reiterc, R.J. Anti-inflammatory actions of melatonin and its metabolites, N1-acetyl-N2-formyl-5-methoxykynuramine (AFMK) and N1-acetyl-5-methoxykynuramine (AMK), in macrophages. J. Neuroimmunol. 2005, 165, 139–149.

Montori, V.M.; Wollan, P.C.; Farmer, A.; Dinneen, S.F. Fish oil supplementation in Type 2 diabetes. Diabetes Care 2000, 23, 1407–1415.

Moore, C.S.; Bryant, S.P.; Mishra, G.D.; Krebs, J.K.; Browning, L.M.; Miller, G.J.; Jebb, S.A. Oily fish reduces plasma triacylglycerols: a primary prevention study in overwright men and women. Nutrition 2006, 22, 1012–1024.

Morris, C.D.; Carson, S. Routine vitamin supplementation to prevent cardiovascular disease: a summary of the evidence for the U.S. Preventive Services Task Force. Ann. Intern. Med. 2003, 139, 56–70.

Mozaffarian, D.; Rimm, E.B. Fish Intake, Contaminants, and Human Health: Evaluating the Risks and the Benefits. JAMA 2006, 296, 1885–1899.

Mune, M.; Meydani, M.; Gong, J.; Fotouhi, N.; Ohtani, H.; Smith, D.; Blumberg, J.B. Effect of dietary fish oil, vitamin E, and probucol on renal injury in the rat. J. Nutr. Biochem. 1999, 10, 539–546.

Nagao, K.; Yanagita, T. Bioactive lipids in metabolic syndrome. Prog. Lipid Res. 2008, 47, 127–146.

Nambi, V.; Ballantyne, C.M. Combination therapy with statins and omega-3 fatty acids. Am. J. Cardio. 2006, 98, 34–38.

Nandhini, A.T.A.; Balakrishnan, S.D.; Anuradha, C.V. Taurine improves lipid profile in rats fed a high fructose-diet. Nutr. Res. 2002, 22, 343–354.

Neschen, S.; Morino, K.; Rossbacher, J.C.; Pongratz, R.; Cline, G.W.; Sono, S.; Gillum, M.; Shulman, G.I. Fish oil regulates adiponectin secretion by a peroxisome proliferator-activated receptor-gannma-dependent mechanism in mice. Diabetes 2006, 55, 924–928.

Nettleton, J.A.; Katz, R. n-3 long-chain polyunsaturated fatty acids in type 2 diabetes: A review. J. Am. Diet. Assoc. 2005, 105, 428–440.

Nieuwehuys, C.M.A.; Feijge, M.A.H.; Vermeer, C.; Hennissen, A.H.H.M.; Beguin, S.; Heemskerk, J.W.M. Vitamin K-dependant and vitamin K-independent hypocoagulant effects of dietary fish oil in rats. Throms. Res. 2001, 104, 137–147.

Nordoy, A.; Bonna, K.H.; Nilsen, H.; Berge, R.K.; Hansen, J.B.; Ingerbretsen, O.C. Effects of Simvastatin and omega-3 fatty acids on plasma lipoproteins and lipid peroxidation in patients with combined hyperlipidaemia. J. Intern. Med. 1998, 243, 163–170.

Nordoy, A.; Hansen, J.B.; Brox, J.; Svensson, B. Effects of Atorvastatin and omega-3 fatty acids on LDL subfractions and postprandial hyperlipemia in patients with combined hyperlipemia. Nutr. Metab.Cardiovasc.Dis. 2001, 11, 7–16.

Okada, T.; Nakai, M.; Maeda, H.; Hosokawa, M.; Sashima, T.; Miyashita, K. Supressive effect of neoxanthin on the differentiation of 3T3-L1 adipose cells. J. Oleo Sci. 2008, 57, 345–351.

Park, S.; Jang, J.S.; Hong, S.M. Long-term consumption of caffeine improves glucose homeostasis by enhancing insulinotropic action through islet insulin/insulin-like growth factor 1 signaling in diabetic rats. Metabolism 2007, 56, 599–607.

Paul, A.G. The roles of PPARs in adipocyte differentiation. Lipid. Res. 2001, 40, 269–281.

Persson, C.; Glimelius, B.; Ronnelid, J.; Nygren, P. Impact of fish oil and melatonin on cachexia in patients with advanced gastrointestinal cancer: A randomized pilot study. Nutrition 2005, 21, 170–178.

Petrie, H.J.; Chown, S.E.; Belfie, L.M.; Duncan, A.M.; McLaren, D.H.; Conquer, J.A.; Graham, T. Caffeine ingestion increases the insulin response to an oral-glucose-tolerance test in obese men before and after weight loss. Am. J. Clin. Nutr. 2004, 80, 22–28.

Pryor, W.A. Vitamin E and heart disease:: Basic science to clinical intervention trials. Free Radic. Biol. Med. 2000, 28, 141–164.

Qi, K.; Fan, C.; Jiang, J.; Zhu, H.; Jiao, H.; Meng, Q.; Deckelbaum, R.J. Omega-3 fatty acid containing diets decrease plasma triglyceride concentrations in mice by reducing endogenous triglyceride synthesis and enhancing the blood clearance of triglyceride-rich particles. Clin. Nutr 2008, 27, 424–430.

Roche, H.M. Fatty acids and the metabolic syndrome. Proc. Nutr. Soc. 2005, 64, 23–29.

Roynette, C.E.; Calder, P.C.; Dupertuis, Y.M.; Pichard, C. n-3 Polyunsaturated fatty acids and colon cancer prevention. Clin. Nutr. 2004, 23, 139–151.

Saify, Z.S.; Ahmed, F.; Akhtar, S.; Arif, M. Biochemical studies of marine fish oil, part II: effects of fish oil and lipid lowering drugs on total cholesterol and total lipid. PJPS 2005, 22, 1–9.

Saify, Z.S.; Ahmed, F.; Akhtar, S.; Siddiqui, S.; Arif, M.; Hussain, S.A.; Mushtaq, N. Biochemical studies on marine fish oil part-1: effects of fish oil and lipid lowering drugs on HDL/LDL cholesterol levels. PJPS 2003, 16, 1–8.

Sánchez-Barceló, E.J.; Cos, S.; Mediavilla, D.; Martínez-Campa, C.; González, A.; Alonso-González, C. Melatonin–estrogen interactions in breast cancer. J. Pineal Res. 2005, 4, 217–222.

Sanders, T.A.B. Marine oil: metabolic effects and role in human nutrition. Proc. Nutr. Soc. 1993, 52, 457–472.

Schacky, C. Omega-3 fatty acids and cardiovascular disease. Curr. Opinion Clinical. Nutr. Metab. Care 2007, 10, 129–135.

Schneider, C. Chemistry and biology of vitamin E. Mol. Nutr. Food. Res. 2005, 49, 7–30.

Schuller-Levis, G.B.; Park, E. Taurine: new implications for an old amino acid. FEMS Microbiol. Lett. 2003, 226, 195–202.

Shiratori, K.; Ohgami, K.; Ilieva, L.; JIN, X.H.; Koyama, Y.; Miyashita, K.; Yoshida, K.; Kase, S.; Ohno, S. Effects of fucoxanthin on lipopolysaccharide-induced inflammation in vitro and in vivo. Exp. Eye. Res. 2005, 81, 271–277.

Shiu, S.Y.W. Towards rational and evidence-based use of melatonin in prostate cancer prevention and treatment. J. Pineal Res. 2007, 43, 1–9.

Shlomo, W.; Yossi, W.; Elena, K.; Nadia, L.; Haim, M.; Riad, A.; Yoav, S.; Joseph, L. Tomato lycopene extract supplementation decreases insulin-like growth factor-I levels in colon cancer patients. Eur. J. Cancer Prev. 2007, 16, 298–303.

Soria, A.; Chicco, A.; D'Alessandro, M.E.; Rossi, A.; Lombardo, Y.B. Dietary fish oil reverse epididymal tissue adiposity, cell hypertrophy and insulin resistance in dyslipemic sucrose fed rat model. J. Nutr. Biochem. 2002, 13, 209–218.

Steinberg, D.; Workshop-participants. Antioxidants in the preventino of human atherosclerosis. Circulation 1992, 85, 2337–2344.

Stephens, N.G.; Parsons, A.; Brown, M.J.; Schofield, P.M.; Kelly, F.; Cheeseman, K.; Mitchinson, M. Randomised controlled trial of vitamin E in patients with coronary disease: Cambridge Heart Antioxidant Study (CHAOS). The Lancet 1996, 347, 781–786.

Svaneborg, N.; Kristensen, S.D.; Hansen, L.M.; Büllow, I.; Husted, S.E.; Schmidt, E.B. The acute and short-time effect of supplementation with the combination of n-3 fatty acids and acetylsali-cylic acid on platelet function and plasma lipids. Thrombo. Res. 2002, 105, 311–316.

Tang, F.Y.; Cho, H.J.; Pai, M.H.; Chen, Y.H. Concomitant supplementation of lycopene and eicosapentaenoic acid inhibits the profileration of human colon cancer cells. J. Nutr. Biochem. 2009, 20, 462–434.

Tang, F.Y.; Shih, C.J.; Cheng, L.H.; Ho, H.J.; Chen, H.J. Lycopene inhibits growth of human colon cancer cells via suppression of Akt signaling pathway. Mol. Nutr. Food. Res. 2008, 52, 646–654.

Taylor, A.J.; Stanek, E.J. Flushing and the HDL-C response to extended-release niacin. J.Clin. Lipidol. 2008, 2, 285–288.

Taylor, A.J.; Sullenberger, L.E.; Lee, H.J.; Lee, J.K.; Grace, K.A. A double-blind, placebo-con-trolled study of extended-release niacin on atherosclerosis progression in secondary prevention patients treated with statins. Circulation 2004, 110, 3512–3517.

Tolan, I.; Ragoobirsingh, D.; Morrison, E.Y.S.A. The effect of capsaicin on blood glucose, plasma insulin levels and insulin binding in dog models. Phytotherapy Res. 2001, 15, 391– 394.

Tsuboyama-Kasaoka, N.; Shozawa, C.; Sano, K.; Kamei, Y.; Kasaoka, S.; Hosokawa, Y.; Ezaki, O. Taurine (2-Aminoethanesulfonic Acid) Deficiency Creates a Vicious Circle Promoting Obesity. Endocrinol. 2006, 147, 3276–3284.

Tsuboyama-Kawaoka, N.; Takahashi, M.; Kim, H.; Ezaki, O. Up-regulation of liver uncoupling protein-2 mRNA by either fish oil feeding or fibrate administration in mice. Biochem. Biophys. re. Commun. 1999, 257, 879–885.

Vanamala, J.; Glagolenko, A.; Yang, P.; Carroll, R.J.; Murphy, M.E.; Newman, R.A.; Ford, J.R.; Branby, L.A.; Chapkin, R.S.; Turner, N.D.; et al. Dietary fish oil and pectin enhance colonocyte apoptosis in part through suppression of $PPAR\ddot{a}/PGE_2$ and elevation of PGE_3. Carcinogenesis 2008, 29, 790–796.

Vecchia, C.L. Mediterranean diet and cancer. Pub. Health Nutr. 2004, 7, 965–968.

Wallace, F.A.; E.A.Miles; P.C.Calder Comparison of the effects of linseed oil and different doses of fish oil on mononuclear cell function in healthy human subjects. British J. Nutr. 2003, 89, 679–689.

Wang, H.; Storlien, L.H.; Huang, X.F. Effects of dietary fat types on body fatness, leptin, and ARC leptin receptor, NPY, and AgRP mRNA expression. Am. J. Physiol. Endocrinol. Metab. 2002, 282, E1352–1359.

Weisberg, S.P.; McCann, D.; Desai, M.; Rosenbaum, M.; Leibel, R.L.; Jr, A.W.F. Obesity is associated with macrophage accumulation in adipose tissue. J. Clin. Invest. 2003, 112, 1796–1808.

Westphal, S.; Borucki, K.; Taneva, E.; Makarova, R.; Luley, C. Aipokines and treatment with niacin. Metab.Clin.Exp. 2006, 55, 1283–1285.

Westphal, S.; Luley, C. Preferential increase in high-molecular weight adiponectin after niacin. Atherosclerosis 2008, 198, 179–183.

Yam, D.; Peled, A.; Shinitzky, M. Duppression of tumor growth and metastasis by dietary fish oil combined with vitamins E and C and cisplatin. Can. Chem. Pharm. 2001, 47, 34–40.

Yang, J.; Zhao, S.-p.; Li, J.; Dong, S.-z. Effect of niacin on adipocyte leptin in hypercholesterolemic rabbits. Cardio. Pathol. 2008, 17, 219–225.

Yusuf, S.; Dagenais, G.; Pogue, J.; Bosch, J.; Sleight, P. Vitamin E supplementation and cardiovascular events in high-risk patients. The Heart Outcomes Prevention Evaluation Study Investigators. N. Engl. J. Med. 2000, 342, 154–160.

Zhang, M.; Bi, L.F.; Fang, J.H.; Su, X.L.; Da, G.L.; Kuwamori, T.; Kagamimori, S. Beneficial effects of taurine on serum lipids in overweight or obese non-diabetic subjects. Amino Acids 2004, 26, 267–271.

·•9•·

Docosahexaenoic Acid Containing Phosphatidylcholine Alleviates Obesity-Related Disorders in Obese Rats

Bungo Shirouchi , Koji Nagao, and Teruyoshi Yanagita
Laboratory of Nutrition Biochemistry, Department of Applied Biochemistry and Food Science, Saga University, Saga, Japan

Introduction

Lifestyle-related diseases, such as hyperlipidemia, arteriosclerosis, diabetes mellitus, and hypertension, are widespread and increasingly prevalent diseases in industrialized countries and contribute to the increase in cardiovascular morbidity and mortality (Kissebah & Krakower, 1994; Formiguera & Canton, 2004). Accompanied by the rapid increase in the number of elderly people, this becomes important not only medically but also socioeconomically. Although the pathogenesis of lifestyle-related diseases is complicated and the precise mechanisms have not been elucidated, obesity has emerged as one of the major cardiovascular risk factors according to epidemiologic studies (Fujioka et al., 1987; Kanai et al., 1990; Nakamura et al., 1994).

Because diet, especially dietary fat, has been recognized as contributing to the development and prevention of obesity, the influence of quantities and qualities of dietary fats on the pathogenesis of obesity-related disorders has been studied (Cunnane et al., 1986; Hill et al., 1993; Nakatani et al., 2003). Differential effects have arisen with respect to individual fatty acids (Nagao et al., 2005; Yanagita et al., 2008; Nagao et al., 2008; Nagao et al., 2010). Omega-3 polyunsaturated fatty acids (n-3 PUFAs), such as eicosapentaenoic acid (EPA; 20:5) and docosahexaenoic acid (DHA; 22:6), are abundant in fish, shellfish, and sea mammals. Evidence from animal models and human studies have suggested that n-3 PUFAs have lipid-lowering effects (Ikeda et al., 1998; Park & Harris, 2003; Gotoh et al., 2009).

Although the majority of dietary fat is triglyceride, it contains approximately 10% of phospholipids (PLs). Growing evidence indicates that dietary PLs have beneficial effects compared with dietary triglyceride (Cohn et al., 2008; Pandey et al., 2008; Yanagita et al., 2008; Nagao et al., 2008; Shirouchi et al., 2008; Shirouchi et al., 2009). For example, phosphatidylcholine (PC), which is a major component of dietary phospholipids, has been reported to improve brain function in animals (Chung et al., 1995; Masuda et al., 1998) and alleviate orotic acid-induced fatty liver in rats (Buang et al., 2005). PLs are composed of hydrophobic (e.g., fatty acid) and hydrophilic (e.g., choline, ethanolamine, serine, or inositol) constituents, and either or both of them could be responsible for the physiological function of dietary PLs. The concept of a "structured-lipid" implies modification of the fatty acid composition and/or its location in the glycerol backbone and improvement of the physical and/or physiological properties of dietary lipids. A broad definition of a "structured-lipid" may include PLs from marine sources, such as fish roe, squid meal, and starfish, which contain abundant EPA and DHA in their fatty acids. In the present study, we investigated the effect of dietary DHA-PC, extracted from salmon roe and containing DHA in its fatty acid composition, on obesity-related disorders in obese Otsuka Long-Evans Tokushima Fatty (OLETF) rats. OLETF rats develop a syndrome with multiple metabolic and hormonal disorders that shares many features with human obesity (Yagi et al., 1997; Takiguchi et al., 1997; Moran et al., 1998; Hida et al., 2000). OLETF rats have hyperphagia because they lack receptors for cholecystokinin and become obese, developing hyperlipidemia, fatty liver, and diabetes.

Effect of DHA-PC on Hepatic Triglyceride Metabolism in Obese Rats

Male OLETF rats were assigned to two groups, with dietary fats composed of a mixture of 5% corn oil + 2% egg-PC, extracted from hen egg-yolk (PC group), and a mixture of 5% corn oil + 2% DHA-PC (DHA-PC group). The fatty acid compositions of egg-PC and DHA-PC are given in Table 9.A.

Fig. 9.1 shows liver weights and hepatic triglyceride levels of OLETF rats after 4 weeks of consuming these diets. Although there was no significant difference in final body weight or food intake between the groups, the DHA-PC diet significantly decreased liver weights and hepatic triglyceride levels of OLETF rats. As a consequence, obesity-induced fatty liver was alleviated by the DHA-PC diet compared with the PC diet. In addition, the DHA-PC diet lowered, but not significantly, serum triglyceride levels (PC group, 135 ± 30; DHA-PC group, 77.0 ± 16.5 mg/dL) consistent with the alleviation of hepatic triglyceride accumulation.

Table 9.A Fatty Acid Composition of Egg-PC and DHA-PC.

	Egg-PC [a]	DHA-PC [b]
fatty acid composition, wt (%)		
14:0	1.4	1.4
16:0	29.9	15.0
16:1	1.4	1.2
18:0	17.5	13.9
18:1	27.9	10.5
18:2	15.4	0.7
20:5ω3	n.d.	12.0
22:5ω3	n.d.	5.2
22:6ω3	n.d.	28.1
Others	7.8	12.0
Total	100	100

[a] Contained 80 % PC, 17 % PE, 3 % Others.

[b] Contained 74 % PC, 13 % PE, 13 % Others.; n.d., not detected.

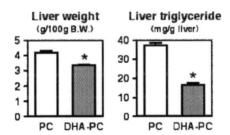

Fig. 9.1. Effect of DHA-PC on liver weights and hepatic triglyceride levels in OLETF rats. Rats were fed diets containing either 5% corn oil + 2% egg-PC (PC group) or 5% corn oil + 2% DHA-PC (DHA-PC group) for 4 weeks. Values are expressed as means ± standard error of six rats. Asterisk shows significant difference at P<0.05. To investigate the regulation of hepatic triglyceride metabolism, we analyzed the effect of the DHA-PC diet on activities of enzymes related to fatty acid synthesis and fatty acid β-oxidation. The activity of fatty acid synthase (FAS), a key enzyme of fatty acid synthesis, was markedly suppressed by the DHA-PC diet, whereas the activities of carnitine palmitoyltransferase (CPT), a rate-limiting enzyme of mitochondrial β-oxidation, and peroxisomal β-oxidation were markedly enhanced by the DHA-PC diet (Fig. 9.2). Therefore, these results suggest that the alleviation of hepatic triglyceride accumulation by the DHA-PC diet was attributable to the suppression of fatty acid synthesis and the enhancement of fatty acid β-oxidation.

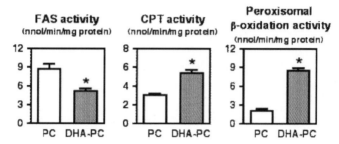

Fig. 9.2. Effect of DHA-PC on activities of hepatic enzyme related to triglyceride metabolism in OLETF rats. Rats were fed diets containing either 5% corn oil + 2% egg-PC (PC group) or 5% corn oil + 2% DHA-PC (DHA-PC group) for 4 weeks. Values are expressed as means ± standard error of six rats. Asterisk shows significant difference at P<0.05.

Effect of DHA-PC on Levels of Hepatic mRNA Related to Lipid Metabolism in Obese Rats

Fig. 9.3 shows mRNA levels of genes related to lipid metabolism in the liver of OLETF rats after 4 weeks of feeding. Expression of lipogenic genes, such as acetyl-CoA carboxylase (ACC) and stearoyl-CoA desaturase 1 (SCD1) was markedly decreased by the DHA-PC diet. Additionally, FAS mRNA levels tended to decrease (30%, data not shown) in rats fed the DHA-PC diet. The expression of lipogenic genes is regulated by sterol regulatory element binding pretein-1c (SREBP-1c), a lipogenic transcriptional factor (Horton et al., 2002). In this study, SREBP-1c mRNA levels were markedly decreased by the DHA-PC diet. Previous studies have shown that DHA has a suppressive effect on SREBP-1 mRNA expression (Fujiwara et al., 2003), which suggests that DHA can decrease fatty acid synthesis through the transcriptional suppression of SREBP-1 signaling. Given our previous results indicating that dietary PC suppresses fatty acid synthesis through the decreased expression of FAS mRNA (Buang et al., 2005), we speculate that DHA enhanced the suppressive effect of PC on fatty acid

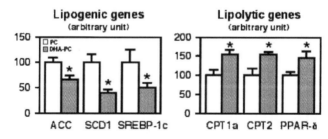

Fig. 9.3. Effect of DHA-PC on levels of hepatic mRNA related to lipid metabolism in OLETF rats. Rats were fed diets containing either 5% corn oil + 2% egg-PC (PC group) or 5% corn oil + 2% DHA-PC (DHA-PC group) for 4 weeks. Values are expressed as means ± standard error of six rats. Asterisk shows significant difference at P<0.05.

synthesis in the liver. On the other hand, expression of lipolytic genes, such as CPT1a and CPT2, was markedly increased by the DHA-PC diet. The expression of lipolytic genes is regulated by ligand-inducing transcription factors called peroxisome proliferator activated receptor-alpha (PPAR-α) (Kota et al., 2005). DHA has been reported to act as a ligand for PPAR and to enhance the gene expression of the lipolytic enzymes (Neschen et al., 2002). In addition, we previously reported that dietary PC enhances fatty acid β-oxidation (Buang et al., 2005). Therefore, we suggest that DHA and PC synergistically enhanced fatty acid β-oxidation. In this study, there was no significant difference in mRNA level of PPAR-α between the groups. Thus, we suggest that DHA-PC might act on the activity of PPAR-α as ligands. The DHA-PC diet markedly increased the mRNA level of PPAR-δ in this study. Recently, it has been demonstrated that PPAR-δ involves lipid and glucose metabolism (Luquet et al., 2005). Hence, we suggest that the increased PPAR-δ mRNA level by the DHA-PC diet might be associated with the normalization of lipid and glucose metabolism in OLETF rats.

Effect of DHA-PC on Adipose Tissue Weights and Serum Adiponectin Levels in Obese Rats

Recent advances in molecular and cell biology have shown that adipose tissue not only stores excess energy in the form of fat, but also secretes physiologically active substances called adipocytokines (Maeda et al., 1997; Matsuzawa, 2006). Among those, adiponectin is the most abundant adipose-specific protein. The expression of adiponectin is reduced in obesity, and blood levels are negatively correlated with visceral fat accumulation (Matsuzawa, 2005; 2006). Several reports indicate that adiponectin enhances glucose uptake, glucose utilization, and fatty acid oxidation and suppresses gluconeogenesis and fatty acid synthesis by activating AMP-activated protein kinase and PPAR-α in the liver and muscle (Yamauchi et al., 2002; 2003; Xu et al., 2003; Zhou et al., 2005). In this study, the DHA-PC diet significantly decreased abdominal white adipose tissue weights (10%, data not shown) and serum glucose levels, whereas it increased serum adiponectin levels of OLETF rats (Fig. 9.4). Therefore, we

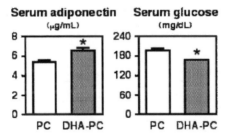

Fig. 9.4. Effect of DHA-PC on serum levels of adiponectin and glucose in OLETF rats. Rats were fed diets containing either 5% corn oil + 2% egg-PC (PC group) or 5% corn oil + 2% DHA-PC (DHA-PC group) for 4 weeks. Values are expressed as means ± standard error of six rats. Asterisk shows significant difference at P<0.05.

suggest that the alleviation of lipid and glucose metabolism by the DHA-PC diet was attributable to the increased serum adiponectin levels.

Conclusions

The present study investigated the effect of DHA-PC on obesity-related disorders in OLETF rats. After 4 weeks of feeding DHA-PC, OLETF rats showed a decrease of abdominal white adipose tissue weights, liver weights and hepatic triglyceride levels. These changes were attributable to the significant suppression of FAS activity and significant enhancement in the activities of CPT and peroxisomal β-oxidation in DHA-PC-fed OLETF rats. Moreover, the DHA-PC diet reduced serum glucose levels concomitant with the increase of serum adiponectin levels. These results show that compared with egg-PC, DHA-PC can prevent or alleviate obesity-related disorders through the suppression of lipogenesis, enhancement of lipolysis, and increase of adiponectin production in OLETF rats (Shirouchi et al., 2007).

Acknowledgment

We thank Takeshi Ohkubo and Hidehiko Hibino (NOF Co., Ltd., Kanagawa, Japan) for providing DHA-PC.

References

Buang, Y.; Wang, Y.M.; Cha, J.Y.; Nagao, K.; Yanagita, T. Dietary phosphatidylcholine alleviates fatty liver induced by orotic acid. Nutrition 2005, 21, 867–873.

Chung, S.Y.; Moriyama, T.; Uezu, E.; Uezu, K.; Hirata, R.; Yohena, N.; Masuda, Y.; Kokubu, T.; Yamamoto, S. Administration of phosphatidylcholine increases brain acetylcholine concentration and improves memory in mice with dementia. J. Nutr. 1995, 125, 1484–1489.

Cohn, J.S.; Wat, E.; Kamili, A.; Tandy, S. Dietary phospholipids, hepatic lipid metabolism and cardiovascular disease. Curr. Opin. Lipidol. 2008, 19, 257–262.

Cunnane, S.C.; McAdoo, K.R.; Horrobin, D.F. ω-3 essential fatty acids decrease weight gain in genetically obese mice. Br. J. Nutr. 1986, 56, 87–95.

Formiguera, X.; Canton, A. Obesity: epidemiology and clinical aspects. Best Pract. Res. Clin. Gastroenterol. 2004, 18, 1125–1146.

Fujioka, S.; Matsuzawa, Y.; Tokunaga, K.; Tarui, S. Contribution of intra-abdominal fat accumulation to the impairment of glucose and lipid metabolism in human obesity. Metabolism 1987, 36, 54–59.

Fujiwara, Y.; Yokoyama, M.; Sawada, R.; Seyama, Y.; Ishii, M.; Tsutsumi, S.; Aburatani, H.; Hanaka, S.; Itakura, H.; Matsumoto, A. Analysis of the comprehensive effects of polyunsaturated fatty acid on mRNA expression using a gene chip. J. Nutr. Sci. Vitaminol. 2003, 49, 125–132.

Gotoh, N.; Nagao, K.; Onoda, S.; Shirouchi, B.; Furuya, K.; Nagai, T.; Mizobe, H.; Ichioka, K.; Watanabe, H.; Yanagita, T.; et al. Effect of three different highly purified n-3 series highly unsaturated fatty acids on lipid metabolism in C57BL/KsJ-db/db Mice. J. Agric. Food Chem. 2009, 57, 11047–11054.

Hida, K.; Wada, J.; Zhang, H.; Hiraguchi, K.; Tsuchiyama, Y.; Shikata, K.; Makino, H. Identification of genes specifically expressed in the accumulated visceral adipose tissue of OLETF rats. J. Lipid Res. 2000, 41, 1615–1622.

Hill, J.O.; Peters, J.C.; Lin, D.; Yakubu, F.; Greene, H.; Swift, L. Lipid accumulation and body fat distribution is influenced by type of dietary fat fed to rats. Int. J. Obes. Relat. Metab. Disord. 1993, 17, 223–236.

Horton, J.D.; Goldstein, J.L.; Brown, M.S. SREBPs: activators of the complete program of cholesterol and fatty acid synthesis in the liver. J. Clin. Invest. 2002, 109, 1125–1131.

Ikeda, I.; Cha, J.Y.; Yanagita, T.; Nakatani, N.; Oogami, K.; Imaizumi, K.; Yazawa, K. Effects of α-linolenic, eicosapentaenoic and docosahexaenoic acids on hepatic lipogenesis and β-oxidation in rats. Biosci. Biotechnol. Biochem. 1998, 62, 675–680.

Kanai, H.; Matsuzawa, Y.; Kotani, K.; Keno, Y.; Kobatake, T.; Nagai, Y.; Fujioka, S.; Tokunaga, K.; Tarui, S. Close correlation of intra-abdominal fat accumulation to hypertension in obese women. Hypertension 1990, 16, 484–490.

Kissebah, A.H.; Krakower, G.R. Regional adiposity and morbidity. Physiol. Rev. 1994, 74, 761–811.

Kota, B.P.; Huang, T.H.; Roufogalis, B.D. An overview on biological mechanisms of PPARs. Pharmacol. Res. 2005, 51, 85–94.

Luquet, S.; Gaudel, C.; Holst, D.; Lopez-Soriano, J.; Jehl-Pietri, C.; Fredenrich, A.; Grimaldi, P.A. Roles of PPAR delta in lipid absorption and metabolism: a new target for the treatment of type 2 diabetes. Biochim. Biophys. Acta. 2005, 1740, 313–317.

Maeda, K.; Okubo, K.; Shimomura, I.; Mizuno, K.; Matsuzawa, Y.; Matsubara, K. Analysis of an expression profile of genes in the human adipose tissue. Gene 1997, 190, 227–235.

Masuda, Y.; Kokubu, T.; Yamashita, M.; Ikeda, H.; Inoue, S. EGG phosphatidylcholine combined with vitamin B12 improved memory impairment following lesioning of nucleus basalis in rats. Life Sci. 1998, 62, 813–822.

Matsuzawa, Y. Adiponectin: identification, physiology and clinical relevance in metabolic and vascular disease. Atheroscler. Suppl. 2005, 6, 7–14.

Matsuzawa, Y. Therapy insight: adipocytokines in metabolic syndrome and related cardiovascular disease. Nat. Clin. Pract. Cardiovasc. Med. 2006, 3, 35–42.

Moran, T.H.; Katz, L.F.; Plata-Salaman, C.R.; Schwartz, G.J. Disordered food intake and obesity in rats lacking cholecystokinin A receptors. Am. J. Physiol. 1998, 274, R618–R625.

Nagao, K.; Yanagita, T. Conjugated fatty acids in food and their health benefits. J. Biosci. Bioeng. 2005, 100, 152–157.

Nagao, K.; Yanagita, T. Bioactive lipids in metabolic syndrome. Prog. Lipid Res. 2008, 47, 127–146.

Nagao, K.; Yanagita, T. Medium-chain fatty acids: Functional lipids for the prevention and treatment of the metabolic syndrome. Pharm Res. 2010, 61, 208–212.

Nakamura, T.; Tokunaga, K.; Shimomura, I.; Nishida, M.; Yoshida, S.; Kotani, K.; Islam, A.H.; Keno, Y.; Kobatake, T.; Nagai, Y.; et al. Contribution of visceral fat accumulation to the development of coronary artery disease in non-obese men. Atherosclerosis 1994, 107, 239–246.

Nakatani, T.; Kim, H.J.; Kaburagi, Y.; Yasuda, K.; Ezaki, O. A low fish oil inhibits SREBP-1 proteolytic cascade, while a high-fish-oil feeding decreases SREBP-1 mRNA in mice liver: relationship toanti-obesity. J. Lipid Res. 2003, 44, 369–379.

Neschen, S.; Moore, I.; Regittnig, W.; Yu, C.L.; Wang, Y.; Pypaert, M.; Petersen, K.F.; Shulman, G.I. Contrasting effects of fish oil and safflower oil on hepatic peroxisomal and tissue lipid content. Am. J. Physiol. Endocrinol. Metab. 2002, 282, E395–E401.

Pandey, N.R.; Sparks, D.L. Phospholipids as cardiovascular therapeutics. Curr. Opin. Invest. Drugs 2008, 9, 281–285.

Park, Y.; Harris, W.S. Omega-3 fatty acid supplementation accelerates chylomicron triglyceride clearance. J. Lipid Res. 2003, 44, 455–463.

Shirouchi, B.; Nagao, K.; Inoue, N.; Ohkubo, T.; Hibino, H.; Yanagita, T. Effect of dietary omega-3 phosphatidylcholine on obesity-related disorders in obese Otsuka Long-Evans Tokushima fatty rats. J. Agric. Food Chem. 2007, 55, 7170–7176.

Shirouchi, B.; Nagao, K.; Inoue, N.; Furuya, K.; Koga, S.; Matsumoto, H.; Yanagita, T. Dietary phosohatidylinositol prevents the development of nonalcoholic fatty liver disease in Zucker (fa/fa) rats. J. Agric. Food Chem. 2008, 56, 2375–2379.

Shirouchi, B.; Nagao, K.; Furuya, K.; Inoue, N.; Inafuku, M.; Nasu, K.; Otsubo, S.; Koga, S.; Matsumoto, H.; Yanagita, T. Effect of dietary phosohatidylinositol on cholesterol metabolism in Zucker (fa/fa) rats. J. Oleo Sci. 2009, 58, 111–115.

Takiguchi, S.; Taketa, Y.; Funakoshi, A.; Miyasaka, K.; Kataoka, K.; Fujimura, Y.; Goto, T.; Kono, A. Disrupted cholecystokinin type-A receptor (CCKAR) gene in OLETF rats. Gene 1997, 197, 169–175.

Xu, A.; Wang, Y.; Keshaw, H.; Xu, L.Y.; Lam, K.S.; Cooper, G.J. The fat-derived hormone adiponectin alleviates alcoholic and nonalcoholic fatty liver diseases in mice. J. Clin. Invest. 2003, 112, 91–100.

Yagi, K.; Kim, S.; Wanibuchi, H.; Yamashita, T.; Yamamura, Y.; Iwao, H. Characteristics of diabetes, blood pressure, and cardiac and renal complications in Otsuka Long-Evans Tokushima Fatty rats. Hypertension 1997, 29, 728–735.

Yamauchi, T.; Kamon, J.; Minokoshi, Y.; Ito, Y.; Waki, H.; Uchida, S.; Yamashita, S.; Noda, M.; Kita, S.; Ueki, K.; et al. Adiponectin stimulates glucose utilization and fatty-acid oxidation by activating AMP-activated protein kinase. Nat. Med. 2002, 8, 1288–1295.

Yamauchi, T.; Kamon, J.; Waki, H.; Imai, Y.; Shimozawa, N.; Hioki, K.; Uchida, S.; Ito, Y.; Takakuwa, K.; Matsui, J.; et al. Globular adiponectin protected ob/ob mice from diabetes and ApoE-deficient mice from atherosclerosis. J. Biol. Chem. 2003, 278, 2461–2468.

Yanagita, T.; Nagao, K. Functional lipids and the prevention of metabolic syndrome. Asia Pac. J. Clin. Nutr. 2008, 17, 189–191.

Zhou, H.; Song, X.; Briggs, M.; Violand, B.; Salsgiver, W.; Gulve, E.A.; Luo, Y. Adiponectin represses gluconeogenesis independent of insulin in hepatocytes. Biochem. Biophys. Res. Commun. 2005, 338, 793–799.

·10··

Health Benefits of Flaxseed

Kelley C. Fitzpatrick

Flax Council of Canada, Winnipeg, Manitoba, Canada

Introduction

Flaxseed, or linseed, (*Linum usitatissimum,* L., subspecies *usitatissimum,* Linaceae) has been used for food and industrial fiber since ancient times (Vaisey-Genser & Morris, 2003). Flaxseed cultivation was reported to date back to around 9,000–8,000 B.C. in the Middle East, Turkey (van Zeiste, 1972), Iran (Hopf, 1983), Jordon (Rollefson et al., 1985), and Syria (Hillman, 1975; 1989). Domestication of flaxseed is dated back to 7000–4500 B.C. (Zohary & Hopf, 2000; Vaisey-Genser & Morris, 2003). About two million metric tons of flaxseed are produced annually with Canada being the main producer (*ca* 33%), followed by China (20%), the United States (16%), and India (11%) (Vaisey-Genser & Morris, 2003).

The terms "flaxseed" and "linseed" are often used interchangeably, although North Americans use "flaxseed" to describe flax when it is eaten by humans and "linseed" to describe flax when it is used for industrial purposes, such as linoleum flooring. In Europe, the term "flaxseed" describes the varieties grown for making linen.

Flaxseed contains lipid (40%), protein (21%), dietary fiber (28%), ash (4%), and other soluble components such as sugars, phenolic acids, and lignans (*ca* 6%). The oil content in flaxseed represents between 29 and 45% of the seed depending on the cultivar, location, and agroclimatic conditions (Oomah & Mazza, 1997; Daun et al., 2003; Wakjira et al., 2004). The main nutritional advantage of flaxseed oil is related to the high level of ALA in the oil (50%–60%). About 20% of the flaxseed is a mucilagenous hull. Flaxseed mucilage is comprised of gum-like polysaccharides containing acidic (54.5% rhamnose and 23.4% galactose) and neutral arabinoxylan (62.8% xylose) (Cui et al., 1994; Warrand et al., 2005). Flaxseed contains about 1%–2% total phenolic compounds (Oomah et al., 1995; Hall & Shultz, 2001), of which the lignan secoisolariciresinol diglucoside (SDG) is a major component. SDG is present in the seed as a mixture of oligomers with hydroxymethylglutaric acid having an average molecular weight of 4000 Da (Kamal-Eldin et al., 2001). A number of bioactivities are claimed for SDG including antioxidant and estrogenic/oestrogenic effects (Adlercreutz et al., 1992; Hutchins & Slavin, 2003), leading to health benefits with respect to cardiovascular diseases, diabetes, and menopause.

Composition

Fatty Acids

Flax has been valued historically for its abundance of fat, which provides a unique mix of fatty acids. It is rich in polyunsaturated fatty acids, particularly alpha-linolenic acid (ALA), the essential omega-3 fatty acid, and linoleic acid (LA), the essential omega-6 fatty acid.

The fatty acid composition of flaxseed, which is cultivar dependent, is dominated by ALA that constitute about 50%–60% of the total fatty acids (Oomah & Mazza, 1997). The relative fatty acid composition of the major fatty acids is palmitic (5.5–6.5%), stearic (2.2%–4.1%), oleic (13.4%–22.2%), linoleic (15.2%–17.4%), and ALA (51.8%–60.4%) (Wakjira et al., 2004; Choo et al., 2007). Linola, a low ALA cultivar developed for commercial vegetable oil market, contains only 3%–4% ALA and 75% linoleic acid (Lukaszewicz et al., 2004).

Nonacylglycerol Constituents

Flaxseed oil contains 0.7%–1.3% of unsaponifiable compounds (Painter & Nesbitt 1943). As with most other vegetable oils, the major unsaponifiable constituents are plant sterols. The total sterol content in flaxseed oil is about 700 mg/100 g, with the main sterols being sitosterol (30%), cycloartenol (29%), campesterol (15%), 24-methylene cycloartanol (9%), Δ5-avenasterol (8.5%), and stigmasterol (5%) (Schwartz et al., 2008). Other sterols, including cholesterol, brassicasterol, campestanol, and sitostanol, are found as minor components (Schwartz et al., 2008).

The major tocopherol in flaxseed oil is gamma-tocopherol (ca 130–575 mg/kg), which is usually accompanied by very small levels of alpha- and deltaocopherols (tr.-1 mg/kg). The carotenoids, β-carotene, lutein, and violaxanthin, are present in flaxseed oil (Pretova & Vojtekova, 1985). Flaxseed oils have chlorophyll content in the range of 8–58 mg/kg, with the level being dependant on seed maturity (Choo et al., 2007). In addition, flaxseed oils contain 127–256 mg/kg of total flavonoids (luteolin equivalents) and 768–3073 mg/kg of total phenolic acids (Choo et al., 2007).

Dietary Fiber

Fiber occurs as structural material in the cell walls of plants and has important health benefits for humans. Dietary fiber consists of nondigestible plant carbohydrates and other materials that are found intact in plants. Whole flax seeds and milled flax are sources of dietary fiber. Functional fiber consists of nondigestible carbohydrates that have been extracted from plants, purified and added to foods and other products (Institute of Medicine, 2002).

Total fiber accounts for about 28% of the weight of full-fat flax seeds. The major fiber fractions in flax consist of cellulose and mucilage gums, which are a type of polysaccharide that becomes viscous when mixed with water or other fluids. Flax mucilage consists of three distinct types of arabinoxylans, which form large aggregates in solution and contribute to its gel qualities (Warrand, et al., 2005). Mucilage levels range from 6 to 8% dry weight (DW), with the content and composition varying with cultivar (Mazza & Biliaderis, 1979; Cui et al., 1994).

Flax also contains sources of insoluble fiber including lignin, a highly-branched fiber found within the cell walls of woody plants, and lignans, which are phytochemicals ("phyto" means "plant"). Flax is a very rich source of a lignan called secoisolariciresinol diglucoside (SDG), which is found in amounts ranging from 1 to 26 mg/g of seed. Total lignan content approximates 380 mg/g and includes SDG, lariciresinol, matairesinol, pinoresinol, and secoisolariciresinol (SECO) (Thompson, et al., 2006). The wide range in SDG content reflects differences in flax cultivars, growing region and method of analysis (Muir, 2006).

Metabolism of Alpha-linolenic Acid

Alpha-linolenic acid (ALA) is found in plants, animals, plankton, and marine species, but flax and perilla seed oils are the richest sources (Morris, 2007). ALA is an essential fatty acid (EFA) that plays an important role in growth and development, reproduction, vision, maintaining healthy skin, maintaining cell structure, the metabolism of cholesterol, and gene regulation. ALA, which is the most commonly consumed omega-3 fatty acid in the typical Western diet (Lanzmann-Petithory, 2001), has also been linked to the prevention and/or amelioration of several chronic conditions including cardiovascular disease, certain cancers, rheumatoid arthritis, and autoimmune disorders.

EFAs are required in the diet, as they cannot be synthesized by humans. The two established EFAs are the omega-6 fatty acid linoleic acid (C18:2n-6, LA) and the omega-3 fatty acid ALA. LA and ALA are components of cellular membranes and act to increase membrane fluidity. These fatty acids are necessary for cell membrane function, as well as for the proper functioning of the brain and nervous system (Davis & Kris-Etherton, 2003; Harper & Jacobson, 2001). ALA is converted to the long-chain omega-3 fatty acids, eicosapentaenoic acid (EPA) and docosapentaenoic acid (DPA), and, to some extent, to docosahexaenoic acid (DHA), which are commonly consumed in fish and fish oil. The conversion of ALA to EPA, DPA, and DHA occurs primarily in the liver in the endoplasmic reticulum and involves a series of elongation enzymes that sequentially add 2-carbon units to the fatty acid backbone and desaturation enzymes that insert double bonds into the molecules (Fig. 10.1). The final conversion of ALA to DHA requires a translocation to the peroxisome for a beta-oxidation reaction. Similarly, LA is converted to long-chain omega-6 fatty acids in particular arachidonic acid (AA), also by the same series of desaturations and elongations that metabolize ALA.

The desaturation and elongation of LA and ALA, as well as the subsequent production of eicosanoids, occurs competitively using the same group of enzymes.

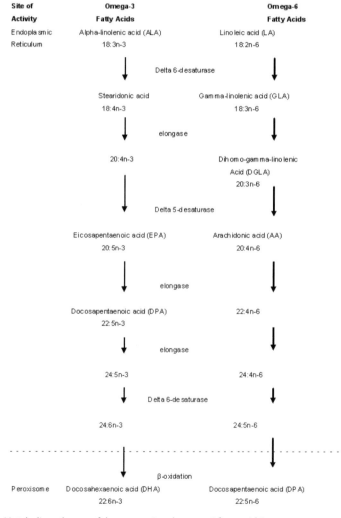

Fig. 10.1. Metabolic pathways of the omega-3 and omega-6 fatty acids[a].

Thus, an excess of one family of fatty acids can interfere with the metabolism of the other, reducing its incorporation into tissue lipids and altering their biological effects (Harper & Jacobson, 2001). Although the rate-limiting enzyme D6-desaturase shows greater substrate specificity for ALA, the overabundance of dietary LA gives this fatty acid a quantitative advantage that is believed to limit the conversion of ALA to EPA in vivo (Burdge & Calder, 2005). The intake of LA in the American diet, which is about 15 g/d, is approximately one order of magnitude greater than the intake of ALA (Arterburn et al., 2006). See Fig. 10.2.

About 96% of dietary ALA appears to be absorbed in the gut (Burdge, 2006). After absorption, ALA has several metabolic fates, including desaturation and elongation to the longer-chain omega-3 polyunsaturated fatty acids (PUFA) (Fig 10.1) or shortening by β-oxidation, which is considered the major catabolic route. In men, 24%–33% of an ingested dose of ALA undergoes β-oxidation (Bretillon et al., 2001; DeLany et al., 2000) compared to 19%–22% in women (Burdge & Wootton, 2002;

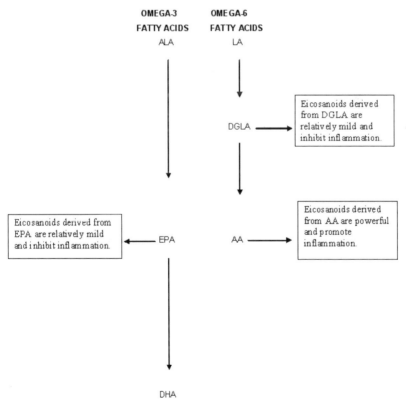

Fig. 10.2. Sources and actions of eicosanoids.

McCloy et al., 2004). The greater β-oxidation of ALA in men probably reflects a larger mass of active tissues such as muscles, heart, liver, and kidney compared with women. Furthermore, the figures may underestimate by about 30% the actual amount of dietary ALA that undergoes β-oxidation due to the trapping of labeled CO_2 in bicarbonate pools (Burdge, 2006). The amount of ingested ALA shunted into the β-oxidation pathway appears to be stable and is not affected by dietary intake. In a study of 14 healthy men aged 40–64 years, the proportion of ALA undergoing β-oxidation did not differ between the men who consumed a diet rich in ALA (10 g/day) versus those who consumed a diet rich in EPA + DHA (1.5 g/day) for 8 weeks (Burdge et al., 2003).

Like other fatty acids, ALA can be stored in adipose tissue. In a typical 75 kg man with a fat mass of 15%, adipose tissue is calculated to contain 79 g of ALA. In a typical 65 kg woman with a fat mass of 23%, adipose tissue is calculated to contain 105 g of ALA (Burdge & Calder, 2005). The greater capacity for ALA storage by women probably reflects their greater fat mass compared with men.

ALA Conversion to Longer Chain Polyunsaturated Fatty Acids

Estimates of the amount of ALA converted to EPA range from 0.2% to over 10% (Burdge & Calder, 2005; Burdge et al., 2002; Cao et al., 2006), with young women showing a conversion rate as high as 21% (Burdge & Wootton, 2002). Conversion of ALA to DPA is estimated to range from 0.13% to 6% (Burdge, 2006; Cao et al., 2006). The conversion rate for young women is on the higher end (6%) (Burdge & Wootton, 2002).

Conversion of ALA to DHA appears to be limited in humans, with most studies showing a conversion rate of about 0.05% (Burdge, 2006; Pawlosky et al., 2001), although one study reported a value of 4% (Emken et al., 1994). A conversion rate of 9% was found in young women (Burdge & Wootton, 2002) and up-regulation of the conversion of EPA to DHA, which may be the result of the actions of estrogen on D6-desaturase and may be of particular importance in maintaining adequate provision of DHA in pregnancy (Williams & Burdge, 2006).

However, stable isotope tracer studies have shown that healthy adults have the ability to synthesize DHA from ALA (Emken et al., 1990; Salem Jr., et al., 1999; Vermunt et al., 2000). The large differences in the rates of ALA conversion may be due to major differences in study methodologies. In addition, some have argued that a failure of ALA to elevate DHA levels in blood compartments does not necessarily mean that DHA concentrations do not increase in tissues (Barceló-Coblijn & Murphy, 2009; Brenna, 2002).

Conversion and conservation of ALA may be efficient in developing neural tissue and in very active tissues such as retina, which actively recycles DHA (Barceló-Coblijn et al., 2005). Further, the conversion of ALA to EPA and DHA appears to be dependent on the tissue and phospholipid class, with significant accumulation of ALA, DPA, and EPA, but not DHA, in the heart and liver phospholipids of rats fed flax oil. Additionally, diets enriched in either flax or fish oil diets increased phospholipid DHA mass in the brains of rats. In all tissues, both oils decreased the AA mass, although the effect was more marked in the fish oil than in the flax oil group. This study was not able to identify the origin of the DHA in the rats fed the flax oil diet, that is, whether it was imported from the blood as preformed DHA or synthesized in the brain from plasma-derived ALA. However, several studies demonstrate minimal elongation and desaturation of ALA in the plasma, with conversion limited to EPA and DPA (Burge et al., 2002; Zhao et al., 2004; de Groot et al., 2004) and thus suggesting that ALA, DPA, and EPA are taken up by the brain and converted to DHA.

In 20 humans who received supplementation with either fish oil (1296 mg EPA and 864 mg DHA/day) or flaxseed oil (3510 mg ALA; 900 mg LA/day) for 8 weeks, erythrocyte membrane EPA and DHA increased 300% and 42%, respectively, following fish oil (Cao et al., 2006). Flaxseed oil supplementation increased erythrocyte membrane EPA to 133% and DPA to 120% of baseline. The rapid conversion between EPA and DPA indicates the possibility that DPA can be a potential storage form for EPA. ALA supplementation enriches EPA and DPA composition in erythrocyte membranes, which may act to sustain a constant supply of EPA to body tissues.

Similarly, flaxseed oil (17 g/day ALA) increased phospholipid ALA (>threefold), EPA (>twofold), and DPA (50%), but did not change DHA, in 21 moderately hyperlipidemic men supplemented for 12 weeks (Hussein et al., 2005).

The relationship between the dietary intake of ALA and changes in membrane EPA is positive and linear over intakes of ALA between 2 and 10 g but strongly influenced by several factors, most significantly being dietary intakes of LA. An early demonstration of the competitive nature of LA and ALA showed that an increase in dietary concentrations of LA causes a decrease in products of omega-3 long chain PUFA and vice versa (Mohrhauer & Holman, 1963). Further, a diet rich in LA can reduce ALA conversion by as much as 40%, with a net reduction in long chain omega-3 fatty acid accumulation of 70%. It is important to note that considerable variability in the conversion rates among individuals has been reported, even when the subjects have similar background diets (Emken, 1995).

In a study of 22 healthy men, an LA rich diet (10.5% energy) reduced the EPA content of plasma phospholipids significantly after 4 weeks compared with a low LA diet (3.8% energy), even though both diets contained the same amount of ALA (1.1% energy) (Liou et al., 2007). Optimal conversion of ALA to n-3 long chain PUFA is expected when the diet is low in both n-6 fatty acids, particularly LA, and in omega-3 long chain PUFA. Additionally, a high intake of LA by pregnant women has been reported to lower EPA and DHA levels in umbilical plasma, suggesting reduced ALA conversion and availability of omega-3 long chain PUFA for the developing fetus (Al et al., 1996).

Maximum conversion of ALA was observed in human hepatoma cells (HepG2) incubated with a mixture of [13C]LA/[13C]ALA at a ratio of 1:1, where 0.7% and

17% of the recovered [13C]ALA was converted to DHA and EPA, respectively (Harnack et al., 2009). Regulative cellular signal transduction pathways involved in conversion were studied through the determination of transcript levels of the genes encoding delta-5 desaturase and delta-6 desaturase, peroxisome proliferator-activated receptor alpha (PPARα), and sterol regulatory element binding protein 1c (SREBP-1c). Gene expression of PPARα, SREBP-1c, and D5D were higher in the presence of ALA solely. When comparing the percentage rates of conversion of [13C]LA and [13C]ALA to AA and EPA/DHA, respectively, the conversion of [13C]ALA to EPA/DHA was higher than the conversion of [13C]LA to AA. This effect might be attributed to differential effects of omega-6 and omega-3 fatty acids on desaturase activity and/or gene expression.

Using stable isotope tracers, the absolute amounts of linoleic acid and ALA were shown to be of greater importance in influencing the conversion of ALA to EPA and DHA than the relative proportions of these fatty acids (Goyens et al., 2006). This study, which employed a randomly controlled intervention design of 258 subjects, examined the optimal ratio of omega-6/omega-3 fatty acids in the United Kingdom diet, using a 6-month, food-based intervention in older men and women. Four dietary treatments with omega-6/omega-3 ratios of 3:1 and 5:1, consisting of either ALA or EPA/DHA or combinations of both, versus a control (omega-6/omega-3 ratio of 10:1) were assessed. The results showed that lowering LA was more effective in promoting the conversion of ALA to EPA, while increasing ALA facilitated conversion of EPA to DHA. The authors argued that a reduction in dietary LA together with an increase in ALA intake would be the most appropriate way to enhance EPA and DHA synthesis from ALA. However, the study also noted no differences between the ALA and EPA plasma phospholipid contents between the two low ratio groups, which support the hypothesis that incorporation and bioconversion of ALA are rather influenced by the ratio of dietary LA/ALA. In additional work by the same group, 29 healthy subjects consumed for 28 days a diet which provided 7% of energy from LA and 0.4% from ALA (Goyens et al., 2005). On day 19, subjects received a single bolus of 30 mg of uniformly labeled [13 C]ALA and for the next 8 days 10 mg twice daily. Nearly 7% of dietary ALA was incorporated into plasma phospholipids. From this pool, 99.8% was converted into EPA and 1% was converted into DPA and subsequently into DHA. The quantification of the separate conversion reactions remains complex.

Most of the investigations of ALA metabolism in human subjects have focused on groups of relatively young, healthy or slightly hypercholesterolemic individuals. Research suggests that age does not seem to be a major determining factor in the metabolism of ALA to EPA. Replacing soybean oil (SO) with perilla oil (PO) was used to increase ALA intake to 3 g/day in 20 Japanese subjects over the age of 60 years (Ezaki et al., 1999). As a result, the omega-6/omega-3 ratio in the diet changed from 4:1 to 1:1. Following 10 months and in comparison to the SO diet, the higher ALA diet resulted in significant increases in serum EPA and DHA from 2.5 to 3.6%

and 5.3 to 6.4%, respectively. These data indicate that in elderly subjects a 3 g/d increase of dietary ALA could increase serum EPA and DHA.

Subjects aged 18–29 or 45–69 years consuming 6 g/day of ALA in the form of ground flaxseed (30 g) or flaxseed oil showed significant increases in plasma ALA and EPA concentrations over a period of 4 weeks (Patenaude et al., 2009). The diets induced no major changes in platelet aggregation, plasma total cholesterol, low-density lipoprotein, or high-density lipoprotein cholesterol levels in any of the groups. However, younger subjects showed a decrease in triacylglycerols (TG) values compared with older subjects. In contrast, studies of a longer duration that administered larger ALA doses to older populations resulted in lower plasma TG concentrations (Djousse et al., 2003; Zhao et al., 2004).

High intakes of EPA and DHA can also block ALA conversion, possibly by signaling that tissue levels of omega-3 fats are adequate. A diet containing more than 12 g of LA per day can reduce ALA conversion (Cunnane et al., 1993). Other factors that influence the rates of ALA conversion include the intake of high levels of dietary cholesterol (Garg et al., 1988; Leikin & Brenner, 1987), saturated fat, oleic acid (Berger et al., 1992; Li et al., 1999), *trans* fatty acids (Houwelingen & Hornstra, 1994) and the ratio of polyunsaturated to saturated fats in the diet (Layne et al., 1996).

Individuals who do not eat fish or fish oils (vegans and non-fish-eating vegetarians and meat eaters) could be at risk of low or inadequate omega-3 status as plasma concentrations are lower in vegetarians and in vegans than in fish eaters (Sanders et al., 1978). Although non-fish-eating meat eaters and vegetarians have much lower intakes of EPA and DHA than do fish eaters, their omega-3 status is higher than would be expected. Using data from a cross-sectional study of 196 meat-eating, 231 vegetarian, and 232 vegan men in the United Kingdom, the proportions of plasma EPA and DHA were found to be lower in the vegetarians and in the vegans than in the meat eaters (Rosell et al., 2005). Only small differences were seen for DPA. Plasma EPA, DPA, and DHA proportions were not significantly associated with the duration of time since the subjects became vegetarian or vegan, which ranged from <1 year to >20 years. In the vegetarians and the vegans, plasma DHA was inversely correlated with plasma LA. When animal foods are excluded from the diet, this data shows that the endogenous production of EPA and DHA results in low but stable plasma concentrations of these fatty acids.

Using data from the EPIC (European Prospective Investigation into Cancer and Nutrition)-Norfolk cohort, Welch et al. (2010) assessed intakes of omega-3 fatty acids in 14,422 men and women aged 39 to 78 with 7 day diary data. Plasma phospholipid fatty acid measures were conducted in a sub-group of 4902 individuals. ALA intake was highest in fish eaters and lowest in vegans, women, and meat eaters. ALA contributed 82% of total dietary omega-3 fatty acids in the whole population, including 80% in fish eaters, 98% in vegetarian men, 99% in vegetarian women and 97% in meat eaters. EPA intakes in meat eaters were only 15% (men) and 18% (women) of those in fish eaters and in vegetarians were 9% (women) and 15% (men) of those in fish eaters. The ratio of EPA/DHA:ALA in plasma phospholipids was 209% higher in vegan men and 184% higher in vegan women than in fish eaters, 14% higher in vegetarian men and 6% higher in vegetarian women than in fish eaters, and 17% and 18% higher in male and female meat eaters, respectively, than in fish eaters, suggesting that statistically estimated conversion may be higher in non-fish eaters than in fish eaters. This is the first large population study to investigate intakes, status, and the precursor-product ratio by using statistical models as surrogate estimates of conversion of ALA to EPA and DHA in different dietary groups.

Significant increases in total plasma ALA and DPA but not in DHA using levels of 32g/day of flaxseed in the diets of type 2 diabetes (Taylor et al., 2010) and in healthy menopausal women fed 40 g/day of flaxseed (Dodin et al., 2008) have also recently been reported.

As summarized here, a number of studies have reported the effects of consuming increased amounts of dietary ALA on the fatty acid composition of plasma or cell lipids and consistently demonstrate that intakes ranging from <5 to approximately 20 g/day result in enhancement of EPA and increased proportions of DPA. Many studies also demonstrate that increased consumption of ALA does not result in increased proportions of DHA in plasma or cell lipids, but some do report a tendency for DHA to increase. What is important to note is the variation in the response between studies, which might reflect differences in the age and gender mix of the subjects studied and variations in background diet (e.g., habitual omega-3 long chain PUFA and LA intakes), as well as differences in the way in which ALA was provided (capsules, oils, margarines, prepared foods), the duration of studies, and differences in the analytical procedures used.

Preliminary data suggest that there are important differences between men and women in their capacity for synthesis of EPA and DHA from ALA, and that this capacity may be affected by physiological state (e.g., pregnancy). The ability to up regulate this pathway during pregnancy may be one of the adaptational mechanisms by which circulating maternal DHA levels are increased and provision of fetal DHA needs are met. This capacity for adaptation may be of particular importance in vegan pregnancies and in multiple and sequential pregnancies, where demand for DHA will be much greater than normal. Clearly, more research in this area is required.

The conversion of ALA to EPA in the body may be physiologically and clinically important. As described in subsequent sections, dietary supplementation with ALA does result in cardiovascular benefits similar to that seen with EPA, but whether these effects are due to its conversion to EPA or another mechanism unrelated to EPA (i.e., improvements in endothelial function, inflammation, lipid changes, or antiarrhythmic effect) require further assessment. The consumption of ALA containing products has important clinical and public health implications because many people do not consume fish or do not have access to fish rich in EPA. Additionally, ALA intake is feasible and realistic through a number of different sources, supplements and foods.

Table 10.A. Comparison of Health Consequences of Diets Rich in Omega-6 Versus Omega-3 Fats.

Consequences of eating a diet rich in omega-6 fats	Benefits of eating a diet rich in omega-3 fats
↑ n-6/n-3 in cell membrane phospholipids	↓ n-6 fatty acids in cell membranes
↑ production of arachidonic acid	↓ n-6/n-3 in cell membrane phospholipids
↑ release of pro-inflammatory eicosanoids derived from arachidonic acid	↓ levels of pro-inflammatory compounds like eicosanoids and cytokines
↑ production of pro-inflammatory cytokines	↓ clumping (aggregation) of blood platelets
↑ expression (activation) of pro-inflammatory genes	↓ expression (activation) of pro-inflammatory genes
↑ biomarkers of inflammation such as C-reactive protein	↓ biomarkers of inflammation such as C-reactive protein
↑ blood viscosity	↓ production of interleukin-10, an anti-inflammatory cytokine
↑ constriction of blood vessels	
↑ oxidative modification of low-density-lipoprotein (LDL) cholesterol	

Sources: Gebauer et al. (2006); Simopoulos (2006)

Flaxseed and Cardiovascular Disease

Cardiovascular disease (CVD) includes all diseases of the blood vessels and circulatory system, such as coronary heart disease (CHD), ischemic heart disease (IHD), myocardial infarction (MI), and stroke. CVD is the leading cause of death in the United States and Canada (Rosamond, 2007; Heart and Stroke Foundation of Canada, 2003). Flaxseed contains ALA, lignans, and fiber, which can positively affect blood lipid levels, blood pressure, endothelial function, and inflammation in CVD.

Alpha-Linolenic Acid and Cardiovascular Disease

Four case-control studies (Lemaitre et al., 2003; Baylin et al., 2003; Baylin et al., 2007; Guallar et al., 1999; Rastogi et al., 2004), one cross-sectional study (Manav et al., 2004), three prevention trials (Pietinen et al., 1997; de Lorgeril et al., 1994; de Lorgeril, et al., 1999; Dolecek, 1992) and three cohort studies (Djoussé et al., 2001; Djoussé et al., 2003a; 2003b; Djoussé et al., 2005a; Ascherio et al., 1996; Mozaffarian et al., 2005; Hu et al., 1999; Albert et al., 2005) found a benefit of ALA-rich diets in lowering the risk of CHD, IHD, nonfatal MI, and stroke. One prevention trial found no change in the estimated 10-year IHD risk but reported a significant decrease in the levels of the pro-thrombotic protein, fibrinogen and the pro-inflammatory C-reactive protein (CRP) following ALA-rich diets (Bemelmans et al., 2002;

Bemelmans et al., 2004). The number of participants in these studies ranged from 233 to 76,283.

Modest intakes of ALA appear to have a significant effect on reducing nonfatal MI (Campos, 2008). A nonlinear inverse relationship between 0.7% adipose tissue ALA and dietary ALA intake of about 1.8 g/day (1/2 teaspoon of flax oil) and risk of nonfatal MI has been observed in a study of 1819 patients who survived an MI and 1817 matching controls. The relationship between ALA and MI was nonlinear; risk did not decrease with intakes above 0.65% energy (1.8 g/day). These observations are significant in that ALA as assessed both by questionnaire and in adipose tissue was associated with reduced risk of MI in a large population. The maximum benefit of ALA was obtained within a realistic and achievable range of intake, and the association between ALA and MI was independent of fish intake.

These results confirm earlier studies in which the ALA content of adipose tissue was reported to be inversely related to risk of MI in one case-control study conducted in Europe and Israel (Guallar et al., 1999) and inversely related to nonfatal acute MI in another study in Costa Rica (Baylin et al., 2003). In the Nurse's Health Study, which involved a 10-year follow-up of 76,283 women with no previously diagnosed CVD, a higher intake of ALA was associated with a lower relative risk of fatal and non-fatal MI (Hu et al., 1999; Albert et al., 2005). A recent evaluation of food balance sheets and CHD outcomes for 11 Eastern European countries revealed that populations that experienced the greatest increase in ALA consumption since 1990 also experienced a substantial decline in CHD mortality. These results were consistent in men and women (Zantonski, 2008). It was noted that the countries that achieved an ALA increase of more than 0.6 g/day between 1990 and 2002 had substantial reductions in CHD risk. It is possible that the partial replacement of oils that do not have ALA, such as sunflower oil, with oils that are rich in ALA could lead to considerable health benefits.

ALA appears to have an important role among populations with low fish intake in part because EPA can inhibit the action of delta-5 and delta-6 desaturase activity. Consistent with this hypothesis, 1 g/day ALA intake has been associated with a 50% lower risk of nonfatal MI among men consuming very low (<100 mg/day) omega-3 long-chain PUFA from fish, but no association was found among those with a higher intake (Mozaffarian et al., 2005). The data assessed was from the Health Professional Follow-up Study, which began in 1986 with a cohort of 45,772 health professionals. These data strongly supports a direct role of ALA consumption in decreasing CHD risk and further indicates that ALA may be of particular importance in sectors of the population that do not eat fatty fish.

The Lyon Diet Heart Study included participants who had previously survived a myocardial infarction compared to an experimental group who consumed a typical Mediterranean-style diet rich in ALA. The control group consumed a typical Western-type diet low in ALA. The results were impressive with a 75% reduction in non-fatal myocardial infarctions and a 70% reduction in total death noted amongst the ALA group in comparison to the control group (de Lorgeril et al., 1994; 1999).

Using the National Health and Nutrition Examination Survey (NHANES), a higher intake of ALA was associated with lower prevalence of peripheral arterial disease (PAD) (Lane et al., 2008). Data from the period 1999 to 2004 from 7203 lower extremity examinations, of which 422 individuals had prevalent PAD (5.9%), were included in the analysis.

The beneficial effects of ALA seen in these studies is most likely not due to changes in serum lipids, as most clinical observations show no effect of flax oil consumption (2 to 20 g/day) consumed for 4 to 20 weeks, on blood total cholesterol (TC) and LDL-cholesterol (LDL-C) levels in normal and hypercholesterolemic subjects (Sanders & Roshanai, 1983; Mantzioris et al., 1994; Kestin et al., 1990; Nestel et al., 1997; Clandinin et al., 1997; Goh et al., 1997; Paschos et al., 2005; Rallidis et al., 2004; Schwab et al., 2006; Singer et al., 1986; Singer et al., 1990).

In a recent study, 86 healthy male and female volunteers completed a 12-week double-blinded, placebo-controlled trial supplemented with two 1 gm capsules of placebo, fish oil, flaxseed oil, or hempseed oil per day for 12 weeks (Kaul et al., 2008). The lipid parameters (TC, high density HDL-C, LDL-C, and TG) did not show any significant differences nor was the oxidative modification of LDL affected. None of the dietary interventions induced any significant change in collagen or thrombin stimulated platelet aggregation or the level of inflammatory markers.

A significant decrease in TC, however, was noted among men who consumed 2 tbsp of flax oil daily for 12 weeks (Wilkinson et al., 2005). HDL-C decreased significantly between 4% and 10% in 4 of the 13 studies (Nestel et al., 1997; Paschos et al., 2005; Rallidis et al., 2004; Wilkinson et al., 2005). TG decreased significantly between 9% and 25% in 3 studies (Schwab et al., 2006; Singer et al., 1986, 1990).

Elderly subjects appear to react favorably to ALA. Goyens and Mensink (2006) studied an ALA-rich diet (6.8 g/day) compared to EPA/DHA-rich diet (1.05 g EPA/day + 0.55 g DHA/day) in 37 mildly hypercholesterolemic subjects, aged between 60 and 78 years. Both EPA/DHA and ALA induced increases in tissue factor pathway inhibitor (TFPI) of 14.6% and 18.3%, respectively. TFPI is a critical inhibitor of tissue factor-induced coagulation, and low levels are involved in the development of deep-vein thrombosis. Additionally, ALA affected concentrations of LDL-C and apoB more favorably than EPA and DHA. Results from well-controlled and larger clinical trials are needed to provide a more conclusive answer regarding the effects of ALA on blood lipids.

Two population studies reported a benefit of ALA in reducing stroke risk. In the Edinburgh Artery Study, significantly lower levels of ALA were found in the red blood cell phospholipids of men and women who had had a stroke compared with participants who had no evidence of the disease (Leng et al., 1999). In the Multiple Risk Factor Intervention Trial (MRFIT), 96 men who had had a stroke were compared with 96 men without stroke who were matched for age. In the multivariate model, each increase of 0.13% in serum ALA level was associated with a 37% decrease in risk of stroke (Simon et al., 1995). After controlling for risk factors of stroke such as smoking and blood pressure, ALA emerged as an independent predictor of stroke risk. In one clinical trial, supplementing the diet with flax oil (1 tbsp providing 8 g ALA/day) for 12 weeks lowered systolic and diastolic blood pressure significantly in middle-aged hypercholesterolemic men compared with a safflower oil group. The magnitude of the effect (5 mmHg) was clinically relevant (Paschos et al., 2007).

Endothelial dysfunction is the earliest detectable stage in the development of atherosclerosis. An increase in systemic arterial compliance (SAC) denoting an improvement in endothelial function, combined with a reduction in mean arterial pressure, was reported among 15 obese adults who ate daily a diet enriched with flax oil (providing 20 g of ALA) for 4 weeks (Sanders & Roshanai, 1983). The increase in SAC with flax oil was similar to that achieved through exercise training. West and coworkers (2005) measured endothelial function by the method of flow-mediated vasodilation (FMD) in 18 healthy adults with type 2 diabetes. FMD was measured before and 4 hours after 3 test meals, each providing 50 g of a specific type of fat—monounsaturated fat (MUFA) obtained from high-oleic safflower and canola oils, the MUFA diet plus EPA and DHA from sardine oil, or the MUFA diet plus ALA from canola oil. In volunteers with high fasting triacylglycerols, meals containing omega-3 fatty acids increased FMD by 50%–80%. Sardine and plant omega-3 fats were equally effective in improving endothelial function as measured by FMD.

Endothelial dysfunction is also characterized by a tendency for leukocytes to adhere to the endothelium in a process controlled by cell adhesion molecules, including E-selectin, vascular cell adhesion molecule type 1 (VCAM-1), and intercellular adhesion molecule type 1 (ICAM-1) (Hwang et al., 1997). A diet rich in ALA significantly decreased VCAM-1, ICAM-1, and E-selectin compared to an average American diet in 23 hypercholesterolemics (Zhao et al., 2004). Consuming 1 tbsp of flax oil daily for 12 weeks reduced VCAM-1 levels by 18.7% in a group of male hypercholesterolemic subjects (Rallidis et al., 2004). These findings suggest that an ALA-rich diet containing flax oil has a beneficial effect on the endothelium.

Anti-inflammatory Effects of ALA

In recent years, medical research has moved toward almost a unifying theory of chronic disease as a consequence of low-grade, chronic inflammation. Inflammation is a controlled, ordered process whereby the body responds to infection or injury. Symptoms of inflammation include redness, swelling, heat and pain. Chronic inflammation is linked with age-related diseases such as CHD, obesity, diabetes and cancer. Agents that exert anti-inflammatory actions are likely to be important in both prevention and therapy of a wide range of human diseases and conditions. An increasing amount of research suggests that the consumption of ALA may provide protection against inflammatory diseases by reducing inflammatory eicosanoids and cytokines.

Pro-inflammatory eicosanoids such as thromboxane A_2 (TXA_2) and leukotriene B_4 (LTB_4) are derived from AA. TXA_2 is one of the most potent promoters of platelet aggregation known (Reiss & Edelman, 2006; Ross, 1999). LTB_4 increases the release of reactive oxygen species and cytokines like tumor necrosis factor α (TNF-α), interleukin 1β (IL-1β), IL-6, and IL-8 (Calder, 2006).

The potential anti-inflammatory effects of ALA could be mediated in part through its conversion to EPA or through direct protective effects. In cell membranes, omega-3 PUFA replace the omega-6 long chain AA, thereby reducing the potential release of AA under basal conditions and during pathophysiological insults which precipitate inflammation. Mechanistically, this would effectively reduce the basal levels of pro-inflammatory eicosanoids.

In a clinical study of healthy men, consumption of 1.75 tbsp of flax oil daily for 4 weeks led to a 30% reduction in the immune cells concentration of TXB_2, which is an inactive metabolite of TXA_2 (Caughey et al., 1996). Concentrations of the pro-inflammatory cytokines TNF-α and IL-1β in immune cells decreased 26% and 28%, respectively. In 64 patients with chronic obstructive pulmonary disease (COPD), serum and sputum LTB_4 levels decreased 32% and 41%, respectively, in those patients who received an ALA-rich nutritional support (1.4% ALA) daily for 24 months compared to those who received a low-ALA nutritional support (0.18% ALA) (Matsuyama et al., 2005). Serum levels of IL-6 decreased 25% in men who consumed 1 tbsp of flax oil daily for 12 weeks (Paschos et al., 2005). The serum levels of TNF-α decreased by 43%, and the production by immune cells of TNF-α, IL-6, and IL-1β decreased between 18% and 22% in hypercholesterolemics who consumed a diet rich in ALA compared with the average American diet (Zhao et al., 2007).

During inflammation the liver releases acute-phase proteins such as CRP and serum amyloid A (SAA) in response to acute injury, infection, malignancy, hypersensitivity reactions, and trauma. CRP and SAA are markers of systemic inflammation and are present in the lesions of atherosclerosis. CRP is an independent risk factor for CVD (Getz, 2005). Consuming flax oil reduced CRP by 48% and serum SAA by 32% in 50 hypercholesterolemic men who consumed 1 tbsp of flax oil daily for 12 weeks (Paschos et al., 2005). In a U.S. study of 23 adults with high blood cholesterol levels, consuming a high-ALA diet based on walnuts, walnut oil, and flax oil resulted in a 75% decrease in CRP levels after 6 weeks (Zhao et al., 2004).

In a recent assessment, inverse relationships were reported between dietary intake of both omega-3 and omega-6 PUFA and serum CRP concentrations in 300 Japanese men aged 21 to 67 years (Poudel-Tandukara et al., 2009). Self-administered diet history questionnaires were used to assess the dietary intake of the preceding month in all subjects. In men, the mean serum CRP concentrations in the highest ALA intake group were 47% lower than that with the lowest intake. Although not statistically significant, serum CRP concentrations tended to decrease with increasing intakes of EPA and DHA in women.

Other Mechanisms of Alpha-linolenic Acid

ALA may lower the risk of fatal or nonfatal MI, which appears to involve an effect on cardiac rhythm. In the Family Heart Study, Djoussé et al. (2005b) found that the higher the dietary ALA intake, the lower the risk of abnormally prolonged repolarization of the heart muscle, an indicator of cardiac arrhythmia. In a clinical study among women referred for elective coronary angiography, ALA content of adipose tissue was positively correlated with 24-hour heart rate variability (HRV), a strong predictor of arrhythmic events and sudden cardiac death (Christensen et al., 2005). Decreased HRV is a strong predictor of SCD and arrhythmic events, and thus this association supports an anti-arrhythmic effect of ALA. ALA may reduce ventricular fibrillation (Ander, 2004), and its cardio-protective effects have also been attributed to improvements in arrhythmia (Vos & Cunnane, 2003).

ALA may also have anti-thrombotic activities. An increase in activated protein C resistance (APC(tm) resistance), which demonstrates increased anticoagulant activity, was noted in 15 healthy male subjects who consumed an ALA-rich diet with 31.5% energy fat diet and approximately 7% energy from PUFA. The high ALA diet had an ALA:LA ratio of 1:1.2 and was compared to a diet with an ALA:LA ratio of 1:21 (LA-rich) (Allman-Farinelli et al., 1999).

The effects of doubling the ALA intake from canola-type rapeseed oil (RSO) was assessed in 42 volunteers for 6 weeks in a parallel design (Seppanen-Laakso et al., 2010). Efficient competitive inhibition by ALA was deemed to be responsible for a decrease in long-chain omega-6 PUFA at 3 weeks. Initial elevated levels of fibrinogen (2.6–3.9 g/l) decreased by 30% (0.95 g/l) at 6 weeks. Fibrinogen, a protein involved in coagulation processes, is found in elevated levels in prothrombotic and proinflammatory states, associated with higher risk of CHD, stroke, diabetes, Alzheimer disease, and dementia. In addition, DHA in plasma phospholipids increased when fibrinogen levels were reduced following the enhancement in ALA intakes.

It has been shown that ALA decreases the nuclear transcription factor $\kappa\beta$, a major transcription factor involved in the regulation of inflammatory genes (Perez-Martinez et al., 2007; Ren & Chung, 2007). Furthermore, ALA inhibits the production of nitric oxide and down-regulates inducible nitric oxide synthase, cyclooxygenase-2, and tumor necrosis factor-R gene expression in murine macrophages (Ren & Chung, 2007).

Overall, ALA appears to protect against cardiovascular diseases by altering the omega-3 fat content of cell membranes (Harper et al., 2006) by improving blood lipids and endothelial function and by exerting significant anti-inflammatory and anti-thrombotic effects (Bloedon & Szapary, 2004). These beneficial effects have been reported with ranges of about 3–20 g of ALA per day (equivalent to 1/2 tsp to 2 $^1/_2$ tbsp of flax oil). In epidemiologic studies, ALA intakes associated with reduced CVD risk averaged about 2 g/day (range = 0.7–6.3 g). A meta-analysis of prospective studies suggested that increasing the intake of ALA by 1.2 g/day decreases the risk of fatal CHD by at least 20% (Brouwer et al., 2004). The Institute of Medicine has set an

Table 10.B. Estimated Dietary Intakes of Major Omega-6 and Omega-3 Fatty Acids in Canada and the United States (g/day)[a].

Country	Omega-6 fatty acid intake (Linoleic acid)	Omega-3 fatty acid intake	
		ALA	Long-chain fatty acids (EPA, DHA and/or DPA)
Europe[b]	15	1–2	0.1–0.5
Canada[c]	8–11	1.3–1.6	0.14–0.24
United States[d]			
Men	18	1.7	0.13
Women	14	1.3	0.10

[a]Abbreviations: ALA, alpha-linolenic acid; DPA, docosapentaenoic acid; DHA, docosahexaenoic acid; EPA, eicosapentaenoic acid.

[b]Values for the general European population (Sanders, 2000).

[c]Values are for pregnant women living in British Columbia and Ontario (Innis et al., 2003; Denomme et al., 2005).

[d]Values are for men and women aged 20–39 years (Gebauer et al., 2006)

Adequate Intake for ALA, based on the median daily intake of healthy Americans who are not likely to be deficient in this nutrient, as shown in Table 10.B (Institute of Medicine, 2002). The Adequate Intake is 1.6 g ALA per day for men and 1.1 g ALA per day for women. Up to 10% of the Adequate Intake for ALA can be provided by EPA and DHA.

More studies are needed to clarify the role of ALA in reducing CVD risk. In particular, there is an urgent need for randomized, controlled clinical trials with good study designs, clearly defined outcomes, appropriate control groups, realistic dietary interventions and thorough statistical analyses (Stark et al., 2008).

Milled Flaxseed and CVD

Several clinical trials have investigated the effects of flaxseed and flaxseed-derived products (flaxseed gums, oil or lignans) on blood lipids; however, the findings have been inconsistent. A recent meta-analysis of 28 studies published between January 1990 and October 2008 found that flaxseed supplementation of 30–40 g/day was associated with a decrease in blood total and LDL-cholesterol concentrations but did not substantially affect HDL cholesterol and triglycerides (Pan et al., 2009). Flaxseed interventions reduced total and LDL cholesterol by 0.10 mmol/L (95% CI: 20.20, 0.00 mmol/L) and 0.08 mmol/L (95% CI: 20.16, 0.00 mmol/L), respectively; significant reductions were observed with whole flaxseed (20.21 and 20.16 mmol/L, respectively) and lignan (20.28 and 20.16 mmol/L, respectively) supplements but not with flaxseed oil. The cholesterol-lowering effects were more apparent in females (particularly postmenopausal women) and in individuals with high initial cholesterol concentrations. Results were influenced by the treatment form of flaxseed, and quality of the study. Conclusions were difficult to draw due to inconsistency in the quality of products, the amount of specific bioactive components in the flaxseed treatments, and their bioavailability.

Patade et al. (2009) have published data which supports the total and LDL-C lowering effects of the daily incorporation of approximately 30 g of flaxseed in the diets of Native American postmenopausal women. Fifty-five mild to moderately hypercholesterolemic women experienced 7% and 10% reductions in total and LDL-C, respectively, following 3 months of supplementation. Bloedon et al. (2008) reported cardiovascular benefits in 62 men and post-menopausal women randomized to consume 40 g/day of ground flaxseed-containing baked products or matching wheat bran products for 10 weeks. Compared to wheat, flaxseed significantly reduced LDL-C at 5 weeks (13%) but not at 10 weeks (7%). Flaxseed reduced lipoprotein α (Lp[a]) by a net of 14% and reduced the homeostatic model assessment of insulin resistance (HOMA-IR) index by 23.7%. A significant reduction in CRP in a subset of hypertensive patients, as well as a significant negative correlation between CRP and serum ALA, was noted. In men, flax reduced HDL-C concentrations by a net of 16% and 9% at 5 and 10 weeks, respectively. The observed reduction in efficacy by 10 weeks could be caused by either a reduction in adherence or by biologic adaptation.

These observations support results from earlier studies in which blood total cholesterol decreased 6%–13% and LDL-cholesterol decreased 9%–18% in healthy young adults (Cunnane et al., 1993; 1995), hypercholesterolemic men and women (Bierenbaum et al., 1993), postmenopausal women (Arjmandi et al., 1998; Lucas et al., 2002), in adults with systemic lupus erythematosus (Clark et al., 1995), and men with prostate cancer (Demark-Wahnefried et al., 2001) fed for 4–12 weeks. HDL cholesterol and triacylglycerol levels were unchanged by diets containing milled flax.

One short-term study found no effect of flax on blood lipids among postmenopausal women who consumed 40 g (5 tbsp) of milled flax daily for 2 months (Lemay et al., 2002). Two long-term interventions of 1 year in duration similarly reported no changes in blood lipids among postmenopausal women (Dodin et al., 2005) or adults with lupus nephritis (Clark et al., 2001) who consumed between $3^3/_4$ tbsp and 5 tbsp (30–40 g) of milled flax daily. In a study using 50 g of partially defatted flax (less than 10% fat by weight) reduced total cholesterol, LDL-C and apo B by 5.5%, 9.7%, and 6%, respectively, and triacylglycerols increased 10% in 29 hypercholesterolemic men and women after 3 weeks (Innis & Elias, 2003). Significant reductions in apo B concentrations of 7.5% have also been reported among postmenopausal women who ate 40 g of milled flax daily for 3 months (Lucas et al., 2002).

Flaxseed has also been demonstrated to reduce Lp(a), an independent predictor of atherosclerosis. In an early study, a 7.4% decrease in Lp(a) over baseline was reported in hypercholesterolemic postmenopausal women consuming 38 g flaxseed per day for 3 months (Arjmandi et al., 1998). Further research by this group showed that flaxseed lowered Lp(a) by approximately 22%, although the mean values did not reach statistical significance (Arjmandi, 2001).

Overall results to date support the observations that milled flax or partially defatted flax decrease total cholesterol, LDL-C, and apo B levels without a significant decrease in HDL-cholesterol. Further, the majority of studies have reported either no change or a non-significant decrease in triacylglycerols. In limited research, flaxseed also reduced Lp(a).

Lignans and CVD

Overview of Lignans

Phytoestrogens including isoflavones, coumestans, flavonoids, and lignans are plant chemicals that can have estrogen-like actions in humans and animals (Martin et al., 2007). Lignans are widely distributed in the plant kingdom, play a role in plant growth, and act as antioxidants in human metabolism (Raffaelli et al., 2002). Flaxseed is the richest source of SDG (7 mg/g or 3.7 mg SECO 2/g), a range of 75 to 800 times more than any other food (Westcott & Muir, 2003). Variation in flaxseed lignan concentrations depend on the variety, location, and crop year (Westcott & Muir, 1996). Whole seed and ground flax typically contain between 0.7% and 1.9% SDG, which is approximately 1 mg/g to 26 mg/g of seed (Muir, 2006).

The lignans SDG, SECO, pinoresinol, lariciresinol, and matairesinol in flax are converted by bacteria in the colon to the mammalian lignans, enterodiol and enterolactone. A simplified diagram showing the conversion of flax lignans to mammalian lignans is given in Fig. 10.3.

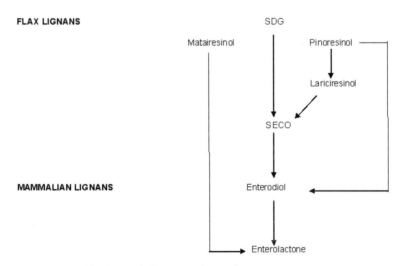

Fig. 10.3. Metabolism of flax lignans by bacteria in the gut[a,b].

Enterodiol and enterolactone can be excreted directly in the feces; absorbed by epithelial cells lining the human colon, conjugated with glucuronic acid or sulfate, and excreted in the feces or enter the circulation (Jansen et al., 2005); or can be absorbed from the gut and transported to the liver, where free forms are conjugated before being released into the bloodstream (Raffaelli et al., 2002). Eventually, enterodiol and enterolactone undergo enterohepatic circulation and are excreted in the urine in conjugated form (Axelson et al., 1982).

The concentration of enterodiol and enterolactone in the feces, blood, and urine is related to the concentration of plants lignans in the diet. Flax consumption increases the blood levels of mammalian lignans (Nesbitt et al., 1999; Morton et al., 1994; Tarpila et al., 2002; Knust et al., 2006) and the excretion of mammalian lignans and/or total lignans in feces (Kurzer et al., 1995) and urine (Nesbitt et al., 1999; Morton et al., 1994; Knust et al., 2006). Consuming a diet supplemented with a lignan/SDG complex derived from flax also increases mammalian lignan excretion in urine (Hallund et al. 2006). The bioavailability of the mammalian lignans can be enhanced by crushing and milling flax (Kuijsten et al., 2005).

Lignans and CVD

One area of active research concerning the health effects of lignans is in the area of CVD in which their anti-oxidant properties may confer benefit by contributing to a reduction in hypercholesterolemia and atherosclerosis (Prasad, 2000). SDG for example acts as an antioxidant for certain free radicals including hydroxyl ion (Prasad, 1997). The antioxidant activity of the flax lignans and metabolites include the ability to exert protective effects against AAPH-induced oxidation (Hosseinian et al., 2007). The antioxidant action of SECO and enterodiol has been reported to be greater than that of vitamin E (Bhathena & Velasquez, 2002).

To date, a number of clinical trials have been conducted using dietary flax and isolated lignans, which suggest that SDG may lower plasma cholesterol concentrations. Several mechanisms involved in the beneficial effects of phytoestrogens on lipid profile have been described, including increased bile acid secretion, promotion of removal of LDL, enhanced thyroid function, and modified hepatic metabolism (Lissin & Cooke, 2000). However, the results have not shown consistent benefit. In most of these studies, the concentration of lignans was not determined. Differences of study designs, subject characteristics, and treatment conditions could confound the outcomes and interpretation of the results (Pan et al., 2009).

One case-control study has reported that men without CHD had higher serum enterolactone, a mammalian lignan, and a 65% lower risk of CHD than men with CHD (Vanharanta et al., 1999). The effects of a lignan complex on CVD risk factors in 22 healthy postmenopausal women who consumed daily a low-fat muffin with or without the lignan complex for 6 weeks has been assessed (Hallund et al., 2006). The lignan complex provided 500 mg of SDG per day, a dose which corresponds to 38–82 g of whole flax seeds. Plasma concentrations of total cholesterol, LDL-C, HDL-C, and triacylglycerols did not differ after the 2 intervention periods.

An 8-week, randomized, double-blind, placebo-controlled study was conducted in 55 hypercholesterolaemic subjects, using treatments of 0 (placebo), 300, or 600 mg/day of SDG extract, resulting in significant reductions in total cholesterol, LDL-C, and glucose concentrations, as well as their percentage decrease from baseline (Zhang et al., 2008). At weeks 6 and 8 in the 600 mg SDG group, the decreases in total cholesterol and LDL-C concentrations were 22.0 to 24.38%, respectively. For the 300 mg SDG group, only significant differences from baseline were observed for decreases in total cholesterol and LDL-C. A substantial effect on lowering concentrations of fasting plasma glucose was also noted in the 600 mg SDG group at weeks 6 and 8, especially in the subjects with high baseline glucose concentrations.

A randomized, double-blind, placebo-controlled, cross-over trial of supplementation with SDG capsules (360mg/day) or placebo for 12 weeks was conducted with 70 diabetic patients (26 men and 44 post-menopausal women) with mild hypercholesterolaemia (Pan et al., 2009). Baseline to follow-up concentrations of CRP increased significantly within the placebo group but were unchanged in the SDG group; a significant difference was observed between treatments. This effect was confined to women but not observed in men. No effects were noted for IL-6 and retinol-binding protein 4 (RBP4) levels, which are associated with insulin resistance and diabetes mellitus. This study suggests that SDG might modulate CRP levels in type 2 diabetics.

Cornish et al. (2009) used a randomized double-blind placebo controlled study design to assess the effects of SDG supplementation during exercise training on a metabolic syndrome composite score and osteoporosis risk in older adults (Cornish et al., 2009). One hundred subjects (>50 years) were randomized to receive SDG (543 mg/day) or placebo while completing a 6 month walking program (30–60 min/day; 5–6 days/week). A composite score of 6 risk factors for metabolic syndrome (fasting glucose, HDL cholesterol, TAG, abdominal adiposity, blood pressure, and inflammatory cytokines) was calculated at baseline and at 6 months. Men taking the placebo increased their metabolic syndrome composite score, but there were no changes in the other groups. There were no differences between groups for change in bone measures, body composition, lipoproteins, or cytokines. Males taking the flaxseed lignan complex reduced metabolic syndrome score relative to men taking placebo, but a similar trend was not seen in females. Flaxseed lignan had no effect on bone mineral density or content, body composition, lipoproteins, glucose, or inflammation.

The limited studies that have been performed to date suggest that SDG have a cholesterol-lowering effect and may modulate CRP levels in type 2 diabetics. These results need to be confirmed by further large clinical trials of longer duration.

Flaxseed Summary—Effects on CVD

Fig. 10.4 summarizes the data from dietary studies and shows the mechanisms by which flaxseed, ALA, fiber and lignans may protect against CVD. Flaxseed influences CVD risk by altering the omega-3 fat content of cell membranes (Harper et al., 2006), by improving blood lipid profile and endothelial function and by exerting antioxidant, anti-inflammatory, anti-thrombotic effects (Bloedon & Szapary, 2004). These effects were achieved with intakes of 28 -40 g/day of milled flaxseed or between 14g and 30g/day of flaxseed oil (3.6–10.8 g of ALA if consumed as seed or about 3–20 g of ALA as oil). In epidemiologic studies, ALA intakes associated with reduced CVD risk averaged about 2 g/day (range = 0.7–6.3 g). A meta-analysis of prospective studies suggested that increasing the intake of ALA by 1.2 g/day—which is equivalent to about 2 tsp of milled flaxseed daily—decreases the risk of fatal CHD by at least 20% (Brouwer et al., 2004). Dietary fiber, ALA, and lignans may all contribute to the lipid-lowering effects of flax. It is not clear whether humans benefit more from consuming isolated components such as pure SDG or the SDG/lignan complex. Mechanistic studies are needed to clarify the component(s) responsible for effects reported. In addition, longer clinical trials which include a larger and equal number of both men and women are needed to fully assess the potential of flaxseed as part of a healthy cardioprotective diet.

Diabetes

Soluble fiber, protein (Velasquez et al., 2003), SDG (Prasad, 2000; Prasad et al., 2000; Prasad, 2001) and ALA (Ghafoorunissa et al., 2005) may affect insulin secretion and activity in maintaining plasma glucose homeostasis. Flaxseed lowers blood glucose in healthy young adults (Cunnane et al., 1995) and hypercholesterolemic postmenopausal women (Lemay et al., 2002). Following an overnight fast, 6 healthy volunteers consumed in random order a test meal containing 50 g of carbohydrate as bread made from milled flaxseed or white flour (Cunnane et al., 1993). The blood glucose response was 28% lower following flaxseed. In the same study, volunteers who consumed flax mucilage gums mixed with glucose showed a 27% decrease in the blood glucose response compared with consuming glucose alone (Cunnane et al., 1993). Glycemic response was improved when healthy volunteers ate bread made with milled flaxseed compared with bread made of regular wheat flour (Dah, et al., 2005).

More recent studies however have yielded conflicting results. Neither milled flaxseed (32 g/day) or flaxseed oil (13 g/day) fed daily for 12 weeks affected glycemic control (as measured by fasting plasma hemoglobin A1c, glucose and insulin) in 34 type 2 diabetics fed for 3 months (Taylor et al., 2010). Both treatments contained equivalent amounts of ALA (7.4 g/day). Both groups had increases in plasma phospholipid n-3 fatty acids (ALA, EPA, and DPA), and the flaxseed oil group had more EPA and DPA in plasma phospholipids compared to the milled flaxseed group. The

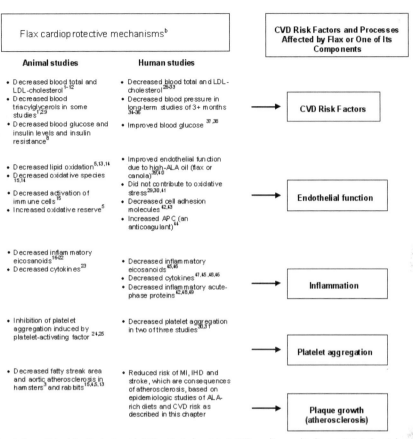

Flax cardioprotective mechanisms[b]		CVD Risk Factors and Processes Affected by Flax or One of Its Components
Animal studies	**Human studies**	
• Decreased blood total and LDL-cholesterol[1-12] • Decreased blood triacylglycerols in some studies[1,2,9] • Decreased blood glucose and insulin levels and insulin resistance[8]	• Decreased blood total and LDL-cholesterol[26-33] • Decreased blood pressure in long-term studies of 3+ months[34-36] • Improved blood glucose[37,28]	**CVD Risk Factors**
• Decreased lipid oxidation[5,13,14] • Decreased oxidative species[15,14] • Decreased activation of immune cells[15] • Increased oxidative reserve[5]	• Improved endothelial function due to high-ALA oil (flax or canola)[39,40] • Did not contribute to oxidative stress[29,30,41] • Decreased cell adhesion molecules[42,43] • Increased APC (an anticoagulant)[44]	**Endothelial function**
• Decreased inflammatory eicosanoids[16-22] • Decreased cytokines[23]	• Decreased inflammatory eicosanoids[45,46] • Decreased cytokines[47,45,48,46] • Decreased inflammatory acute-phase proteins[42,48,49]	**Inflammation**
• Inhibition of platelet aggregation induced by platelet-activating factor[24,25]	• Decreased platelet aggregation in two of three studies[30,31]	**Platelet aggregation**
• Decreased fatty streak area and aortic atherosclerosis in hamsters[3] and rabbits[15,4,5,13]	• Reduced risk of MI, IHD and stroke, which are consequences of atherosclerosis, based on epidemiologic studies of ALA-rich diets and CVD risk as described in this chapter	**Plaque growth (atherosclerosis)**

[a]Abbreviations: ALA, alpha-linolenic acid; APC, activated protein C; CVD, cardiovascular disease; IHD, ischemic heart disease; LDL-cholesterol, low-density lipoprotein cholesterol; MI, myocardial infarction.

[b] Sources:

1. Bhathena et al., 2002.
2. Bhathena et al., 2003.
3. Lucas et al., 2004.
4. Prasad et al., 1998.
5. Prasad, 1999.
6. Garg et al., 1989.
7. Vijaimohan et al., 2006.
8. Morise et al., 2004.
9. Morise et al., 2005.
10. Yang et al., 2005.
11. Huang & Horrobin, 1987.
12. Pellizzon et al., 2007.
13. Prasad, 2005.
14. Kinniry et al., 2006.
15. Prasad, 1997b.
16. Croft et al., 1984.
17. Fritsche & Johnston, 1989.

18. Hubbard et al., 1994.
19. Ingram et al., 1995.
20. Magrum & Johnston, 1983.
21. Marshall & Johnston, 1982.
22. Weiler et al., 2002.
23. Morris et al., 1991.
24. Hall et al., 1993.
25. Vas Dias et al., 1982.
26. Cunnane et al., 1993.
27. Arjmandi et al., 1998.
28. Lucas et al., 2002.
29. Cunnane et al., 1995.
30. Bierenbaum et al., 1993.
31. Clark et al., 1995.
32. Demark-Wahnefried et al., 2001.
33. Wilkinson et al., 2005.

34. Dodin et al., 2005.
35. Paschos et al., 2007.
36. Spence et al., 2003.
37. Lemay et al., 2002.
38. Dahl et al., 2005.
39. Nestel et al., 1997.
40. West et al., 2005.
41. Finnegan et al., 2003.
42. Zhao et al., 2004.
43. Rallidis et al., 2004.
44. Allman-Farinelli et al., 1999.
45. Caughey et al., 1996.
46. Matsuyama et al., 2005.
47. Zhao et al., 2007.
48. Paschos et al., 2005.
49. Ferrucci et al., 2006.

Fig. 10.4. Mechanisms by which animal and clinical studies support a role for flax in preventing cardiovascular disease[a].

control group experienced a 4% weight gain compared to baseline, while both flax groups had constant body weights during the study period. The soluble fiber or SDG content of the milled flaxseed diet may have benefits beyond the parameters assessed in this study, such as postprandial glucose response. A component of both flaxseed and oil may have an inhibitory effect on weight gain and warrants further investigation.

Sixty-eight type 2 diabetic patients with mild hypercholesterolemia completed a randomized, double-blind, placebo-controlled, cross-over trial supplemented with flaxseed-derived lignan capsules (360 mg lignan per day) or placebo for 12 weeks, separated by an 8-week wash-out period (Pan et al., 2007). The lignan supplement significantly improved glycemic control as measured by HbA1c compared to placebo; however, no significant changes were observed in fasting glucose and insulin concentrations, insulin resistance and blood lipid profiles. Further studies are needed to validate these findings and explore the efficacy of lignans on type 2 diabetes. Therefore, more studies are needed to clarify the efficiency of lignans in the long-term glycemic control, especially since it is not completely clear whether the relatively small reduction of HbA1c observed in this study is, in fact, real or due to chance.

Further research described earlier by this group examined the effects of supplementation with SDG capsules (360mg/day) in 70 diabetic patients (26 men and 44 post-menopausal women) with mild hypercholesterolemia (Pan et al., 2009). As noted, baseline to follow-up concentrations of CRP increased significantly within the placebo group, but were unchanged in the SDG group; a significant difference was observed between treatments. This observation suggests that SDG might modulate CRP levels in type 2 diabetics.

The effects of flaxseed gum incorporated in wheat flour chapattis on glycemic control in 60 patients with type 2 diabetes has been evaluated (Thakur et al., 2009). Six wheat flour chapattis containing flaxseed gum (5 g) were consumed daily for 3 months. Significant reductions in fasting blood sugar, total cholesterol, and LDL-C were reported.

As with CVD prevention, the role of dietary fiber, ALA, lignans, and soluble fiber may all contribute to improvements in glycemic control as noted in some studies. Again, it is not clear whether benefits are derived from isolated components such as fiber or SDG and further research is necessary in this regard.

Hormone Metabolism

The most significant components in flaxseed as related to its effects on hormone metabolism are the lignans and the mammalian lignans (enterodiol and enterolactone). Depending on their concentration and other factors, these phytoestrogens can act like weak estrogens by binding to the estrogen receptor on cell membranes. Following menopause, they act as estrogen antagonists by preventing estrogens from binding to the receptors (Benassayag et al., 2002).

Lignans have anti-cancer and antiviral effects, influence gene expression (activation) and may protect against estrogen-related diseases such as osteoporosis (Martin et al., 2007; Raffaelli et al., 2002; Benassayag et al., 2002) and breast cancer risk (Touillaud et al., 2007); as well as prostate cancer (Hedelin et al., 2006) in men. The mammalian lignans stimulate the synthesis of sex hormone-binding globulin (SHBG) (Adlercreutz, 1995), which binds sex hormones and reduces their circulation in the bloodstream, thus decreasing their biologic activity. The mammalian lignans also inhibit the activity of aromatase, an enzyme involved in the production of estrogens (Wang, 2002) which may protect against breast cancer (Brooks & Thompson, 2005).

Flaxseed has been investigated for effects on conditions related to estrogen metabolism. In recent years a phase 2 pilot study on the effect of 6 weeks of 40 g/day milled flaxseed consumption on hot flash scores in women not wishing to receive estrogen therapy was evaluated (Pruthi et al., 2007). Eligibility included 14 hot flashes per week for at least 1 month and 30 women were enrolled. The mean decrease in hot flash scores after flaxseed was 57% (median decrease was 62%). The mean reduction in daily hot flash frequency was 50% (median reduction 50%), from 7.3 hot flashes to 3.6. This study suggests that flaxseed decreases hot flash activity in women not taking estrogen therapy, a reduction greater than what would be expected with placebo.

Earlier research found similar results. In 18 premenopausal women with normal menstrual cycles, eating 10 g of milled flaxseed daily for 3 months lengthened the luteal phase of the menstrual cycle (Phipps et al., 1993). Milled flaxseed fed at a level of 25 g daily for 2 weeks, vaginal cell maturation was stimulated in 25 postmenopausal women, suggesting an estrogenic effect of flaxseed on the reproductive tract (Wilcox et al., 1990). However, several other clinical studies lasting 2–12 weeks have reported no effect of consuming 10–40 g (1+ to 5 tbsp) of milled flaxseed daily on blood levels of estradiol, estrone, follicle-stimulating hormone or luteinizing hormone in young women of reproductive age (Phipps et al., 1993) or in postmenopausal women (Brooks et al., 2004; Wilcox et al., 1990; Arjmandi et al., 1998; Lemay et al., 2002; Lucas et al., 2002).

Twenty-five menopausal women with mild symptoms either ate 40 g of milled flaxseed daily or took an oral estrogen-progesterone hormone replacement (0.625 mg conjugated estrogens per day) for 2 months (Lemay et al., 2002). After a 2-month period free of treatment, each group crossed over to the other intervention for 2 months. Flaxseed was as effective as hormone replacement therapy in improving mild menopause symptoms. The Kupperman index was used to measure the 11 most common menopausal complaints in this study. In two other studies flaxseed consumption reduced the severity of menopausal symptoms but the severity scores did not differ from those of the placebo (Lewis et al., 2006; Dodin et al., 2005).

The effects of partially defatted ground flaxseed on the climacteric symptoms and endometrial thickness of postmenopausal women has recently been tested using a double-blind, placebo-controlled, randomized clinical trial (Simbalista et al., 2010). Twenty patients and 18 in the control group who had been postmenopausal for 1–10 years, respectively, consumed 2 slices of bread containing 25 g of flaxseed (46 mg lignans) or wheat bran (1 mg lignans; control) every day for 12 consecutive weeks. The Kupperman index (KMI) was used to measure the 11 most common menopausal complaints. Both groups had significant, but similar, reductions in hot flashes and KMI after 3 months of treatment. Endometrial thickness was not affected in either group.

These results require validation in larger, placebo-controlled trials. The weak estrogenic properties identified in flaxseed seem to be the most likely mechanism for its effectiveness in reducing hot flash activity.

Cancer

Breast cancer is a hormone-sensitive cancer, in which tumor growth is influenced by sex hormones, particularly estrogen. Other hormone-sensitive cancers include those of the endometrium and prostate. Breast tumors containing receptors for estrogen are estrogen receptor positive (ER-positive or ER+); tumors without estrogen receptors are ER negative (ER-negative or ER(). Women with ER+ tumors are more likely to respond to hormone therapy than women whose tumors are ER((American Cancer Society. http://documents.cancer.org/104.00/104.00.pdf. 2007).

Breast Cancer

Animal Studies

Milled flaxseed decreased tumor incidence, number and size when fed to carcinogen-treated rats at the initiation (Serraino & Thompson, 1991), promotion (Serraino & Thompson, 1992), and late stages of mammary cancer (Thompson et al., 1996). In the cancer initiation stage, feeding milled flaxseed or defatted flaxseed to rats resulted in lower levels of cell proliferation and fewer nuclear aberrations in mammary gland tissue compared with the basal diet (Serraino & Thompson, 1991). Although not lethal events, nuclear aberrations are considered an early warning sign of cancer. In this rat study, the effects of flax on cell proliferation and nuclear aberrations were due to both its ALA and lignan content.

Feeding milled flaxseed slowed the tumor growth rate in mice implanted with an ER(human breast cancer cell line (Dabrosin et al., 2002) and decreased final tumor weight and volume in mice implanted with an ER+ human breast cancer cell line (Chen et al., 2004). Feeding rats milled flaxseed at dietary levels of 2.5% and 5% decreased the volume of established mammary tumors by more than 70% (Thompson et al., 1996).

Milled flaxseed also enhanced the effects of tamoxifen in nude mice (Chen et al., 2004). This study is noteworthy because it was designed to compare the effects of milled flaxseed and tamoxifen, alone and in combination, on the growth of mammary tumors in conditions of high versus low blood levels of estrogen. The study design mimicked the case of premenopausal women, who have high blood levels of estrogen, and postmenopausal women, who have low circulating estrogen. Tamoxifen has been the leading anticancer drug used in breast cancer treatment for women whose breast tumors are ER+ (Conte & Frassoldati, 2007). In this mouse study, milled flaxseed inhibited the growth of human ER+ breast cancer in mouse mammary glands. The combination of milled flax and tamoxifen had a greater inhibitory effect on tumor growth than tamoxifen treatment alone.

Milled flaxseed inhibited the growth and spontaneous metastasis of ER-human breast cancer cells in mice (Dabrosin et al., 2002; Chen et al., 2002; Wang et al., 2005). A recent study showed that in ovariectomized, athymic mice injected with human estrogen receptor (ER) positive breast cancer cells (MCF-7), milled flaxseed and SDG significantly decreased the palpable tumor size and induced significantly higher apoptosis (Chen et al., 2009). Both milled flaxseed and SDG significantly decreased mRNA expressions of Bcl2, cyclin D1, pS2, ERalpha, and ERbeta, epidermal growth factor receptor, and insulin-like growth factor receptor. Milled flaxseed also reduced human epidermal growth factor receptor 2 mRNA and SDG decreased phospho-specific mitogen-activated protein kinase expression. A lignan-rich flaxseed hull did not significantly reduce these biomarkers.

Feeding flaxseed oil to rats or mice slowed tumor growth (Rao et al., 2000), decreased the number of tumors (Cameron et al., 1989), decreased tumor diameter and weight, and increased survival time (Fritsche & Johnston, 1990). Feeding flaxseed oil also decreased metastasis in nude mice (Wang et al., 2005; Chen et al., 2006). Metastasis to lung decreased by [16%; to lymph nodes by (52%; and to other organs such as liver, bones and kidney by more than 90% (Wang et al., 2005)]. Flaxseed oil decreased cell proliferation (measured by the Ki-67 labeling index) by 26.5% and increased cell apoptosis (cell death) by 60% compared with the basal diet in this study. These actions interfered with cancer processes.

SDG derived from flaxseed has been shown to inhibit mammary tumor growth at the early promotion stage. Feeding rats purified SDG after treatment with a cancer-causing agent produced a 37% decrease in the number of tumors (Thompson et al. 1996) and decreased new and total tumor volume but had no effect on the volume of established tumors (Thompson et al., 1996). Thus, SDG may exert a stronger effect on new tumor development, whereas milled flaxseed and flaxseed oil appear to exert their effects at later stages of tumor development (Thompson et al., 1996).

SDG inhibited metastasis in two studies of implanted ER-human breast cancer cells in nude mice (Wang et al., 2005; Chen et al., 2006). Feeding SDG decreased metastasis to lung, lymph nodes, and other organs, but its effects were enhanced when it was combined with flax oil, where the combination significantly decreased total metastasis by (43% (Wang et al., 2005). The researchers concluded that the

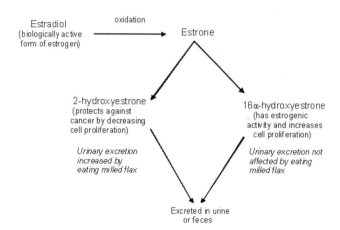

Fig. 10.5. Metabolism of estradiol and the effects of flax consumption on two estrogen metabolites.

inhibitory effects of milled flaxseed were mediated by both the oil and lignan (SDG) components.

Based on the findings of current studies, the initiation stage appears to be affected by milled flaxseed, as shown in Fig. 10.5, whereas the promotion stage and metastasis are affected by milled flaxseed, flaxseed oil, and purified SDG.

Human Studies

Several randomized, controlled clinical trials have investigated the effects of flaxseed on factors related to breast cancer risk in postmenopausal women. After menopause, most estrogen is derived by the conversion of plasma androstenedione to estrone in adipose tissue. Some estrone is metabolized to estradiol. Free estradiol, unbound to sex hormone-binding globulin (SHBG) is hypothesized to be the most biologically active fraction in the breast. Estrone sulfate is an abundantly circulating estrogen that serves as a reservoir for the more biologically active estrogens (Thomas et al., 1997). Epidemiologic studies have indicated that elevated serum levels of total and free estradiol, estrone, estrone sulfate, and lower levels of SHBG after menopause each substantially increase the risk of breast cancer (Dorgan et al., 1997). Further, estrone can be converted to 2-hydroxyestrone and 16α-hydroxyestrone, the latter of which enhances the actions of estrogen and increases cell proliferation (Brooks et al., 2004). Individuals who produce more 16α-hyroxyestrone may have an increased risk of breast cancer (Modugno et al., 2006).

Consuming 7.5 g of flaxseed per day for 6 weeks and 15 g of flaxseed for an additional 6 weeks resulted in modest, non-statistically significant declines in serum levels of testosterone, estrone, and estradiol but not SHBG in a group of postmenopausal women (Sturgeon et al., 2008). In the subset of overweight women, a mean reduction of 6.5 pg/ml for estrone was statistically significant. The results of this study suggest that dietary flaxseed may modestly lower serum levels of sex steroid hormones, especially in overweight/obese women.

Twenty-five g of flaxseed fed as a muffin was compared to a soy muffin (made with 25 g of soy flour) or a placebo muffin (made with whole wheat flour) for 16 weeks in the diets of 46 postmenopausal women (Brooks et al., 2004). The urinary concentration of 2-hydroxyestrone increased significantly in the flax group, whereas the urinary concentration of 16α-hydroxyestrone did not. Urinary excretion of these two estrogen metabolites did not change in the placebo and soy groups. In a study of 28 postmenopausal women, consumption of 10 g of milled flaxseed daily for 7 weeks increased significantly the urinary excretion of 2-hydroxyestrone (Haggans et al., 1999). In both studies, flaxseed increased the ratio of 2-hydroxyestrone to 16α-hydroxyestrone.

A randomized, controlled clinical trial assessed the effects of flaxseed on tumor biological markers in 32 postmenopausal women with newly diagnosed breast cancer (Thompson et al., 2005). After confirmation by biopsy of breast cancer and before surgical excision, the women were randomized to eat daily either a muffin containing 25 g of milled flaxseed or a placebo muffin and consumed the muffins for between 32 and 39 days. Women who ate the flaxseed exhibited a significant reduction in cell proliferation (measured by the Ki-67 labeling index), increased cell apoptosis and decreased c-erbB2 expression. Expression of c-erbB2 is associated with aggressive types of breast cancer and a greater potential for metastasis. Although the study consisted of a small number of subjects, its findings suggest that flaxseed has promise as an adjunct diet therapy in breast cancer treatment.

Prostate Cancer

Prostate cancer is the most frequently diagnosed cancer in men (American Cancer Society, 2007. http://www.cancer.org). Like breast cancer, it is hormone-sensitive and in the early stages of development, tumor growth is influenced by the sex hormones estrogen and testosterone and their active metabolites (McCann et al., 2005; Coffey, 2001).

One case-control study conducted among 1,499 Swedish men with prostate cancer and 1,130 controls found that a high intake of foods rich in phytoestrogens, including lignans, was associated with a decreased risk of prostate cancer (Hedelin et al., 2006). In this large scale population, the diet assessment instrument used contained phytoestrogen-rich foods like flax, berries, nuts, peanuts, beans, sunflower seeds and soy as part of a typical diet.

In a pilot study, 25 men with prostate cancer who were awaiting surgery consumed 30 g of milled flaxseed daily for a month as part of a low-fat diet (Demark-Wahnefried et al., 2001). Prostate cancer cell proliferation decreased and apoptotic death of cancer cells increased compared with a matched historic control group. Total serum prostate-specific antigen (PSA) levels did not change, although total testosterone and the free androgen index both decreased significantly between baseline and surgery. In a related pilot study, total serum PSA and the cell proliferation rate decreased significantly after 6 months of flaxseed supplementation (30 g/day) in men scheduled for a repeat prostate biopsy (Demark-Wahnefried et al., 2004).

A more extensive assessment of flaxseed and prostate cancer has recently been published by this same group (Demark-Wahnefried et al., 2008). A multisite, randomized controlled trial design was used to test the effects of low-fat and/or flaxseed supplemented diets on the biology of the prostate and other biomarkers. Prostate cancer patients (n=161) scheduled at least 21 days before prostatectomy were randomly assigned to consume their normal diet (control); a flaxseed-supplemented diet (30 g/day), a low-fat diet (<20% total energy), or a flaxseed-supplemented, low-fat diet, for an average of 30 days. Proliferation rates were significantly lower among men assigned to the flaxseed arms. Median Ki-67-positive cells/total nuclei ratios were 1.66 (flaxseed-supplemented diet) and 1.50 (flaxseed-supplemented, low-fat diet) versus 3.23 (control) and 2.56 (low fat diet). These results suggest that flaxseed consumption is associated with biological alterations that may be protective for prostate cancer.

A flaxseed lignan extract containing 33% SDG was evaluated for its ability to alleviate lower urinary tract symptoms (LUTS) in 78 subjects with Benign Prostatic Hyperplasia (BPH) (Zhang et al., 2008). A randomized, double-blind, placebo-controlled clinical trial with repeated measurements was conducted over a 4-month period using treatment dosages of 0 (placebo), 300, or 600 mg/day SDG. For the 0, 300, and 600 mg/day SDG groups, respectively, the International Prostate Symptom Score (IPSS) decreased 3.67 ± 1.56, 7.33 ± 1.18, and 6.88 ± 1.43 compared to baseline, the Quality of Life score (QOL score) improved by 0.71 ± 0.23, 1.48 ± 0.24, and 1.75 ± 0.25 compared to placebo and baseline, and the number of subjects whose LUTS grade changed from "moderate/severe" to "mild" increased by 3, 6, and 10 compared to baseline. Maximum urinary flows insignificantly increased 0.43 ± 1.57, 1.86 ± 1.08, and 2.7 ± 1.93 mL/second, and post-voiding urine volume decreased insignificantly by 29.4 ± 20.46, 19.2 ± 16.91, and 55.62 ± 36.45 mL. The observed decreases in IPSS and QOL score were correlated with the concentrations of plasma total lignans, SECO, ED, and EL. In this study, SDG improved LUTS in BPH subjects, and the therapeutic efficacy appeared comparable to that of commonly used intervention agents of α1A-adrenoceptor blockers and 5 α-reductase inhibitors.

By comparison, another study among 29 men diagnosed with prostate cancer and scheduled for surgery found no effect of consuming a bread made with soy grits + flax on PSA levels, free PSA, testosterone and sex-hormone-binding globulin compared with a bread containing only soy grits or a bread made with pearled wheat (Dalais et al., 2004).

The anticancer effects of flaxseed appear to be due to hormone and non-hormone-related actions. Flaxseed lignans exert hormone-related actions by competing with estrogen and testosterone for binding to their respective receptors and by inhibiting the enzyme aromatase, which converts androgens into estrogen (Saarinen et al., 2003; Brodie et al., 2006).

Non-hormone-related actions include decreasing nuclear aberrations and genetic damage (Serraino & Thompson, 1991; Trentin et al., 2004); cell proliferation (Serraino & Thompson, 1991; Chen et al., 2004); the production of prostaglandin E_2 (PGE_2), an eicosanoid that enhances cell proliferation and metastasis (Fritsche & Johnston, 1990); the production of growth factors like vascular endothelial growth factor (VEGF) that promote the building of new capillaries to supply the tumor (a process called angiogenesis) (Dabrosin et al., 2002); and the production of insulin-like growth factor 1 (IGF-1), which promotes tumor growth (Chen et al., 2002). Flaxseed also increases cell apoptosis (Chen et al., 2004), which inhibits cell proliferation and tumor growth.

ALA and Prostate Cancer

Some epidemiologic studies have suggested that ALA is associated with an increased risk of prostate cancer. Greater levels of ALA were found in tissues taken from men with prostate cancer than in tissues taken from men with benign prostatic hyperplasia (BPH) (Christensen et al., 2006). Two other studies, however, found that ALA concentrations were lower in men with prostate cancer compared with controls, especially when the tumors were advanced (Freeman et al., 2000). The most recent analysis found no correlation between ALA concentration in prostate tissue and locally advanced disease (Freeman et al., 2004). These conflicting study findings may be due to different population ages, sample sizes and methods for confirming and classifying prostate carcinoma.

Five case-control and 5 prospective cohort studies examined the relationship between dietary ALA intake and risk of prostate cancer. Three of the case-control studies (Andersson et al., 1996; Bairati et al., 1998; Bidoli et al., 2005) found no evidence of an association between dietary ALA and prostate cancer risk, while two (De Stéfani et al., 2000; Ramon et al., 2000) reported that ALA intake was associated with increased prostate cancer risk. Among the cohort studies, two found no association (Koralek et al., 2006; Laaksonen et al., 2004) and one found an inverse relationship (Schuurman et al., 1999) between ALA intake and prostate cancer risk. Three analyses of data from the Health Professionals Follow-up Study (Giovannucci et al., 1993; Giovannucci et al., 2007; Leitzmann et al., 2004) found a positive association between dietary ALA and risk of advanced prostate cancer.

Regulations for Alpha-linolenic Acid

United States

GRAS

In the United States, a substance can be legally added to foods if it is a food additive and hence approved for use, or if it is deemed to be generally recognized as safe (GRAS). A substance can be confirmed as GRAS through a formal rulemaking process or be designated GRAS status informally through its long history of safe use outside the United States. Whole and milled flaxseed received a no objection from the FDA for use in foods up to 12% flaxseed on August 24, 2009 (United States food and Drug Adminstration. 2009. http://www.fda.gov/Food/ FoodIngredientsPackaging/ GenerallyRecognizedasSafeGRAS/).

Nutrient Content Claims

On May 16, 2004, the U.S. Food and Drug Administration approved a petition to establish Daily Values (DV) and Nutrient Content Claims for ALA, EPA, and DHA. There was no upper-limit established for omega-3 consumption in the form of ALA as there was for EPA and DHA (U.S. Food and Drug Administration, 2010). The FDA action established a minimum recommended Daily Value for ALA at 1,300 mg per day. Recommended Daily Values for EPA and DHA were provided at 130 mg per day. Only "high" claims can be made for EPA and DHA, while ALA declarations can include "high", "good" and "more".

When making a claim, the omega-3 fatty acid in the product must be identified as ALA omega-3.

The packaging label claims that can be made for foods and in dietary supplements are as follows:

"contains _____ mg of ALA per serving, which is _____% of the Daily value for ALA (1.3g)"

"contains _____ mg of ALA per serving. The Daily Value for ALA is 1.3g"

For *"more"* claims: *"___% more of the Daily Value of ALA per serving than [reference food]. This product contains ___mg ALA omega-3 per serving, which is ___% of the Daily Value for ALA omega-3 (1.3g). [Reference Food] contains ___mg ALA omega-3 per serving".*

Nutrient Content Claim	Amount of ALA-omega-3 (per serving)
"high", "rich" or "excellent" source of ALA omega-3	260 mg
"good" source of ALA omega-3	130 mg
"more" ALA omega-3 (than reference food)	130 mg

Structure/Function (S/F) Claims

S/F claims for food and dietary supplement products describe the effect that the product has on the normal structure or function of the body (U.S. Food and Drug

Administration, 2010a). These need not be pre-approved by the FDA, but they must be true and not misleading to the consumer. Products containing flax oil qualify for S/F claims such as *"Now with Omega-3 ALA To Help Support A Healthy Heart."*

Health Claims

Health claims allow a statement on a food label which describes the relationship between a substance and a reduction in risk of a disease (U.S. Food and Drug Administration, 2010a). In 2003 the FDA launched the Consumer Health Information for Better Nutrition Initiative for foods and supplements. The result of the initiative saw the FDA allow the use of *qualified health claims* (U.S. Food and Drug Administration, 2010b). These claims do not have sufficient scientific agreement in support of them and would be required to have a qualifying statement detailing the limitations of the evidence in support of the claim. Since 2004, producers of foods containing omega-3 from fish and algal oils are able to make a qualified health claim on the benefits of their products:

"Supportive but not conclusive research shows that consumption of EPA and DHA Omega3 fatty acids may reduce the risk of coronary heart disease."

The claim does not include ALA because this omega-3 was not included in the petition reviewed by the FDA as it was submitted by fish oil companies.

Canada

In Canada, flax is regulated as a food and not as a food additive. There is presently no regulation that limits the level of flax in foods, although the Health Protection Branch of Health Canada has specified a guideline for use of flax in foods, indicating it has no objection to the use of flax at the level of 8% or lower on a dry-weight basis in baking (or 4% in dry cereal). Foods in Canada are sold containing flax in the range of 12%.

Nutrient Content Claims

Similar to the U.S., nutrient content claims in Canada are statements or expressions which describe, directly or indirectly, the level of a nutrient in a food or a group of foods. Food manufacturers may claim the nutrient content of specific nutrients if the food meets criteria described by regulations of the Canadian Food Inspection Agency (CFIA). The label can state the food is "low", "high", "a good source" or other descriptor of a particular nutrient. In order to make these claims the food must meet the specifications set out in the table in section B.01.513 of the Food and Drug Regulations (Department of Justice Canada, 2010).

Health Canada requires the following guidelines to be met ("Guide" refers to the CFIA "Nutrition Labeling Guide") for foods claiming omega-3 fatty acids. The claims in quotation marks in column 1 are those which are permitted by the *Food and Drug Regulations*. The reference amounts are found in Part D, Schedule M of the *Food and Drug Regulations* (Canadian Food Inspection Agency, 2010).

Omega-3 and Omega-6 Polyunsaturated Fatty Acid Claims

Column 1 Claim	Column 2 Conditions— Food	Column 3 Conditions—Label or Advertisement	FDR Reference
a) Source of omega-3 polyunsaturated fatty acids "source of omega-3 polyunsaturated fatty acids" "contains omega-3 polyunsaturated fatty acids" "provides omega-3 polyunsaturated fatty acids" Note: "polyunsaturated fatty acids" may be substituted with "polyunsaturated fat" or "polyunsaturates" in the above claims	The food contains: (a) 0.3 g or more of omega-3 polyunsaturated fatty acids per reference amount and serving of stated size; or (b) 0.3 g or more of omega-3 polyunsaturated fatty acids per 100 g, if the food is a prepackaged meal.	Must comply with the general requirements for nutrient content claims—see 7.5 of the *Guide* Nutrition Facts table must include a declaration of omega-3 polyunsaturated fatty acids, omega-6 polyunsaturated fatty acids, and monounsaturated fatty acids Nutrition Facts table required on products otherwise exempted by B.01.401(2)*(a)* and *(b)* When used in an advertisement, must comply with the requirements for advertisements—see 7.11 of the *Guide*	[B.01.402 (3) and (4)] [B.01.401(3)*(e)*(ii)] Table following B.01.513, item 25
b) Source of omega-6 polyunsaturated fatty acids "source of omega-6 polyunsaturated fatty acids" "contains omega-6 polyunsaturated fatty acids" "provides omega-6 polyunsaturated fatty acids" Note: "polyunsaturated fatty acids" may be substituted with "polyunsaturated fat" or "polyunsaturates" in the above claims	The food contains: (a) 2 g or more of omega-6 polyunsaturated fatty acids per reference amount and serving of stated size; or (b) 2 g or more of omega-6 polyunsaturated fatty acids per 100 g, if the food is a prepackaged meal.	See conditions set out for item a) of this table.	Table following B.01.513, item 26

Health Claims

A health claim is a statement or representation that states, suggests or implies that a relation exists between a food or component of that food and health, and of disease reduction. There are no health claims allowed in Canada for any edible oil or seed containing omega-3 fatty acids.

European Union

The European Food Safety Authority (EFSA) was established in January 2002, following a series of food crises in the late 1990s, as an independent source of scientific advice and communication on risks associated with the food chain. EFSA's responsibility includes food and feed safety, nutrition and health claim guidance, animal health and welfare, plant protection and plant health. In all these fields, EFSA provides objective and independent science-based advice and clear communication grounded in the most up-to-date scientific information and knowledge.

Under Article 13 of Regulation (EC) No 1924/20063, a general health claim is permissible on products containing ALA (EFSA, 2009) related to blood cholesterol. The target population is defined as the general population. In order to bear the claim a food should contain at least 15% of the proposed labeling reference intake value of 2 g ALA per day. The allowable claim is "contributes to healthy blood cholesterol level/helps to maintain normal cholesterol level/maintenance of normal blood cholesterol level" or "Alpha-linolenic acid contributes to maintenance of normal blood cholesterol concentrations."

A second claim is permissible under Article 14 of Regulation (EC) No 1924/20061 and is specific to either ALA or LA and a relationship to the normal growth and development of children (EFSA, 2008). The target population for the health claim is children from 1 to 12 years of age.

References

Adlercreutz, H.; Mousavi, Y.; Clark, J.; Hockerstedt, K. ; Hämäläinen, E.; Wähälä, K.; Mäkelä, T.; Hase, T. Dietary phytoestrogens and cancer: in vitro and in vivo studies. J. Steroid Biochem. Mol. Biol. 1992, 41, 331–337.

Adlercreutz, H. Phytoestrogens: epidemiology and a possible role in cancer protection. Environ. Health Perspect. 1995, 103 (suppl 7), 103–112.

Al, M.D.M.; Badart-Smook, A.; Houwelingen, A.; Hasaart, T.; Hornstra, G. Fat intake of women during normal pregnancy: Relationship with maternal and neonatal essential fatty acid status. J. Am. Coll. Nutr. 1996, 15, 49–55.

Albert, C.M.; Oh, K.; Whang, W.; Manson, J.E.; Chau, C.E.; Stampfer, M.; Willett, W.; Hu, F.B.. Dietary Alpha-linolenic Acid intake & risk of sudden cardiac death & coronary heart disease. Circulation 2005, 112, 3232–3238.

Allman-Farinelli, M.A.; Hall, D.; Kingham, K.; et al. Comparison of the effects of two low fat diets with different alpha-linolenic:linoleic acid ratios on coagulation and fibrinolysis Atherosclerosis. 1999, 142, 1, 159–168.

American Cancer Society. Breast cancer: detailed guide. 2007. Available at http://documents. cancer.org/104.00/104.00.pdf. Accessed 12 May 2007.

American Cancer Society. Cancer Facts & Figures 2007. Atlanta: American Cancer Society; 2007. Available at http://www.cancer.org. Accessed 30 June 2007.

Ander, B.P.; Weber, A.R.; Rampersad, P.P.; Gilchrist, J.; Pierce, G.; Lukas, A.. Dietary flaxseed protects against ventricular fibrillation induced by ischemia-reperfusion in normal and hypercholesterolemic rabbits. J. Nutr. 2004, 134, 3250–3256.

Andersson, S.-O.; Wolk, A.; Bergström, R.; et al. Energy, nutrient intake and prostate cancer risk: a population-based case-control study in Sweden. Int. J. Cancer. 1996, 68, 716–722.

Arjmandi, B.H.; Khan, D.A.; Juma, S.; et al. Whole flaxseed consumption lowers serum LDL-cholesterol and lipoprotein(a) concentrations in postmenopausal women. Nutr. Res. 1998, 18, 1203–1214.

Arjmandi, B.H. The Role of Phytoestrogens in the Prevention and Treatment of Osteoporosis in Ovarian Hormone Deficiency. J. Am. Coll. Nutr. 2001, 20, 5, 398S–402S.

Arterburn, L.M.; Hall, E.B.;; Oken, H. Distribution, interconversion, and dose response of n-3 fatty acids in humans. Am. J. Clin. Nutr. 2006, 83(suppl), 1467S–76S.

Ascherio, A.; Rimm, E.B.; Giovannucci, E.L.; et al. Dietary fat and risk of coronary heart disease in men: Cohort follow up study in the United States. Br. Med. J. 1996. 313, 84–90.

Axelson, M.; Sjövall, J.; Gustafsson, B.E.; Setchell, K.D.R. Origin of lignans in mammals and identification of a precursor from plants. Nature 1982, 298. 659–660.

Bairati, I.; Meyer, F.; Fradet, Y.; Moore, L. Dietary fat and advanced prostate cancer. J. Urol. 1998, 159, 1271–1275.

Barceló-Coblijn, G.; Murphy, E.J. Alpha-linolenic acid and its conversion to longer chain n-3 fatty acids: Benefits for human health and a role in maintaining tissue n-3 fatty acid levels. Progress in Lipid Research. 2009, 48, 355–374.

Barceló-Coblijn, G.; Collison, L.W.; Jolly, C.A.; et al. Dietary a-Linolenic acid increases brain but not heart and liver docosahexaenoic acid levels. Lipids. 2005, 40, 787–798.

Baylin, A.; Kabagambe, E.K.; Ascherio, A.; Spiegelman, D.; Campos, H. Adipose tissue α-linolenic acid and nonfatal acute myocardial infarction in Costa Rica. Circulation 2003, 107, 1586–1591.

Baylin, A.; Ruiz-Narvaez, E.; Kraft, P.; Campos, H.. α-Linoleic acid, Δ^6-desaturase gene polymorphism, and the risk of nonfatal myocardial infarction. Am. J. Clin. Nutr. 2007, 85, 554–560.

Bemelmans, W.J.E.; Lefrandt, J.D.; Feskens, E.M.J.; van Haelst, P.L.; Broer, J.; Meyboom-de Jong, B.; May, J.F.; Cohen, T.; Smit, A. Increased alpha-linolenic acid intake lowers C-reactive protein, but has no effect on markers of atherosclerosis. Eur. J. Clin. Nutr. 2004, 58, 1083–1089.

Bemelmans, W.J.E.; Broer, J.; Feskens, E.J.M.; Smit, A.; Muskiet, F.; Lefrandt, J.; Bom, V.; May, J.; Meyboom-de Jong, B.. Effect of an increased intake of α-linolenic acid and group nutritional education on cardiovascular risk factors: the Mediterranean Alpha-linolenic Enriched Groningen Dietary Intervention (MARGARIN) study. Am. J. Clin. Nutr. 2002, 75, 221–227.

Benassayag, C.; Perrot-Applanat, M.; Ferre, F. Phytoestrogens as modulators of steroid action in target cells. J. Chromatogr. B. 2002, 777, 233–248.

Berger, A.; Gershwin, M.E.; German, J.B.. Effects of various dietary fats on cardiolipin acyl composition during ontogeny of mice. Lipids. 1992, 27, 605–612.

Bhathena, S.J.; Velasquez, M.T. Beneficial role of dietary phytoestrogens in obesity and diabetes. Am. J. Clin. Nutr. 2002, 76, 1191–1201.

Bhathena, S.J.; Ali, A.A.; Mohamed, A.I.; et al. Differential effects of dietary flaxseed protein and soy protein on plasma triglyceride and uric acid levels in animal models. J. Nutr. Biochem. 2002, 13, 684–689.

Bhathena, S.J.; Ali, A.A.; Haudenschild, C.; et al. Dietary flaxseed meal is more protective than soy protein concentrate against hypertriglyceridemia and steatosis of the liver in an animal model of obesity. J. Am. Coll. Nutr. 2003, 22, 157–164.

Bidoli, E.; Talamini, R.; Bosetti, C.; et al. Macronutrients, fatty acids, cholesterol and prostate cancer risk. Ann. Oncol. 2005, 16, 152–157.

Bierenbaum, M.L.; Reichstein, R.; Watkins, T.R. Reducing atherogenic risk in hyperlipemic humans with flax seed supplementation: a preliminary report. J. Am. Coll. Nutr. 1993, 12, 501–504.

Bloedon, L.T.; Shilpa Balikai, B.S.; Chittams, J.; Cunnane, S.C.; Berlin, J.A.; Rader, D.J.; Szapary, P.O. Flaxseed and cardiovascular risk factors: Results from a double blind, randomized, controlled clinical trial. J. Amer. College of Nutr. 2008, 27, 1, 65–74.

Bloedon, L.T.; Szapary, P.O. Flaxseed and cardiovascular risk. Nutr. Rev. 2004, 62, 18–27.

Brenna, J.T. Efficiency of conversion of alpha-linolenic acid to long chain n-3 fatty acids in man. Curr. Opin. Clin. Nutr. Metab. Care. 2002, 5, 127–132.

Bretillon, L; Chardigny, J.M.; Sébédio, J.L.; Noel, J.; Scrimgeour, C.; Fernie, C.; Loreau, O.; Gachon, P.; Beaufrere, B. Isomerization increases the postprandial oxidation of linoleic acid but not α-linolenic acid in men. J. Lipid Res. 2001, 42, 995–997.

Brodie, A; Sabnis, G.; Jelovac, D. Aromatase and breast cancer. J. Steroid Biochem. Mol. Biol. 2006, 102, 97–102.

Brooks, J.D.; Thompson, L.U. Mammalian lignans and genistein decrease the activities of aromatase and 17β-hydroxysteroid dehydrogenase in MCF-7 cells. J. Steroid Biochem. Mol. Biol. 2005, 94, 461–467.

Brooks, J.D.; Ward, W.E.; Lewis, J.E.; et al. Supplementation with flaxseed alters estrogen metabolism in postmenopausal women to a greater extent than does supplementation with an equal amount of soy. Am. J. Clin. Nutr. 2004, 79, 318–325.

Brouwer, I.A.; Katan, M.B.; Zock, P.L. Dietary α-linolenic acid is associated with reduced risk of fatal coronary heart disease, but increased prostate cancer risk: a meta-analysis. J. Nutr. 2004, 134, 919–922.

Burdge, G.C. Metabolism of α-linolenic acid in humans. Prostaglandins Leukot. Essent. Fatty Acids. 2006, 75, 161–168.

Burdge, G.C.; Wootton, S.A. Conversion of α-linolenic to eicosapentaenoic, docosapentaenoic and docosahexaenoic acids in young women. Br. J. Nutr. 2002, 88, 411–420.

Burdge, G.C.; Finnegan, Y.E.; Minihane, A.M.; et al. Effect of altered dietary n-3 fatty acid intake upon plasma lipid fatty acid composition, conversion of [^{13}C]α-linolenic acid to longer-chain fatty acids and partitioning towards β-oxidation in older men. Br. J. Nutr. 2003, 90, 311–321.

Burdge, G.C.; Calder, P.C. Conversion of α-linolenic acid to longer-chain polyunsaturated fatty acids in human adults. Reprod. Nutr. Dev. 2005, 45, 581–597.

Burdge, G.C.; Jones, A.E.; Wootton, S.A. Eicosapentaenoic and docosapentaenoic acids are the principal products of α-linolenic acid metabolism in young men. Br. J. Nutr. 2002, 88, 355–363.

Calder, P.C. n-3 Polyunsaturated fatty acids, inflammation, and inflammatory diseases. Am. J. Clin. Nutr. 2006, 83 (suppl), 1505S–1519S.

Cameron, E.; Bland, J.; Marcuson, R. Divergent effects of omega-6 and omega-3 fatty acids on mammary tumor development in C3H/Heston mice treated with DMBA. Nutr. Res. 1989, 9, 383–393.

Campos, H.; Baylin, A.; Willett, W.C. α-Linolenic Acid and Risk of Nonfatal Acute Myocardial Infarction, Circulation. 2008, 118, 339–345.

Canadian Food Inspection Agency, Guide to Food Labelling and Advertising Chapter 7–Nutrient Content Claims. http://www.inspection.gc.ca/english/fssa/labeti/guide/ch7be.shtml.).

Cao, J.; Schwichtenberg, K.A.; Hanson, N.Q.; Tsai, M.Y. Incorporation and clearance of Omega-3 fatty acids in erythrocyte membranes and plasma phospholipids. Clinical Chemistry 2006, 52, 12, 2265–2272.

Caughey, G.E.; Mantzioris, E.; Gibson, R.A.; Cleland, L.G.; James, M.J. The effect on human tumor necrosis factor α and interleukin 1β production of diets enriched in n-3 fatty acids from vegetable oil or fish oil. Am. J. Clin. Nutr. 1996, 63, 116–122.

Chen, J.; Hui, E.; Ip, T.; Thompson, L.U. Dietary flaxseed enhances the inhibitory effect of tamoxifen on the growth of estrogen-dependent human breast cancer (MCF-7) in nude mice. Clin. Cancer Res. 2004, 10, 7703–7711.

Chen, J.; Stavro, P.M.; Thompson, L.U. Dietary flaxseed inhibits human breast cancer growth and metastasis and downregulates expression of insulin-like growth factor and epidermal growth factor receptor. Nutr. Cancer. 2002, 43, 187–192.

Chen, J.; Wang, L.; Thompson, L.U. Flaxseed and its components reduce metastasis after surgical excision of solid human breast tumor in nude mice. Cancer Lett. 2006, 234, 168–175.

Chen, J.; Saggar, J.K.; Corey, P.; Thompson, L.U. Flaxseed and pure secoisolariciresinol diglucoside, but not flaxseed hull, reduce human breast tumor growth (MCF-7) in athymic mice. J. Nutr. 2009, 139, 11, 2061–2066.

Christensen, J.H.; Fabrin, K.; Borup, K.; et al. Prostate tissue and leukocyte levels of n-3 polyunsaturated fatty acids in men with benign prostate hyperplasia or prostate cancer. BJU Int. 2006, 97, 270–273.

Christensen, J.H.; Schmidt, E.B.; Mølenberg, D.; Toft, E. Alpha-linolenic acid and heart rate variability in women examined for coronary artery disease. Nutr. Metab. Cardiovasc. Dis. 2005, 15, 345–351.

Choo, W.; Birch, J.; Dufour, J.P. Physicochemical and quality characteristics of cold-pressed flaxseed oils. J. Food Composition Anal. 2007, 20, 202–211.

Clandinin, M.T.; Foxwell, A.; Goh, Y.K.; Layne, K.; Jumpsen, J. Omega-3 fatty acid intake results in a relationship between the fatty acid composition of LDL cholesterol ester and LDL cholesterol content in humans. Biochim. Biophys. Acta 1997, 1346, 247–252.

Clark, W.F.; Parbtani, A.; Huff, M.W.; et al. Flaxseed: a potential treatment for lupus nephritis. Kidney Int. 1995, 48, 475–480.

Clark, W.F.; Kortas, C.; Heidenheim, A.P.; et al. Flaxseed in lupus nephritis: a two-year nonplacebo-controlled crossover study. J. Am. Coll. Nutr. 2001, 20, 143–148.

Clavel, T.; Borrmann, D.; Braune, A.; et al. Occurrence and activity of human intestinal bacteria involved in the conversion of dietary lignans. Anaerobe 2006, 12, 140–147.

Coffey, D.S. Similarities of prostate and breast cancer: evolution, diet, and estrogens. Urol. 2001, 57(suppl 4A), 31–38.

Conte, P.; Frassoldati, A. Aromatase inhibitors in the adjuvant treatment of postmenopausal women with early breast cancer: putting safety issues into perspective. Breast J. 2007, 13, 28–35.

Cornish, S.M.; Chilibeck, P.D.; Paus-Jennsen, L.; Biem, H.J.; Khozani, T.; Senanayake, V.; Vatanparast, H.; Little, J.P.; Whiting, S.J.; Pahwa, P. A randomized controlled trial of the effects of flaxseed lignan complex on metabolic syndrome composite score and bone mineral in older adults. Appl. Physiol. Nutr. Metab. 2009, 34, 89–98.

Croft, K.D.; Beilin, L.J.; Vandongen, R.; Mathews, E. Dietary modification of fatty acid and prostaglandin synthesis in the rat. Biochim. Biophys. Acta 1984, 795, 196–207.

Cui, W.; Mazza, G.; Biliaderis, C. Chemical structure, molecular size distributions and rheological properties of flaxseed gum. J. Agric. Food Chem. 1994, 42, 1891–1895.

Cunnane, S.C.; Hamadeh, M.J.; Liede, A.C.; et al. Nutritional attributes of traditional flaxseed in healthy young adults. Am. J. Clin. Nutr. 1995, 61, 62–68.

Cunnane, S.C.; Ganguli, S.; Menard, C.; et al. High α-linolenic acid flaxseed (Linum usitatissimum): Some nutritional properties in humans. Br. J. Nutr. 1993. 69, 443–453.

Dabrosin, C.; Chen, J.; Wang, L.; Thompson, L.U. Flaxseed inhibits metastasis and decreases extracellular vascular endothelial growth factor in human breast cancer xenografts. Cancer Lett. 2002, 185, 31–37.

Dahl, W.J.; Lockert, E.A.; Cammer, A.L.; Whiting, S.J. Effects of flax fiber on laxation and glycemic response in healthy volunteers. J. Med. Food 2005, 8, 508–511.

Dalais, F.S.; Meliala, A.; Wattanapenpaiboon, N.; et al. Effects of a diet rich in phytoestrogens on prostate-specific antigen and sex hormones in men diagnosed with prostate cancer. Urol. 2004, 64, 510–515.

Daun, J.K.; Barthet, V.J.; Chornick, T.L.; Duguid, S. Structure, composition and variety development of flaxseed. Flaxseed in Human Nutrition, 2nd ed. L.U. Thompson and S.C. Cunnane Eds, AOCS: Champaign, IL, 2003, 1–40.

Davis, B.C.; Kris-Etherton, P.M. Achieving optimal essential fatty acid status in vegetarians: current knowledge and practical implications. Am. J. Clin. Nutr. 2003, 78(Suppl): 640S–660S.

de Groot, R.H.M.; Hornstra, G.; van Houwelingen, A.C.; Roumen, F. Effect of a-Linolenic acid supplementation during pregnancy on maternal and neonatal polyunsaturated fatty acid status and pregnancy outcome. Am .J. Clin. Nutr. 2004, 79, 251–260.

DeLany, J.P.; Windhauser, M.M.; Champagne, C.M.; Bray, G.A. Differential oxidation of individual dietary fatty acids in humans. Am. J. Clin. Nutr. 2000, 72, 905–911.

De Lorgeril, M.; Salen, P.; Martin, J-L.; Monjaud, I.; Delaye, J.; Mamelle, N. Mediterranean Diet, Traditional risk factors, and the rate of cardiovascular complications after myocardial infarction: Final report of the Lyon Diet Heart Study. Circulation. 1999, 99, 779–785.

De Lorgeril, M.; Renaud, S.; Mamelle, N.; Salen, P.; Martin, J.; Monjaud, I.; Guidollet, J.; Touboul, P.; Delaye, J. Mediterranean alpha-linolenic acid-rich diet in secondary prevention of coronary heart disease. Lancet. 1994, 343: 1454–1459.

Demark-Wahnefried, W.; Robertson, C.N.; Walther, P.J.; et al. Pilot study to explore effects of low-fat, flaxseed-supplemented diet on proliferation of benign prostatic epithelium and prostate-specific antigen. Urol. 2004, 63, 900–904.

Demark-Wahnefried, W.; Polascik, T.J.; George, S.L.; Switzer, B.R.; Madden, J.F.; Ruffin IV, M.T.; Snyder, D.C.; Owzar, K.; Hars, V.; Albala, D.M.; et al. . Flaxseed supplementation (not dietary fat restriction) reduces prostate cancer proliferation rates in men presurgery. Cancer Epidemiol. Biomarkers Prev. 2008, 17, 12, 3577–3587.

Demark-Wahnefried, W.; Price, D.T.; Polascik, T.J.; et al. Pilot study of dietary fat restriction and flaxseed supplementation in men with prostate cancer before surgery: exploring the effects on hormonal levels, prostate-specific antigen, and histopathologic features. Urology 2001, 58, 47–52.

Denomme, J.; Stark, K.D.; Holub, B.J. Directly quantitated dietary (n-3) fatty acid intakes of pregnant Canadian women are lower than current dietary recommendations. J. Nutr. 2005, 135, 206–211.

Department of Justice Canada, 2010. Food and Drug Regulations. (http://laws.justice.gc.ca/eng/F-27/index.html) Accessed November 23, 2010.

De Stéfani, E.; Deneo-Pellegrini, H.; Boffetta P.; et al. α-Linolenic acid and risk of prostate cancer: a case-control study in Uruguay. Cancer Epidemiol. Biomarkers Prev. 2000, 9, 335–338.

Djoussé, L.; Pankow, J.S.; Eckfeldt, J.H.; Folsom, A.R.; Hopkins, P.N.; Province, M.A.; Hong, Y.; Ellison, R.C.. Relationship between dietary linolenic acid and coronary artery disease in the national Heart, Lung, and Blood Institute Family Heart Study. Am. J. Clin. Nutr. 2001, 74, 612–619.

Djoussé, L.; Folsom, A.R.; Province, M.A.; Hunt, S.C.; Ellison, R.C. Dietary linolenic acid and carotid artherosclerosis: the National Heart, Lung, and Blood Institute Family Heart Study. Am. J. Clin. Nutr. 2003a, 77, 819–825.

Djoussé, L.; Hunt, S.C.; Arnett, D.K.; Province, M.A.; Eckfeldt, J.H.; Ellison, R.C. Dietary linolenic acid is inversely associated with plasma triacylglycerol: the National Heart, Lung, and Blood Institute Family Heart Study. Am. J. Clin. Nutr. 2003b, 78, 1098–1102.

Djoussé, L.; Arnett, D.K.; Carr, J.; et al. Dietary linolenic acid is inversely associated with calcified atherosclerotic plaque in the coronary arteries: the National Heart, Lung, and Blood Institute Family Heart Study. Circulation. 2005a, 111, 2921–2926.

Djoussé, L.; Rautaharju, P.M.; Hopkins, P.N.; et al. Dietary linolenic acid and adjusted QT and JT intervals in the National Heart, Lung, and Blood Institute Family Heart Study. J. Am. Coll. Cardiol. 2005b, 45, 1716–1722.

Dodin, S.; Cunnane, S.C.; Mâsse, B.; et al. Flaxseed on cardiovascular disease markers in healthy menopausal women: a randomized, double-blind, placebo-controlled trial. Nutr. 2008, 24. 23–30.

Dodin, S.; Lemay, A.; Jacques, H.; et al. The effects of flaxseed dietary supplement on lipid profile, bone mineral density, and symptoms in menopausal women: a randomized, double-blind, wheat germ placebo-controlled clinical trial. J. Clin. Endocrinol. Metab. 2005, 90, 1390–1397.

Dolecek, T.A. Epidemiological evidence of relationships between dietary polyunsaturated fatty acids and mortality in the Multiple Risk Factor Intervention Trial. Pro. Soc. Exp. Biol. Med. 1992, 200, 177–182.

Dorgan, J.F.; Longcope, C.; Stephenson Jr, H.E.; Falk, R.T.; Miller, R.; et al. Serum sex hormones are related to breast cancer risk in postmenopausal women. Environ Health Perspect 1997, 105, (3 Suppl). 583–585,

EFSA, 2008. Scientific Opinion of the Panel on Dietetic Products, Nutrition and Allergies on a request from Unilever PLC/NV on α-linolenic acid and linoleic acid and growth and development of children. EFSA Journal. 2008, 783, 1–9.

EFSA, 2009. Panel on Dietetic Products, Nutrition and Allergies (NDA); Scientific Opinion on the substantiation of health claims related to alpha-linolenic acid and maintenance of normal blood cholesterol concentrations (ID 493) and maintenance of normal blood pressure (ID 625). EFSA Journal. 2009, 7, 9, 1252.

Emken E.A.; Adlof, R.O.; Rakof, H.; et al. Metabolism in vivo of deuterium labelled linolenic and linoleic acids in humans. Biochem. Soc. Trans. 1990, 18, 766–769.

Emken, E.A.; Adlof, R.O.; Gulley, R.M.. Dietary linoleic acid influences desaturation and acylation of deuterium-labeled linoleic and linolenic acids in young adult males. Biochim. Biophys. Acta 1994, 1213, 277–288.

Emken, E.A. Influence of linoleic acid on conversion of linolenic acid to omega-3 fatty acids in humans. In: Proceedings from the Scientific Conference on Omega-3 Fatty Acids in Nutrition, Vascular Biology, and Medicine. American Heart Association, Dallas, TX, 1995, 9–18.

Ezaki, O.; Takahashi, M.; Shigematsu, T.; et al. Long-term effects of dietary alpha-linolenic acid from perilla oil on serum fatty acids composition and on the risk factors of coronary heart disease in Japanese elderly subjects, J. Nutr. Sci. Vitaminol. (Tokyo). 1999, 45, 759–772.

Ferrucci, L.; Cherubini, A.; Bandinelli, S.; et al. Relationship of plasma polyunsaturated fatty acids to circulating inflammatory markers. J. Clin. Endocrinol. Metab. 2006, 91, 439–446.

Finnegan, Y.E.; Minihane, A.M.; Leigh-Firbank, E.C.; et al. Plant- and marine-derived n-3 polyunsaturated fatty acids have differential effects on fasting and postprandial blood lipid concentrations and on the susceptibility of LDL to oxidative modification in moderately hyperlipidemic subjects. Am. J. Clin. Nutr. 2003, 77, 783–795.

Freeman, V.L.; Meydani, M.; Yong, S.; et al. Prostatic levels of fatty acids and the histopathology of localized prostate cancer. J. Urol. 2000, 164, 2168–2172.

Freeman, V.L.; Meydani, M.; Hur, K.; Flanigan, R.C. Inverse association between prostatic polyunsaturated fatty acid and risk of locally advanced prostate carcinoma. Cancer. 2004, 101, 2744–2754.

Fritsche, K.L.; Johnston, P.V. Effect of dietary α-linolenic acid on growth, metastasis, fatty acid profile and prostaglandin production of two murine mammary adenocarcinomas. J. Nutr. 1990, 120, 1601–1609.

Fritsche, K.L.; Johnston, P.V. Modulation of eicosanoid production and cell-mediated cytotoxicity by dietary α-linolenic acid in BALB/c mice. Lipids 1989, 24, 305–311.

Garg, M.L.; Wierzbicki, A.A.; Thomson, A.B.R.; Clandinin, M.T. Dietary cholesterol and/or n-3 fatty acid modulate delta 9 – desaturase activity in rat liver microsomes. Biochim. Biophys. Acta. 1988, 962, 330–336.

Garg, M.L.; Wierzbicki, A.A.; Thomson, A.B.R.; Clandinin; M.T. Dietary saturated fat level alters the competition between α-linolenic and linoleic acid. Lipids 1989, 24, 334–339.

Gebauer, S.K.; Psota, T.L.; Harris, W.S.; Kris-Etherton, P.M. n-3 Fatty acid dietary recommendations and food sources to achieve essentiality and cardiovascular benefits. Am. J. Clin. Nutr. 2006, 83, 1526S–1535S.

Getz, G.S. Immune function in atherogenesis. J. Lipid Res. 2005, 46, 1–10.

Ghafoorunissa, I.A.; Natarajan, S.. Substituting dietary linoleic acid with α-linolenic acid improves insulin sensitivity in sucrose fed rats. Biochim. Biophys. Acta. 2005, 1733, 67–75.

Giovannucci, E.; Rimm, E.B.; Colditz, G.A.; et al. A prospective study of dietary fat and risk of prostate cancer. J. Natl. Cancer Inst. 1993, 85, 1571–1579.

Giovannucci, E.; Liu, Y.; Platz, E.A.; et al. Risk factors for prostate cancer incidence and progression in the Health Professionals Follow-up Study. Int. J. Cancer. 2007, 121, 7, 1571–1578.

Goh, Y.K.; Jumpsen, J.A.; Ryan, E.A.; Clandinin, M.T. Effect of ω3 fatty acid on plasma lipids, cholesterol and lipoprotein fatty acid content in NIDDM patients. Diabetologia. 1997. 40: 45–52.

Goyens, P.L.L.; Spilker, M.E.; Zock, P.L.; et al. Compartmental modeling to quantify alpha-linolenic acid conversion after longer term intake of multiple tracer boluses. J.Lipid Res. 2005, 46, 1474–1483.

Goyens, P.P.L.; Mensink, R.P. Effects of alpha-linolenic acid versus those of EPA/DHA on cardiovascular risk markers in healthy elderly subjects. Eur. J. Clin. Nutr. 2006, 60, 978–984.

Goyens, P.L.L.; Spilker, M.E.; Zock, P.L.; Katan, M.; Mensink, R. Conversion of α-linolenic acid in humans is influenced by the absolute amounts of α-linolenic acid and linoleic acid in the diet and not by their ratio. Am. J. Clin. Nutr. 2006, 84, 44–53.

Guallar, E.; Aro, A.; Jiménez, F.J.; Martin-Moreno, J.M.; Salminen, I.; vant Veer, P.; Kardinaal, A.F.M.; Gomez-Aracena, J.; Martin, B.; Kohlmeier, L.; et al. Omega-3 fatty acids in adipose tissue and risk of myocardial infarction: The EURAMIC Study. Arterioscler. Thromb. Vasc. Biol. 1999, 19, 1111–1118.

Haggans, C.J.; Hutchins, A.M.; Olson, B.A.; et al. Effect of flaxseed consumption on urinary estrogen metabolites in postmenopausal women. Nutr. Cancer. 1999, 33, 188–195.

Hall, A.V.; Parbtani, A.; Clark, W.F.; et al. Abrogation of MRL/lpr lupus nephritis by dietary flaxseed. Am. J. Kidney Dis. 1993, 22, 326–332.

Hall III, C.; Shultz, K. Phenolic antioxidant interactions. In Abstracts of the 92nd American Oil Chemists Society Annual Meeting and Expo, 2001, S88.

Hallund, J.; Ravn-Haren, G.; Bügel, S.; et al. A lignan complex isolated from flaxseed does not affect plasma lipid concentrations or antioxidant capacity in healthy postmenopausal women. J. Nutr. 2006, 136, 112–116.

Harnack, K.; Andersen, G.; Somoza, V. Quantitation of alpha-linolenic acid elongation to eicosapentaenoic and docosahexaenoic acid as affected by the ratio of n6/n3 fatty acids. Nutrition & Metabolism. 2009, 6, 8.

Harper, C.R.; Jacobson, T.A.. The fats of life. Arch. Intern. Med. 2001, 161, 2185–2192.

Harper, C.R.; Edwards, M.J.; DeFilipis, A.P.; Jacobson, T.A. Flaxseed oil increases the plasma concentrations of cardioprotective (n-3) fatty acids in humans. J. Nutr. 2006, 136, 83–87.

Heart and Stroke Foundation of Canada. The growing burden of heart disease and stroke in Canada 2003. http://www.heartandstroke.ca.

Hedelin, M.; Klint, Å.; Chang, E.T.; et al. Dietary phytoestrogen, serum enterolactone and risk of prostate cancer: the Cancer Prostate Sweden Study (Sweden). Cancer Causes Control 2006, 17, 169–180.

Hillman, G. The plant remains from Tell Abu Hureyra: a preliminary report. Proc. Prehist. Soc. 1975, 41, 70–73.

Hillman, G.C.; Colledge, S.M.; Harris, D.R.. Plant-food economy during the Epipalaeolithic period at Tell Abu Hureyra, Syria: dietary diversity, seasonality, and modes of exploitation. Foraging and Farming: the Evolution of Plant Exploitation. D.R. Harris and G.H. Hillman, Eds. Unwin and Hyman: London, 1989, 240–268.

Hopf, M. Jericho plant remains. Excavations at Jericho Vol. 5. K.M. Kenyon and T.A. Holland, Eds. British School of Archaeology in Jerusalem: London, 1983, 576–621.

Hosseinian, F.S.; Muir, A.D.; Westcott, N.D.; Krol, E.S. AAPH mediated antioxidant reactions of secoisolariciresinol and SDG. Organ Biomolec. Chem. 2007, 5, 644–654.

Houwelingen, A.C.; Hornstra, G. Trans fatty acids in early human development. World Rev. Nutr. Diet. 1994, 75,175–178.

Hu, F.B.; Stampfer, M.J.; Manson, J.E.; Rimm, E.B.; Wolk, A.; Colditz, G.A.; Hennekens, C.H.; Willett, W.A. Dietary intake of α-linolenic acid and risk of fatal ischemic heart disease among women. Am. J. Clin. Nutr. 1999, 69, 890–897.

Huang, Y.S.; Horrobin, D.F. Effect of dietary cholesterol and polyunsaturated fats on plasma and liver lipids in guinea pigs. Ann. Nutr. Metab. 1987, 31, 18–28.

Hubbard, N.E.; Chapkin, R.S.; Erickson, K.L. Effect of dietary linseed oil on tumoricidal activity and eicosanoid production in murine macrophages. Lipids 1994, 29, 651–655.

Hussein, N.; Ah-Sing, E.; Wilkinson, P.; et al. Long-chain conversion of [13C]linoleic acid and alpha-linolenic acid in response to marked changes in their dietary intake in men. J. Lipid Res. 2005, 46, 269–280.

Hutchins, A.M.; Slavin, J.L. Effects of flaxseed on sex hormone metabolism. Flaxseed in Human Nutrition, 2nd ed. L.U. Thompson and S.C. Cunnane, Eds. AOCS: Champaign, IL, 2003, 126–149.

Hwang, S.J.; Ballantyne, C.M.; Sharrett, A.R.; Smith, L.; Davis, C.; Gotto Jr., A.; Boerwinkle, E. Circulating adhesion molecules VCAM-1, ICAM-1, and E-selectin in carotid atherosclerosis and incident coronary heart disease cases: the Atherosclerosis Risk in Communities (ARIC) study. Circulation 1997, 96, 4219–4225.

Ingram, A.J.; Parbtani, A., Clark, W.F.; et al. Effects of flaxseed and flax oil diets in a rat-5/6 renal ablation model. Am. J. Kidney Dis. 1995, 25, 320–329.

Innis, S.M.; Elias, S.L. Intakes of essential n-6 and n-3 polyunsaturated fatty acids among pregnant Canadian women. Am. J. Clin. Nutr. 2003. 77, 473–478.

Institute of Medicine. Dietary Reference Intakes for Energy, Carbohydrate, Fiber, Fat, Fatty Acids, Cholesterol, Protein, and Amino Acids, National Academies Press: Washington, DC, 2002, 7-1- 7-69 (dietary fiber), 8-1-8-97 (fat and fatty acids).

256 ● K.C. Fitzpatrick

Jansen, G.H.E.; Arts, I.C.W.; Nielen, M.W.F.; et al. Uptake and metabolism of enterolactone and enterodiol by human colon epithelial cells. Arch. Biochem. Biophys. 2005, 435, 74–82.

Kamal-Eldin A.; Peerlkamp, N.; Johnsson, P.; Andersson, R.; Andersson, R.E.; Lundgren, L.; Åh, P. -3-methyl glutaric acid residues. Phytochemistry 2001, 58, 587–590.

Kaul, N.; Kreml, R.; Austria, J.A.; et al. A comparison of fish oil, flaxseed oil and hempseed oil supplementation on selected parameters of cardiovascular health in healthy volunteers. J. Am. Coll. Nutr. 2008, 27,1, 51–58.

Kestin, M; Clifton, P.; Belling, G.B.; Nestel, P.J.. n-3 Fatty acids of marine origin lower systolic blood pressure and triglycerides but raise LDL cholesterol compared with n-3 and n-6 fatty acids from plants. Am. J. Clin. Nutr. 1990, 51, 1028–1034.

Kinniry, P.; Amrani, Y.; Vachani, A.; et al. Dietary flaxseed supplementation ameliorates inflammation and oxidative tissue damage in experimental models of acute lung injury in mice. J. Nutr. 2006, 136, 1545–1551.

Knust, U.; Spiegelhalder, B.; Strowitzki, T.; Owen; R.W.. Contribution of linseed intake to urine and serum enterolignan levels in German females: a randomized controlled intervention trial. Food Chem. Toxicol. 2006, 44, 1057–1064.

Koralek, D.O.; Peters, U.; Andriole, G.; et al. A prospective study of dietary alpha-linolenic acid and the risk of prostate cancer (United States). Cancer Causes Control 2006, 17, 783–791.

Kuijsten, A.; Arts, I.C.W.; van't Veer, P.; Hollman; P.C.H.. The relative bioavailability of enterolignans in humans is enhanced by milling and crushing of flaxseed. J. Nutr. 2005, 135, 2812–2816.

Kurzer, M.S.; Lampe, J.W.; Martini, M.C.; Adlercruetz, H. Fecal lignan and isoflavonoid excretion in premenopausal women consuming flaxseed powder. Cancer Epidemiol. Biomarkers Prev. 1995, 4, 353–358.

Laaksonen, D.E.; Laukkanen, J.A.; Niskanen, L.; et al. Serum linoleic and total polyunsaturated fatty acids in relation to prostate and other cancers: a population-based cohort study. Int. J. Cancer 2004, 111, 444–450.

Lane, J.S.; Magno, C.P. ; Lane, K.T.; et al. Nutrition impacts the prevalence of peripheral arterial disease in the United States. J. Vasc. Surg. 2008, 48, 897–904.

Lanzmann-Petithory, D. Alpha-Linolenic acid and cardiovascular diseases. J. Nutr. Health Aging. 2001, 5, 79–183.

Layne, K.S.; Goh, Y.K.; Jumpsen, J.A.; Ryan, E.A.; Chow, P.; Clandinin, M.T. Normal subjects consuming physiological levels of 18:3(n-3) and 20:5(n-3) from flaxseed or fish oils have characteristic differences in plasma lipid and lipoprotein fatty acid levels. J. Nutr. 1996, 126, 2130–2140.

Leiken, A.I.; Brenner, R.R. Cholesterol-induced microsomal changes modulate desaturase activities. Biochim. Biophys. Acta. 1987, 922, 294–303.

Leitzmann, M.F.; Stampfer, M.J.; Michaud, D.S.; et al. Dietary intake of n-3 and n-6 fatty acids and the risk of prostate cancer. Am. J. Clin. Nutr. 2004, 80, 204–216.

Lemaitre, R.; King, I.; Mozaffarian, D.; Kuller, L.; Tracy, R.; Siscovick, D. N-3 polyunsaturated fatty acids, fatal ischemic heart disease, and nonfatal myocardial infarction in older adults: the cardiovascular study. Am. J. Clin. Nutr. 2003, 77, 319–325.

Lemay, A.; Dodin, S.; Kadri, N.; et al. Flaxseed dietary supplement versus hormone replacement therapy in hypercholesterolemic menopausal women. Obstet. Gynecol. 2004, 100, 495–504.

Leng, G.C.; Taylor, G.S.; Lee, A.J.; Fowkes, F.G.; Horrobin, D. Essential fatty acids and cardiovascular disease: the Edinburgh Artery Study. Vasc. Med. 1999, 4, 219–226.

Lewis, J.E.; Nickell, L.A.; Thompson, L.U.; et al. A randomized controlled trial of the effect of dietary soy and flaxseed muffins on quality of life and hot flashes during menopause. Menopause 2006, 13, 631–642.

Li, D.; Mann, N.J.; Sinclair, A.J. Comparison of n-3 polyunsaturated fatty acids from vegetable oils, meat and fish in raising platelet eicosapentaenoic acid levels in humans. Lipids. 1999, 34, S309.

Liou, Y.A.; King, D.J.; Zibrik, D.; Innis, S.M. Decreasing linoleic acid with constant α-linolenic acid in dietary fats increases (n-3) eicosapentaenoic acid in plasma phospholipids in healthy men. J. Nutr. 2007, 137, 945–952.

Lissin, L.W.; Cooke, J.P. Phytoestrogens and cardiovascular health. J Am Coll Cardiol 2000, 35, 1403–1410.

Lucas, E.A.; Wild, R.D.; Hammond, L.J.; et al. Flaxseed improves lipid profile without altering biomarkers of bone metabolism in postmenopausal women. J. Clin. Endocrinol. Metab. 2002, 87, 1527–1532.

Lucas, E.A.; Lightfoot, S.A.; Hammond, L.J.; et al. Flaxseed reduces plasma cholesterol and atherosclerotic lesion formation in ovariectomized Golden Syrian hamsters. Atherosclerosis 2004,173, 223–229.

Magrum, L.J.; Johnston, P.V. Modulation of prostaglandin synthesis in rat peritoneal macrophages with ω-3 fatty acids. Lipids 1983, 18, 514–521.

Manav, M.; Su, J.; Hughes, K.; Lee, H.P.; Ong, C.N. ω-3 Fatty acids and selenium as coronary heart disease risk modifying factors in Asian Indian and Chinese males. Nutr. 2004, 20, 967–973.

Mantzioris, E.; James, M.J.; Gibson, R.A.; Cleland, L.G.. Dietary substitution with an α-linolenic acid rich vegetable oil increases eicosapentaenoic acid concentrations in tissues. Am. J. Clin. Nutr. 1994, 59, 1304–1309.

Marshall, L.A.; Johnston, P.V. Modulation of tissue prostaglandin synthesizing capacity by increased ratios of dietary alpha-linolenic acid to linoleic acid. Lipids 1982, 17, 905–913.

Martin, J.H.J.; Crotty, S.; Warren, P.; Nelson, P.N. Does an apple a day keep the doctor away because a phytoestrogen a day keeps the virus at bay? A review of the anti-viral properties of phytoestrogens. Phytochemistry 2007, 68, 266–274.

Matsuyama, W.; Mitsuyama, H.; Watanabe, M.; Oonakahara, K.; Higashimoto, I.; Osame, M.; Arimura, K. Effects of omega-3 polyunsaturated fatty acids on inflammatory markers in COPD. Chest. 2005, 128, 3817–3827.

Mazza, G.; Biliaderis, C.G. Functional properties of flaxseed mucilage. J. Food Sci. 1979, 54, 1392–1305.

McCann, M.J.; Gill, C.I.R.; McGlynn, H.; Rowland, I.R. Role of mammalian lignans in the prevention and treatment of prostate cancer. Nutr. Cancer 2005, 52, 1–14.

McCloy, U.; Ryan, M.A.; Pencharz, P.B.; Ross, R.; Cunnane, S. A comparison of the metabolism of eighteen-carbon ^{13}C-unsaturated fatty acids in healthy women. J. Lipid Res. 2004, 45, 474–485.

Modugno, F.; Kip, K.E.; Cochrane, B.; et al. Obesity, hormone therapy, estrogen metabolism and risk of postmenopausal breast cancer. Int. J. Cancer 2006, 118, 1292–1301.

Mohrhauer, H.; Holman, R.T. The effect of dose level of essential fatty acids upon fatty acid composition of the rat liver. J. Lipid Res. 1963, 6, 494–497.

Morise, A.; Sérougne, C.; Gripois, D.; et al. Effects of dietary alpha linolenic acid on cholesterol metabolism in male and female hamsters of the LPN strain. J. Nutr. Biochem. 2004, 15, 51–61.

Morise, A.; Mourot, J.; Riottot, M.; et al. Dose effect of alpha-linolenic acid on lipid metabolism in the hamster. Reprod. Nutr. Dev. 2005, 45, 405–418.

Morris, D. Flax Nutrition Primer. Flax Council of Canada. www.flaxcouncil.ca. 2007.

Morris, D.D.; Henry, M.M.; Moore, J.N.; Fischer, J.K. Effect of dietary α-linolenic acid on endotoxin-induced production of tumor necrosis factor by peritoneal macrophages in horses. Am. J. Vet. Res. 1991, 52, 528–532.

Morton, M.S.; Wilcox, G.; Wahlqvist, M.L.; Griffiths, K. Determination of lignans and isoflavonoids in human female plasma following dietary supplementation. J. Endocrinol. 1994, 142, 251–259.

Mozaffarian, D.; Ascherio, A.; Hu, F.B.; Stampfer, M.J.; Willett, W.C.; Siscovick, D.S.; Rimm, E.B. Interplay between different polyunsaturated fatty acids and risk of coronary heart Disease in men. Circulation. 2005, 111, 157–164.

Muir, A.D. Flax lignans–analytical methods and how they influence our understanding of biological activity. J. AOAC Int. 2006, 89, 1147–1157.

Nesbitt, P.D.; Lam, Y.; Thompson, L.U. Human metabolism of mammalian lignan precursors in raw and processed flaxseed. Am. J. Clin. Nutr. 1995, 69, 549–555.

Nestel, P.J.; Pomeroy, S.E.; Sasahara, T.; Yamashita, T.; Liang, Y.; Dart, A.; Jennings, G.; Abbey, M.; Cameron, J. Arterial compliance in obese subjects is improved with dietary plant n-3 fatty acid from flaxseed oil despite increased LDL oxidizability. Arterioscler. Thromb. Vasc. Biol. 1997, 17, 1163–1170.

Oomah, B.D.; Mazza, G. Effect of dehulling on chemical composition and physical properties of flaxseed. Lebensm. Wiss. U. Technol. 1997, 30, 135–140.

Oomah, B.D.; Kenaschuk, E.; Cui, W.; Mazza, G. Variation in the composition of water-soluble polysaccharides in flaxseed. J. Agric. Food Chem. 1995, 43, 1484–1488.

Painter, E.P.; Nesbitt, L. L. Fat acid composition of linseed oil from different varieties of flaxseed. Oil and Soap. 1943, 20, 208–211.

Pan, A.; Yu, D.; Demark-Wahnefried, W.; Franco, O.H.; Lin; X. Meta-analysis of the effects of flaxseed interventions on blood lipids. Am. J. Clin. Nutr. 2009, 90, 288–297

Pan, A.; Demark-Wahnefried, W.; Ye, X.; Yu, Z.; Li, H.; Qi, Q.; Sun, J.; Chen, Y.; Chen, X.; Liu, Y.; et al. Effects of a flaxseed-derived lignan supplement on C-reactive protein, IL-6 and retinol-binding protein 4 in type 2 diabetic patients. Br. J. Nutr. 2009, 101, 1145–1149

Pan, A.; Sun, J.; Chen, Y.; Ye, X.; Li, H.; Yu, Z.; Wang, Y.; Gu, W.; Zhang, X.; Chen, X.; et al. Effects of a flaxseed-derived lignan supplement in Type 2 diabetic patients: A randomized, double-Blind, cross-over trial. PLoS ONE. Nov. 2007, 11, e1148.

Paschos, G.K.; Yiannakouris, N.; Rallidis, L.S.; Davies, I.; Griffin, B.A.; Panagiotakos, D. Apolipoprotein E Genotype in dyslipidemic patients and response of blood lipids and inflammatory markers to alpha-linolenic acid. Angiology. 2005, 56, 49–60.

Paschos, G.K.; Magkos, F.; Panagiotakos, D.B.; Votteas, V.; Zampelas, A. Dietary supplementation with flaxseed oil lowers blood pressure in dyslipidaemic patients. Eur. J. Clin. Nutr. 2007, 31, 1–6.

Patade, A.; Devareddy, L.; Lucas, E.A. ; et al. Flaxseed lowers total and LDL cholesterol concentrations in Native American postmenopausal women. J. Women's Health. 2009, 17, 3, 355–366.

Patenaude, A.; Rodriguez-Leyva, D.; Edel, A.L.; et al. Bioavailability of alpha-linolenic acid from flaxseed diets as a function of the age of the subject. Eur. J. Clin. Nutr. 2009, 63, 1123–1129.

Pawlosky, R.J.; Hibbeln, J.R.; Novotny, J.A.; Salem Jr., N. Physiological compartmental analysis of α-linolenic acid metabolism in adult humans. J. Lipid Res. 2001, 42, 1257–1265.

Pellizzon, M.A.; Billheimer, J.T.; Bloedon, L.T.; et al. Flaxseed reduces plasma cholesterol levels in hypercholesterolemic mouse models. J. Am. Coll. Nutr. 2007, 26, 66–75.

Perez-Martinez, P.; Lopez-Miranda, J.; Blanco-Colio, L.; et al. The chronic intake of a Mediterranean diet enriched in virgin olive oil, decreases nuclear transcription factor kappaB activation in peripheral blood mononuclear cells from healthy men. Atherosclerosis. 2007, 194, e141– e146.

Phipps, W.R.; Martini, M.C.; Lampe, J.W.; et al. Effect of flax seed ingestion on the menstrual cycle. J. Clin. Endocrinol. Metab. 1993, 77, 1215–1219.

Pietinen, P.; Ascherio, A.; Korhonen, P.; Hartman, A.; Willett, W.; Albanes, D.; Virtamo, J. Intake of fatty acids and risk of coronary heart disease in a cohort of Finnish men: the Alpha-Tocopherol, Beta-Carotene Cancer Prevention Study. Am. J. Epidemiol. 1997, 145, 876–887.

Poudel-Tandukara, K.; Nanria, A.; Matsushitaa, Y.; et al. Dietary intakes of α-linolenic and linoleic acids are inversely associated with serum C-reactive protein levels among Japanese men. Nutr. Res. 2009, 29, 363–337.

Prasad, K. Antioxidant activity of secoisolariciresinol diglucoside-derived metabolites, secoisolariciresinol, enterodiol, and enterolactone. Int. J. Angiol. 2000, 9, 220–225.

Prasad, K. Hydroxyl radical-scavenging property of secoisolariciresinol diglucoside (SDG) isolated from flax-seed. Mol. Cell. Biochem. 1997, 168, 117–123.

Prasad, K. Hypocholesterolemic and antiatherosclerotic effect of flax lignan complex isolated from flaxseed. Atherosclerosis 2005, 179, 269–275.

Prasad, K. Oxidative stress as a mechanism of diabetes in diabetic BB prone rats: Effects of secoisolariciresinol diglucoside (SDG). Mol. Cell. Biochem. 2000, 209, 89–96.

Prasad, K. Reduction of serum cholesterol and hypercholesterolemic atherosclerosis in rabbits by secoisolariciresinol diglucoside isolated from flaxseed. Circulation 1999, 99, 1355–1362.

Prasad, K.; Mantha, S.V.; Muir, A.D.; Westcott, N.D. Protective effect of secoisolariciresinol diglucoside against streptozotocin-induced diabetes and its mechanism. Mol. Cell. Biochem. 2000, 206, 141–150.

Prasad, K. Secoisolariciresinol diglucoside from flaxseed delays the development of type 2 diabetes in Zucker rat. J. Lab. Clin. Med. 2001, 138, 32–39.

Prasad, K.; Mantha, S.V.; Muir, A.D.; Westcott, N.D.. Reduction of hypercholesterolemic atherosclerosis by CDC-flaxseed with very low alpha-linolenic acid. Atherosclerosis 1998, 136, 367–375.

Prasad, K. Dietary flax seed in prevention of hypercholesterolemic atherosclerosis. Atherosclerosis 1997b, 132, 69–76.

Pretóvá, A.; Vojteková, M. Chlorophylls and carotenoids in flax embryos during embryogenesis. Photosynthetica 1985, 19, 194–197.

Pruthi, S.; Thompson, S.L.; Novotny, P.J.; Barton, D.L.; Kottschade, L.A.; Tan, A.D.; Sloan, J.A.; Loprinzi, C.L.. Pilot evaluation of flaxseed for the management of hot flashes. J. Soc. Int. Oncol. 2007, 5, 3, 106–112.

Raffaelli, B.; Hoikkala, A.; Leppälä, E.; Wähälä, K.. Enterolignans. J. Chromatogr. 2002, B 777, 29–43.

Rallidis, L.S.; Pascho,s G.; Papaioannou, M.L.; Liakos, G.; Panagiotakos, D.; Anastasiadis, G.; Zampelas, A. The effect of diet enriched with α-linolenic acid on soluble cellular adhesion molecules in dyslipidaemic patients. Atherosclerosis. 2004, 174, 127–132.

Ramon, J.M.; Bou, R.; Romea, S. ; et al. Dietary fat intake and prostate cancer risk: A case-control study in Spain. Cancer Causes Control 2000, 11, 679–685.

Rao, G.N.; Ney, E.; Herbert, R.A. Effect of melatonin and linolenic acid on mammary cancer in transgenic mice with c-neu breast cancer oncogene. Breast Cancer Res. Treat. 2000, 64, 287–296.

Rastogi, T.; Reddy, K.S.; Vaz, M.; Spiegelman, D.; Prabhakaran, D.; Willett, W.; Stampfer, M.; Ascherio, A. Diet and risk of ischemic heart disease in India. Am. J. Clin. Nutr. 2004, 79, 582–592.

Reiss, A.B.; Edelman, S.D. Recent insights into the role of prostanoids in atherosclerotic vascular disease. Curr. Vasc. Pharmacol. 2006, 4, 395–408.

Ren, J.; Chung, S.H. Anti-inflammatory effect of alpha-linolenic acid and its mode of action through the inhibition of nitric oxide production and inducible nitric oxide synthase gene expression via NF-kappaβ and nitrogen-activated protein kinase pathways. J. Agric. Food Chem. 2007, 55, 5073–5080.

Rollefson, G.O.; Simmons, A.H.; Donaldson, M.L.; Gillespie, W.; Kafafi, Z.; Kohler-Rollefson, I.U.; McAdam, E.; Ralston, S.; Tubb, M. Excavation at the Pre-Pottery Neolithic B village of 'Ain Ghazal (Jordan). Mitteilungen der Deuschen Orient-Gesellschaft zu Berlin 1985, 117, 69–116.

Rosamond, W. for the Writing Group Members. Heart disease and stroke statistics – 2007 update. A report from the American Heart Association Statistics Committee and Stroke Statistics Subcommittee. Circulation. 2007, 115, e69–e171.

Rosell, M.S.; Lloyd-Wright, Z.; Appleby, P.N.; Sanders, T.A.; Allen, N.E.; Key, T.J. Long-chain n–3 polyunsaturated fatty acids in plasma in British meat-eating, vegetarian, and vegan menAm. J. Clin. Nutr. 2005, 82, 327–334.

Ross, R. Atherosclerosis—an inflammatory disease. N. Engl. J. Med. 1999, 340, 115–126.

Saarinen, N.; Mäkelä, S.; Santti, R. Mechanism of anticancer effects of lignans with a special emphasis on breast cancer. Flaxseed in Human Nutrition, 2nd ed. L.U. Thompson and S.C. Cunnane, Eds. AOCS: Champaign, IL, 2003, 223–231.

Salem Jr., N; Pawlosky, R.; Wegher, B.; Hibbeln, J. In vivo conversion of linoleic acid to arachidonic acid in human adults. Prostaglandins Leukot. Essent. Fatty Acids. 1999, 60, 407–410.

Sanders, T.A.; Ellis, F.R.; Dickerson, J.W.. Studies of vegans: the fatty acid composition of plasma choline phosphoglycerides, erythrocytes, adipose tissue, and breast milk, and some indicators of susceptibility to ischemic heart disease in vegans and omnivore controls. Am. J. Clin. Nutr. 1978, 31, 805–813.

Sanders, T.A.B.; Roshanai, F. The influence of different types of $\omega 3$ polyunsaturated fatty acids on blood lipids and platelet function in healthy volunteers. Clin. Sci. 1983, 64, 91–99.

Sanders, T.A.B. Polyunsaturated fatty acids in the food chain in Europe. Am. J. Clin. Nutr. 2000, 71, 176S–178s.

Schuurman, A.G.; van den Brandt, P.A.; Dorant, E.; et al. Association of energy and fat intake with prostate carcinoma risk: results from the Netherlands Cohort Study. Cancer 1999, 86, 1019–1027.

Schwab, U.S.; Callaway, J.C.; Erkkilä, A.T.; Gynther, J.; Uusitupa, M.; Jarvinen, T. Effects of hempseed and flaxseed oils on the profile of serum lipids, serum total and lipoprotein lipid concentrations and haemostatic factors. Eur. J. Nutr. 2006, 45, 470–477.

Schwartz, H.; Ollilainen, V.; Piironen, V.; Lampi, A.M. Tocopherol, tocotrienol and plant sterol contents of vegetable oils and industrial fats. J. Food Comp. Anal. 2008, 21, 152–161.

Seppanen-Laakso, T.; Laakso, I.; Lehtimaki, T.; et al. Elevated plasma fibrinogen caused by inadequate a-linolenic acid intake can be reduced by replacing fat with canola-type rapeseed oil. Prostaglandins, Leukotrienes and Essential Fatty Acids. 2010, 83, 45–54.

Serraino, M.; Thompson, L.U. The effect of flaxseed supplementation on early risk markers for mammary carcinogenesis. Cancer Lett. 1991, 60, 135–142.

Serraino, M.; Thompson, L.U. The effect of flaxseed supplementation on the initiation and promotional stages of mammary tumorigenesis. Nutr. Cancer 1992, 17, 153–159.

Simbalista, R.L.; Sauerbronn, A.V.; Aldrighi, J.M.; Areas; J. A.G.. Consumption of a flaxseed-rich food is not more effective than a placebo in alleviating the climacteric symptoms of postmenopausal women. J. Nutr. 2010, 140, 293–297.

Simon, J.A.; Fong, J.; Bernert, J.T.; Browner, W.S. Serum fatty acids and the risk of stroke. Stroke 1995, 26, 778–782.

Simopoulos, A.P. Evolutionary aspects of diet, the omega-6/omega-3 ratio and genetic variation: nutritional implications for chronic diseases. Biomed. Pharmacother. 2006, 60, 502–507.

Singer, P.; Berger, I.; Wirth, M.; Goedicke, W.; Jaeger, W.; Voigt, S. Slow desaturation and elongation of linoleic and α-linolenic acids as a rationale of eicosapentaenoic acid-rich diet to lower blood pressure and serum lipids in normal, hypertensive and hyperlipemic subjects. Prostaglandins Leuko. Med. 1986, 24, 173–193.

Singer, P.; Wirth, M.; Berger, I.. A possible contribution of decrease in free fatty acids to low serum triglyceride levels after diets supplemented with n-6 and n-3 polyunsaturated fatty acids. Atherosclerosis 1990, 83,167–175.

Spence, J.D.; Thornton, T.; Muir, A.D.; Westcott, N.D.. The effect of flax seed cultivars with differing content of α-linolenic acid and lignans on responses to mental stress. J. Am. Coll. Nutr. 2003, 22, 494–501.

Stark, A.H.; Crawford, M.; Reifen, R. Update on Alpha-linolenic acid. Nutr. Rev. 2008, 66, 326–332.

Sturgeon, S.R.; Heersink, J.L.; Volpe; S.L.; Bertone-Johnson, E.R.; Puleo, E.; Stanczyk, F.Z.; Sabelawski, S.; Wähälä, K.; Kurzer, M.S.; Bigelow, C. Effect of dietary flaxseed on serum levels of estrogens and androgens in postmenopausal women. Nutr. Cancer. 2008, 60(5), 612–618.

Tarpila, S.; Aro, A.; Salminen, I.; et al. The effect of flaxseed supplementation in processed foods on serum fatty acids and enterolactone. Eur. J. Clin. Nutr. 2002, 56, 157–165.

Taylor, C.G.; Noto, A.D.; Stringer, D. M.; Froese, S.; Malcolmson, L. Dietary milled flaxseed and flaxseed oil improve n-3 fatty acid status and do not affect glycemic control in individuals with well-controlled Type 2 diabetes. J. Amer. Coll. Nutr. 2010, 29(1), 72–80.

Thakur,G.; Mitra, A.; Pal, K.; Rousseau, D. Effect of flaxseed gum on reduction of blood glucose and cholesterol in type 2 diabetic patients. Int. J. Food Sci. Nutr. 2009, 1, 11.

Thomas, H.V.; Reeves, G.K.; Key, T.J. Endogenous estrogen and postmenopausal breast cancer: a quantitative review. Cancer Causes Control 1997, 8, 922–928.

Thompson, L.U.; Seidl, M.M.; Rickard, S.E.; et al. Antitumorigenic effect of a mammalian lignan precursor from flaxseed. Nutr. Cancer 1996, 26, 159–165.

Thompson, L.U.; Rickard, S.E.; Orcheson, L.J.; Seidl, M.M. Flaxseed and its lignan and oil components reduce mammary tumor growth at a late stage of carcinogenesis. Carcinogenesis 1996, 17, 1373–1376.

Thompson, L.U.; Chen, J.M. ; Li, T. ; et al. Dietary flaxseed alters tumor biological markers in postmenopausal breast cancer. Clin. Cancer Res. 2005, 11, 3828–3835.

Thompson, L.U. ; Boucher, B.A. ; Liu, Z. ; et al. Phytoestrogen content of foods consumed in Canada, including isoflavones, lignans, and coumestan. Nutr. Cancer. 2006 54: 184–201.

Touillaud, M.S.; Thiébaut, A.C.M.; Fournier, A. ; et al. Dietary lignan intake and postmenopausal breast cancer risk by estrogen and progesterone receptor status. J. Natl. Cancer Inst. 2007, 99, 475–486.

Trentin, G.A.; Moody, J.; Torous, D.K.; et al. The influence of dietary flaxseed and other grains, fruits and vegetables on the frequency of spontaneous chromosomal damage in mice. Mutat. Res. 2004, 551, 213–222.

United States food and Drug Adminstration. Agency Response Letter GRAS Notice No. GRN 000280 CFSAN/Office of Food Additive Safety. http://www.fda.gov/Food/ FoodIngredientsPackaging/GenerallyRecognizedasSafeGRAS/). August, 2009.

United States Food and Drug Administration. Food and Nutrition. (http://www.fda.gov/Food/ GuidanceComplianceRegulatoryInformation/GuidanceDocuments/FoodLabelingNutrition/ FoodLabelingGuide/ucm064908.htm) 2010a.

United States Food and Drug Administration. Food Labeling and Nutrition. http://www.fda. gov/Food/GuidanceComplianceRegulatoryInformation/GuidanceDocuments/ FoodLabeling Nutrition/ucm073332.htm) 2010b.

Vaisey-Genser, M.; Morris, D. Introduction: history of the cultivation and uses of flaxseed. Flax: The Genus Linum. A. Muir and N. Westcott, Eds. Taylor and Francis: London, 2003, 1–21.

Vanharanta, M.; Voutilainen, S.; Lakka, T.A.; et al. Risk of acute coronary events according to serum concentrations of enterolactone: a prospective population-based case-control study. Lancet 1999, 354, 2112–2115.

Van Zeiste, W. Palaeobotanical results in the 1970 seasons at Cayonu, Turkey. Helinium 1972, 12, 3–19 (Cited by Zohary and Hopf, 1993).

Vas Dias, F.W.; Gibney, M.J.; Taylor, T.G. The effect of polyunsaturated fatty acids of the n-3 and n-6 series on platelet aggregation and platelet and aortic fatty acid composition in rabbits. Atherosclerosis 1982, 43, 245–257.

Velasquez, M.T.; Bhathena, S.J., Ranich, T.; et al. Dietary flaxseed meal reduces proteinuria and ameliorates nephrophathy in an animal mode of type II diabetes mellitus. Kidney Int. 2003, 64, 2100–2107.

Vermunt, S.H.; Mensink, R.P.; Simonis, M.M.; Hornstra, G. Effects of dietary alpha linolenic acid on the conversion and oxidation of 13C-alpha-linolenic acid. Lipids. 2000, 35, 137–142.

Vijaimohan, K.; Jainu, M. ; Sabitha, K.E.; et al. Beneficial effects of alpha linolenic acid rich flaxseed oil on growth performance and hepatic cholesterol metabolism in high fat diet fed rats. Life Sci. 2006, 79, 448–454.

Vos, E.; Cunnane, S.C. α-Linolenic acid, linoleic acid, coronary artery disease, and overall mortality (letter). Am. J. Clin. Nutr. 2003, 77, 521–522.

Wakjira, A.; Labuschagne, M. T.; Hugo, A. Variability in oil content and fatty acid composition of Ethiopian and introduced cultivars of linseed. J. Sci. Food Agric. 2004, 84, 601–607.

Wang, L.-Q. Mammalian phytoestrogens: enterodiol and enterolactone. J. Chromatogr. B 2002, 777, 289–309.

Wang, L.; Chen, J.; Thompson, L.U. The inhibitory effect of flaxseed on the growth and metastasis of estrogen receptor negative human breast cancer xenografts is attributed to both its lignan and oil components. Int. J. Cancer 2005, 116, 793–798.

Warrand, J.; Michaud, P.; Picton, L.; Muller, G.; Courtois, B.; Ralainirina, R.; Courtois, J. Structural investigations of the neutral polysaccharide of Linum usitatissimum L. seeds mucilage. Int. J. Biol. Macromol. 2005, 35, 121–125.

Weiler, H.; Kovacs, H.; Nitschmann, E.; et al. Elevated bone turnover in rat polycystic kidney disease is not due to prostaglandin E$_2$. Pediatr. Nephrol. 2002, 17, 795–799.

Welch, A.A.; Shakya-Shrestha, S.; Lentjes, M.A.H.; et al. Dietary intake and status of n-3 polyunsaturated fatty acids in a population of fish-eating and non-fish-eating meat-eaters, vegetarians, and vegans and the precursor-product ratio of alpha-linolenic acid to long-chain n23 polyunsaturated fatty acids: results from the EPIC-Norfolk. Am. J. Clin. Nutr. 2010, 92, 1040–1051.

West, S.G.; Hecker, K.D.; Mustad, V.A.; Nicholson, S.; Schoemer, S.; Wagner, P.; Hinderliter, A.; Ulbrecht, J.; Ruey, P.; Kris-Etherton, P. Acute effects of monounsaturated fatty acids with and without omega-3 fatty acids on vascular reactivity in individuals with type 2 diabetes. Diabetologia 2005, 48, 113–122.

Westcott, N.D.; Muir, A.D. Variation in the concentration of the flax seed lignan concentration with variety, location and year. Proc 56th Flax Institute of the United States Conference, Flax Inst. of the United States: Fargo, N.D., 1996, 77–80.

Westcott, N.D.; Muir, A.D. Chemical studies on the constituents of Linuum spp. Flax: the genus Linum. A.D. Muirand N.D. Westcott, Eds. Taylor & Francis: London, 2003, 55–73.

Wilcox, G.; Wahlqvist, M.L.; Burger, H.G.; Medley, G. Oestrogenic effects of plant foods in postmenopausal women. Br. Med. J. 1990, 301, 905–906.

Wilkinson, P.; Leach, C.; Ah-Sing, E.E.; Eric, E.; Hussain, N.; Miller, G.; Millward, D.; Griffin, B. Influence of α-linolenic acid and fish-oil on markers of cardiovascular risk in subjects with an atherogenic lipoprotein phenotype. Atherosclerosis 2005, 181, 115–124.

Williams, C.M.; Burdge, G. Long-chain n-3 PUFA: plant v. marine sources. Proceedings of the Nutrition Society. 2006, 65, 42–50.

Yang, L.; Leung, K.Y.; Cao, Y.; et al. α-Linolenic acid but not conjugated linolenic acid is hypocholesterolaemic in hamsters. Br. J. Nutr. 2005, 93, 433–438.

Zatonski, W.; Campos, H.; Willett, W. Rapid declines in coronary heart disease mortality in Eastern Europe are associated with increased consumption of oils rich in alpha-linolenic acid. Eur J. Epidemiol. 2008. 23 (1): 3–10.

Zhang, W.; Wang, X.; Liu, Y.; Tian, H.; Flickinger, B.; Empie, M.W.; Sun, S.Z. Effects of Dietary Flaxseed Lignan Extract on Symptoms of Benign Prostatic Hyperplasia. J. Med. Food. 2008. 11(2) 207–214.

Zhang, W.; Wang, X.; Liu, Y.; et al. Dietary flaxseed lignan extract lowers plasma cholesterol and glucose concentrations in hypercholesterolaemic subjects. Br. J. Nutr. 2008, 99, 1301–1309.

Zhao, G.; Etherton, T.D.; Martin, K.R.; Vanden Heuvel, J.; Gillies, P.; West, S.; Kris-Etherton, P. Dietary α-linolenic acid reduces inflammatory and lipid cardiovascular risk factors in hypercholesterolemic men and women. J. Nutr. 2004, 134, 2991–2997.

Zhao, G.; Etherton, T.D.; Martin, K.R.; Gillies, P.; West, S.; Kris-Etherton, P. Dietary alpha-linolenic acid inhibits proinflammatory cytokine production by peripheral blood mononuclear cells in hypercholesterolemic subjects. Am. J. Clin. Nutr. 2007, 85, 385–391.

Zohary, D.; Hopf, M. Oil and fibre crops. Domestication of Plants in the Old World, 3rd ed. D. Zohary and M. Hopf, Eds. Oxford University Press: Oxford, 2000, 125–132.

·11·

Fish Oil and Aggression

Kei Hamazaki[1], Hidekuni Inadera[1], and Tomohito Hamazaki[2]

[1]Department of Public Health, Faculty of Medicine, University of Toyama and [2]Department of Clinical Sciences, Institute of Natural Medicine, University of Toyama

Introduction

Fish oils have long been suggested to protect the heart from ischemic heart disease and fatal arrhythmia (Tziomalos et al., 2007; Jacobson, 2007). Recently, they have also been suggested to protect the heart in a literal sense (Hibbeln et al., 2006; Sinclair et al., 2007). Actually, fish oils are well-known to ameliorate depression, but not many investigators have been working on effects of fish oils on aggression or hostility. Hostile behavior is often rooted in anger and usually directed against a person. If anger and hostility refer to feelings and attitudes, aggression implies a further step in the sense that it includes the appearance of behaviors that may be destructive, harmful, or punitive when directed to other people or objects (Ramirez et al., 2006). Barefoot et al. (1983) followed 255 medical students who completed the Minnesota Multiphase Personality Inventory (Cook & Medley, 1954) while in medical school for 25 years, and examined the relationship between hostility scores and subsequent health status. Those with hostility scores above the median were nearly 7 times more likely to be dead by age 50 than those with hostility scores at or below the median. Here we discuss the effects of fish oil on aggression. The number of related papers is rather limited, so we will take aggression in a broader sense including hostility, oppositional behavior, violence, etc.

The First Reports that Suggested the Relationship Between n-3 Fatty Acid Deficiency and Aggression

According to the experiments of the research group of Okuyama (Yamamoto et al., 1987), rats on an α-linolenic acid- (α-LNA) rich diet performed better at a brightness-discrimination test than did n-3 fatty-acid deficient rats. In their test either a bright light or a dim light was presented randomly. If a rat touched or pressed the lever under a bright light, the rat obtained a food pellet (correct response), whereas no pellets were given to the responses under a dim light (incorrect response).

Interestingly, the total number of correct responses were essentially the same between the two groups of rats, but the incorrect number in α-LNA-deficient animals was greater than that of α-LNA-fed rats. It may be that α-LNA (or its metabolized material docosahexaenoic acid [DHA]) was necessary for full development of learning ability or that deficiency in α-LNA (and thus, DHA) did not affect learning ability but made animal behavior more aggressive. Rats in the α-LNA-deficient group might have understood the whole situation, but those rats could not resist pressing the lever because of their internal demand to obtain food pellets. If the latter was the case, their learning ability might be normal; the difference only consisted in the degree of aggression. The paper by Yamamoto et al. (1987) was the very first step in our research on aggression and n-3 fatty acids.

Weidner et al. (1992) measured hostility in their 5-year study (The Family Heart Study). Hostility in those participants who ate a typical American diet did not significantly change at the end of the study, but in those who consumed a low-fat, high carbohydrate diet (including fish), hostility was significantly reduced. This was probably the first human experiment relating fish oils and hostility, but unfortunately the amount of fish consumption was not available.

Some of Our Trials with Regard to Fish Oil and Aggression

Nearly 15 years ago we performed a double-blind study using students as subjects. Forty-one students were allocated to either a control (n=22) or DHA group (n=19) in a double-blind manner. Subjects of the DHA group were asked to take DHA-rich fish oil (1.5–1.8 g DHA/d) for 3 months. Those of the control group took a soybean oil-based control oil. At the start and end of the study, aggression of the subjects was measured with the picture frustration (PF) Study (Rosenzweig, 1978). There was a stressor component at the end of the study. A few days after the second (last) PF Study, either the final or most important term exams started for all of the participants. Therefore, they were likely to be stressed while busy preparing for the exams around the last PF Study. Aggression in the control group increased because of the presence of the stressor, but it remained unchanged in the DHA group (Fig. 11.1). There were highly significant differences in changes in aggression between the two groups (p=0.003). This study indicated a possibility that stressor-enhanced aggression might be controlled by prior administration of DHA (Hamazaki et al., 1996).

We tried to answer if aggression was controlled by fish oil also in schoolchildren (Itomura et al., 2005). A placebo-controlled double-blind study with 166 schoolchildren 9–12 years of age was performed. The subjects of the fish oil group (n=83) took fish-oil-fortified foods (3600 mg DHA + 840 mg eicosapentaenoic acid [EPA]/week for 3 months). The rest (the controls, n=83) took control foods. Physical aggression assessed by Hostility-Aggression Questionnaire for Children in girls increased significantly (13 to 15: median, n=42) in the control group and did not change (13 to 13,

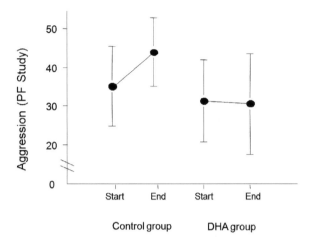

Fig. 11.1. The effects of DHA-rich fish oil on aggression in students. Forty-one students were randomly allocated into either the control or DHA groups in a double-blind fashion. After 3 months of intervention, aggression was measured with PF. Study increased in the control group (p=0.002) because of the presence of stressor (important term or final exams), whereas it was stable in the DHA group (intergroup difference: p=0.003).

n=43) in the fish oil group with a significant intergroup difference (p=0.008). The changes in physical aggression scores over time and those of the ratio of EPA/arachidonic acid in RBC (ΔEPA/AA) were significantly correlated in girls who agreed to blood collection (r=–0.53, p=0.01, n=23). On the contrary, there were no significant changes in physical aggression in boys. Impulsivity of girls assessed by parents/guardians using the diagnostic criteria for attention deficit/hyperactivity disorder of DSM-IV was reduced in the fish oil group (1 to 0) with a significant (p=0.008) intergroup difference from the control group (1 to 1).

A placebo-controlled double-blind study was performed with 40 AD/HD children 6–12 years of age who were mostly without medication. Subjects of the DHA group (n=20) took the same active foods as shown above (the schoolchildren study) for 2 months, whereas controls (n=20) took indistinguishable control foods without fish oil. Unfortunately, we could not find any improvement of AD/HD-related symptoms (Hirayama et al., 2004), but the sum of the hostility scores rated by parents and teachers was significantly reduced in the DHA group (Hamazaki & Hirayama, 2004).

Overview of Intervention Studies

We summarized the results of intervention studies that investigated the effects of fish oils on aggression (in a broader sense as described above). Table 11.A contains the results published during these 6 years. With regard to data published before 2003 and observational studies, please see our review (Hamazaki & Hamazaki, 2008).

Table 11.A. Overview of Intervention Studies on the Effects of n-3 Fatty Acids on Aggression Published for the Last 5 Years[a].

Authors and year of publication	n	Participants and intervention period	Supplementation (per day)	Outcome	Significance
Stevens et al., 2003	50	AD/HD children (age: 6–13 years) 4 months	DHA[b] 480mg EPA[c] 80mg AA[d] 40mg γ-LNA[e] 96mg	Significant reduction in aggressive behaviour according to parents' assessment.	p=0.02 compared with placebo group
Bradbury et al., 2004	30	Moderately stressed university staff (mean age: 41–44) 6 weeks	DHA 1500mg EPA 360mg	"Significant reduction in Perceived Stress Scale compared with no-treatment group, but not with placebo group."	p<0.05 compared with no-treatment group
Hamazaki & Hirayama, 2004	40	AD/HD children (age: 6–12 years) 2 months	DHA 510mg EPA 120mg	The sum of aggression scores assessed by both parents and teachers indicated that aggression was significantly reduced in DHA group.	Intra-group p=0.01 Inter-group p=0.001
Fontani et al., 2005	49	"Healthy 49 adults (15 men and 34 women, mean age=33)" 35 days	DHA 1.6g EPA 800mg others 400mg	Supplementation with n-3 fatty acids significantly increased vigor and decreased the other mood states (including anger) on the POMS[f] analysis.	Fish oil group vigor p<0.0001 anger p<0.001 Control group p=ns
Richardson & Montgomery, 2005	117	Children with developmental coordination disorder (age: 5–12 years) 3 months and another 3 months for crossover	EPA 558mg DHA 174mg γ-LNA 60mg vitamin E 9.6mg	"Omega-3 and -6 (n=50) decreased opposition scores (one of the CTRS-L[g] subscales) compared with the control (olive oil, n=52)."	Group comparison p<0.02

Study	N	Subjects	Dose	Results	Outcome
Itomura et al., 2005	166	Elementary school children (age: 9–12 years) 3 months	DHA 510mg EPA 120mg	Physical aggression in girls was increased in the control group compared with the DHA group with HAQ-C[h].	p=0.008 in girls p=ns in boys
Hallahan B et al., 2007	49	Repeated self-harm patients (age: 16–64) 3 months	EPA 1.2g DHA 900mg	"At 12 weeks, fish oil did not improve scores for aggression and impulsivity; however, significantly improved scores for depression, suicidality and daily stresses."	aggression: p=ns
Amminger et al., 2007	12	Children with autism (age: 5–17 years) 6 weeks	DHA 840mg EPA 700mg vitmin E 7mg	"Supplementation with n-3 fatty acids appeared to decrease stereotypy and hyperactivity with >0.7 effect size, but did not change irritability measured with ABC[i]."	irritability: p=ns
Sinn & Bryan, 2007	104	Children with AD/HD-related symptoms (age: 7–12). 15 weeks and another 15 weeks for crossover	EPA 558mg DHA 174mg γ-LNA 60mg vitamin E 9.6mg	PUFAs for 15 weeks decreased opposition scores on the CPRS-R[j] compared with the placebo (palm oil). Multivitamin mixture did not have any influence.	p<0.01 on ANCOVAs
Buydens-Branchey & Branchey, 2008; Buydens-Branchey et al., 2008	24	Male substance abusers with aggressive behavior and problems with the law 3 months	EPA 2250mg DHA 500mg Other n-3s 250mg + vitamin E	Supplementation with 3g n-3 significantly decreased anger scale scores (modified POMS). Scores remained decreased for another 3 months after n-3 discontinuation.	Anger scores: p=0.025 Correlation(–) with DHA: p=0.037
Hamazaki et al., 2008	233	Elementary school children (age: 9–14 years) in Indonesia 3 months	DHA 650mg EPA 100mg	No changes in aggression/hostility were detected with HAQ-C.	p=ns

All studies were placebo-controlled double-blind trials. [a] Please see our review (Hamazaki & Hamazaki, 2008) for studies published before 2004; cross-sectional and animal studies were also summarized there. [b] docosahexaenoic acid, [c] eicosapentaenoic acid, [d] arachidonic acid, [e] linolenic acid, [f] Profile of Mood States, [g] Conners'Teacher Rating Scales, Long Versionn, [h] Hostility-Aggression Questionnaire for Children, [i] Aberrant Behavior Checklist, [j] Conners Parent Rating Scales-Revised.

Some intervention studies showed that n-3 fatty acids reduced aggression, and the others did not (Table 11.A). The presence of stressors might be important for obtaining positive results of fish oils on aggression. We conducted two very similar intervention studies with university students—one with a formidable stressor as shown above (Hamazaki et al., 1996) and the other without any stressor (Hamazaki et al., 1998). The results were very different. The study with the stressor showed significant effects (Hamazaki et al., 1996), but the study without any stressor did not show aggression-controlling effects of DHA at all (Hamazaki et al., 1998). In some study subjects, chronic unfavorable conditions or environment-like disease might work as a stressor.

How n-3 Fatty Acids Work

Serotonin and Aggression

The serotonergic neuron system is probably the most important factor for aggression (Hibbeln et al., 2006; Garland & Hallahan, 2006). An inverse relationship between a lifelong aggression history and 5-hydroxyindolacetic acid (5-HIAA, the major metabolite of serotonin) concentrations in the cerebrospinal fluid (CSF) was shown in a group of 26 military men with no history of major psychiatric illness (Brown et al., 1979). Since then, the association between low concentrations of CSF 5-HIAA and aggressive behavior has been repeatedly pointed out in other studies (Virkkunen et al., 1994; Higley et al., 1996). The relationship between central serotonin and aggressive behavior is further strengthened by intervention studies controlling central serotonin functions with amino acid mixtures designed to raise or lower tryptophan availability and, thus, to raise or lower brain serotonin synthesis (Pihl et al., 1995). Lowered tryptophan levels were associated with increased aggression. There are quite a few interventional studies using serotonin specific reuptake inhibitors (SSRIs); reduction of impulsive aggressive behavior was reduced with fluoxetine, an SSRI, in patients with personality disorder (Coccaro et al., 1990). Moreover, Knutson et al. (1998) found that an SSRI reduced hostility of normal volunteers in a double-blind trial.

Serotonin and n-3 Fatty Acids

Olsson et al. (1998) reported that a diet low in n-3 fatty acids decreased serotonin and 5-HIAA concentrations in rats. Kodas et al. (2004) found that deficits in fenfluramine-induced serotonin release in the rat hippocampus could be normalized when dietary n-3 fatty acid fortification was initiated. All these facts indicate that supplemental n-3 fatty acids fortify the serotonergic function. The membrane fatty acid composition of neurons can affect the metabolism of serotonin by regulating tryptophan hydroxylase (Mandell, 1984), monoamine oxidase (Delion et al., 1997), and the serotonergic reuptake pump (Block & Edwards, 1987). In rats, an n-3 fatty-acid-deficient diet resulted in a 44% increase in 5-HT_{2A} receptor density in the frontal cortex (Delion et al., 1996).

In observational studies, Hibbeln et al. (1998a) found that higher plasma concentrations of DHA and AA predicted higher concentrations of CSF 5-HIAA among healthy volunteers, but plasma concentrations of DHA were inversely correlated with CSF 5-HIAA concentrations among early-onset alcoholics, who are at risk for aggressive behavior. It was also reported that violent subjects had significantly lower concentrations of CSF 5-HIAA than nonviolent subjects matched for their severity of alcohol dependence and that plasma DHA concentrations were inversely correlated with CSF 5-HIAA among those violent subjects (Hibbeln et al., 1998b). Consequently, the relationship between plasma DHA levels and CSF 5-HIAA might be different depending on whether study subjects are healthy or early-onset alcoholic, or normal or violent.

Noradrenalin and Other Factors and n-3 Fatty Acid

The sympathetic nervous system tone is known to be enhanced in subjects with high hostility scores (Williams, 1994). Short-term (1 month) intervention studies with fish oil did not support the idea that fish oils reduced plasma noradrenaline (NA) concentrations (Hughes et al., 1991; Mills et al., 1990); however, our intervention studies for 2 months (Sawazaki et al., 1999; Hamazaki et al., 2005) and one performed by Singer et al. (1990) for 36 weeks showed that fish oil reduced plasma NA levels. Changes in central NA levels by n-3 fatty acids might explain the relationship between those fatty acids and aggression.

Inflammatory markers, c-aminobutyric acid, neuroactive steroids, endocannabinoids, brain-derived neurotrophic factors, etc. might be other possible parameters that link both n-3 fatty acids and aggression (Hamazaki & Hamazaki, 2008).

Conclusion

Unfortunately, the study of human behavior is completely different from studying medicines that control hypertension. There are no simple parameters like blood pressure in behavior studies. However, many studies including animal experiments appear to indicate a general direction that fish oils (or at least treating n-3 fatty acid deficiency) modulate aggression. Serotonergic neurons are probably the mainstay of the mechanism as to how n-3 fatty acids work. Further investigations are needed. Finally we would like to introduce an interesting cross-country relationship between seafood consumption and mortality from homicide. Homicide is the ultimate deed of aggression, and it is the easiest outcome to assess with very little diagnostic fluctuations. There was a significant inverse correlation between homicide rates and seafood consumption across countries (Hibbeln, 2001). Here we conclude as follows: please keep away from those who do not remember when they last ate fish! It is literally of life and death importance.

References

Amminger, G.P.; Berger, G.E.; Schäfer, M.R.; Klier, C.; Friedrich, M.H.; Feucht, M. Omega-3 fatty acids supplementation in children with autism: a double-blind randomized, placebo-controlled pilot study. Biol. Psychiatry 2007, 61, 551–553. Epub 2006 Aug 22.

Barefoot, J.C.; Dahlstrom, W.G.; Williams Jr., R.B. Hostility, CHD incidence, and total mortality: a 25-year follow-up study of 255 physicians. Psychosom. Med. 1983, 45, 59–63.

Block, E.R.; Edwards, D. Effect of plasma membrane fluidity on serotonin transport by endothelial cells. Am. J. Physiol. 1987, 253(5 Pt 1), C672–C678.

Bradbury, J.; Myers, S.P.; Oliver, C. An adaptogenic role for omega-3 fatty acids in stress; a randomised placebo-controlled double-blind intervention study (pilot) [ISRCTN22569553]. Nutr. J. 2004, 3, 20.

Brown, G.L.; Goodwin, F.K.; Ballenger, J.C.; Goyer, P.F.; Major, L.F. Aggression in humans correlates with cerebrospinal fluid amine metabolites. Psychiatry Res. 1979, 1, 131–139.

Buydens-Branchey, L.; Branchey, M. Long-chain n-3 polyunsaturated fatty acids decrease feelings of anger in substance abusers. Psychiatry Res. 2008, 157, 95–104.

Buydens-Branchey, L.; Branchey, M.; Hibbeln, J.R. Associations between increases in plasma n-3 polyunsaturated fatty acids following supplementation and decreases in anger and anxiety in substance abusers. Prog. Neuropsychopharmacol. Biol. Psychiatry 2008, 32, 568–575.

Coccaro, E.F.; Astill, J.L.; Herbert, J.L.; Schut, A.G. Fluoxetine treatment of impulsive aggression in DSM-III-R personality disorder patients. J. Clin. Psychopharmacol. 1990, 10, 373–375.

Cook, W.; Medley, D. Proposed hostility and pharisaic-virtue scales for the MMPI. J. Appl. Psychol. 1954, 238, 414–418.

Delion, S.; Chalon, S.; Guilloteau, D.; Besnard, J.C.; Durand, G. alpha-Linolenic acid dietary deficiency alters age-related changes of dopaminergic and serotoninergic neurotransmission in the rat frontal cortex. J. Neurochem. 1996, 66, 1582–1591.

Delion, S.; Chalon, S.; Guilloteau, D.; Lejeune, B.; Besnard, J.C.; Durand, G. Age-related changes in phospholipid fatty acid composition and monoaminergic neurotransmission in the hippocampus of rats fed a balanced or an n-3 polyunsaturated fatty acid-deficient diet. J. Lipid. Res. 1997, 38, 680–689.

Fontani, G.; Corradeschi, F.; Felici, A.; Alfatti, F.; Migliorini, S.; Lodi, L. Cognitive and physiological effects of Omega-3 polyunsaturated fatty acid supplementation in healthy subjects. Eur. J. Clin. Invest. 2005, 35, 691–699.

Garland, M.R.; Hallahan, B. Essential fatty acids and their role in conditions characterized by impulsivity. Int. Rev. Psychiatry 2006, 18, 99–105.

Hallahan, B.; Hibbeln, J.R.; Davis, J.M.; Garland, M.R. Omega-3 fatty acid supplementation in patients with recurrent self-harm. Single-centre double-blind randomised controlled trial. Br. J. Psychiatry 2007, 190, 118–122.

Hamazaki, K.; Itomura, M.; Huan, M.; Nishizawa, H.; Sawazaki, S.; Tanouchi, M.; et al. Effect of omega-3 fatty acid-containing phospholipids on blood catecholamine concentrations in healthy volunteers: a randomized, placebo-controlled, double-blind trial. Nutrition 2005, 21, 705–710.

Hamazaki, K.; Syafruddin, D.; Tunru, I.S.; Azwir, M.F.; Asih, P.B.; Sawazaki, S.; Hamazaki, T. The effects of docosahexaenoic acid-rich fish oil on behavior, school attendance rate and malaria infection in school children –a double-blind, randomized, placebo-controlled trial in Lampung, Indonesia. Asia Pac. J. Clin. Nutr. 2008, 17, 258–263.

Hamazaki, T.; Sawazaki, S.; Itomura, M.; Asaoka, E.; Nagao, Y.; Nishimura, N.; et al. The effect of docosahexaenoic acid on aggression in young adults a placebo-controlled double-blind study. J. Clin. Invest. 1996, 97, 1129–1133.

Hamazaki, T.; Sawazaki, S.; Nagao, Y.; Kuwamori, T.; Yazawa, K.; Mizushima, Y.; et al. Docosa-hexaenoic acid does not affect aggression of normal volunteers under nonstressful conditions a randomized, placebo-controlled, double-blind study. Lipids 1998, 33, 663–667.

Hamazaki, T.; Hirayama, S. The effect of docosahexaenoic acid-containing food administration on symptoms of attention-deficit/hyperactivity disorder –a placebo-controlled double-blind study. Eur. J. Clin. Nutr. 2004, 58, 838.

Hamazaki, T.; Hamazaki, K. Fish oils and aggression or hostility. Prog. Lipid Res. 2008, 47, 221–232.

Hibbeln, J.R.; Linnoila, M.; Umhau, J.C.; Rawlings, R.; George, D.T.; Salem Jr., N. Essential fatty acids predict metabolites of serotonin and dopamine in cerebrospinal fluid among healthy control subjects, and early- and late-onset alcoholics. Biol. Psychiatry 1998a, 44, 235–242.

Hibbeln, J.R.; Umhau, J.C.; Linnoila, M.; George, D.T.; Ragan, P.W.; Shoaf, S.E.; et al. A replica-tion study of violent and nonviolent subjects: cerebrospinal fluid metabolites of serotonin and dopamine are predicted by plasma essential fatty acids. Biol. Psychiatry 1998b, 44, 243–249.

Hibbeln, J.R. Seafood consumption and homicide mortality. A cross-national ecological analysis. World Rev. Nutr. Diet 2001, 88, 41–46.

Hibbeln, J.R.; Ferguson, T.A.; Blasbalg, T.L. Omega-3 fatty acid deficiencies in neurodevelop-ment, aggression and autonomic dysregulation: opportunities for intervention. Int. Rev. Psychia-try 2006, 18, 107–118.

Higley, J.D.; Mehlman, P.T.; Poland, R.E.; Taub, D.M.; Vickers, J.; Suomi, S.J.; et al. CSF testos-terone and 5-HIAA correlate with different types of aggressive behaviors. Biol. Psychiatry 1996, 40, 1067–1082.

Hirayama, S.; Hamazaki, T.; Terasawa, K. Effect of docosahexaenoic acid-containing food admin-istration on symptoms of attention-deficit/hyperactivity disorder—a placebo-controlled double-blind study. Eur. J. Clin. Nutr. 2004, 58, 467–473.

Hughes Jr., G.S.; Ringer, T.V.; Francom, S.F.; Caswell, K.C.; DeLoof, M.J.; Spillers, C.R. Effects of fish oil and endorphins on the cold pressor test in hypertension. Clin. Pharmacol. Ther. 1991, 50, 538–546.

Itomura, M.; Hamazaki, K.; Sawazaki, S.; Kobayashi, M.; Terasawa, K.; Watanabe, S.; et al. The effect of fish oil on physical aggression in schoolchildren –a randomized, double-blind, placebo-controlled trial. J. Nutr. Biochem. 2005, 16, 163–171.

Jacobson, T.A. Beyond lipids: the role of omega-3 fatty acids from fish oil in the prevention of coronary heart disease. Curr. Atheroscler. Rep. 2007, 9, 145–153.

Knutson, B.; Wolkowitz, O.M.; Cole, S.W.; Chan, T.; Moore, E.A.; Johnson, R.C.; et al. Selective alteration of personality and social behavior by serotonergic intervention. Am. J. Psychiatry 1998, 155, 373–379.

Kodas, E.; Galineau, L.; Bodard, S.; Vancassel, S.; Guilloteau, D.; Besnard, J.C.; et al. Serotonin-ergic neurotransmission is affected by n-3 polyunsaturated fatty acids in the rat. J. Neurochem. 2004, 89, 695–702.

Mandell, A.J. Non-equilibrium behavior of some brain enzyme and receptor systems. Ann. Rev. Pharm. Toxicol. 1984, 24, 237–274.

Mills, D.E.; Mah, M.; Ward, R.P.; Morris, B.L.; Floras, J.S. Alteration of baroreflex control of forearm vascular resistance by dietary fatty acids. Am. J. Physiol. 1990, 259, R1164–R1171.

Olsson, N.U.; Shoaf, S.; Salem Jr., N. The effect of dietary polyunsaturated fatty acids and alcohol on neurotransmitter levels in rat brain. Nutr. Neurosci. 1998, 1, 133.

Pihl, R.O.; Young, S.N.; Harden, P.; Plotnick, S.; Chamberlain, B.; Ervin, F.R. Acute effect of altered tryptophan levels and alcohol on aggression in normal human males. Psychopharmacology (Berl) 1995, 119, 353–360.

Ramirez, J.M.; Andreu, J.M. Aggression, and some related psychological constructs (anger, hostility, and impulsivity) Some comments from a research project. Neurosci. Biobehav. Rev. 2006, 30, 276–291.

Richardson, A.J.; Montgomery, P. The Oxford-Durham study: a randomized, controlled trial of dietary supplementation with fatty acids in children with developmental coordination disorder. Pediatrics 2005, 115, 1360–1366.

Rosenzweig, S. Rosenzweig picture-frustration study Basic manual. Rana House: St. Louis, MO, 1978.

Sawazaki, S.; Hamazaki, T.; Yazawa, K.; Kobayashi, M. The effect of docosahexaenoic acid on plasma catecholamine concentrations and glucose tolerance during long-lasting psychological stress: a double-blind placebo-controlled study. J. Nutr. Sci. Vitaminol. 1999, 45, 655–665.

Sinclair, A.J.; Begg, D.; Mathai, M.; Weisinger, R.S. Omega 3 fatty acids and the brain: review of studies in depression. Asia Pac. J. Clin. Nutr. 2007, 16(Suppl. 1), 391–397.

Singer, P.; Melzer, S.; Goschel, M.; Augustin, S. Fish oil amplifies the effect of propranolol in mild essential hypertension. Hypertension 1990, 16, 682–691.

Sinn, N.; Bryan, J. Effect of supplementation with polyunsaturated fatty acids and micronutrients on learning and behavior problems associated with child ADHD. J. Dev. Behav. Pediatr. 2007, 28, 82–91.

Stevens, L.; Zhang, W.; Peck, L.; Kuczek, T.; Grevstad, N.; Mahon, A.; et al. EFA supplementation in children with inattention, hyperactivity, and other disruptive behaviors. Lipids 2003, 38, 1007–1021.

Tziomalos, K.; Athyros, V.G.; Mikhailidis, D.P. Fish oils and vascular disease prevention: an update. Curr. Med. Chem. 2007, 14, 2622–2628.

Virkkunen, M.E.; Rawlings, R.; Tokola, R.; Poland, R.E.; Guidotti, A.; Nemeroff, C.; et al. CSF biochemistries, glucose metabolism, and diurnal activity rhythms in alcoholic, violent offenders, fire setters, and healthy volunteers. Arch. Gen. Psychiatry 1994, 1, 20–27.

Weidner, G.; Connor, S.L.; Hollis, J.F.; Connor, W.E. Improvements in hostility and depression in relation to dietary change and cholesterol lowering. The family heart study. Ann. Intern. Med. 1992, 117, 820–823.

Williams, R.B. Neurobiology, cellular and molecular biology, and psychosomatic medicine. Psychosom. Med. 1994, 56, 308–315.

Yamamoto, N.; Saitoh, M.; Moriuchi, A.; Nomura, M.; Okuyama, H. Effect of dietary α-linolenate/1inoleate balance on brain lipid compositions and learning ability of rats. J. Lipid Res. 1987, 28, 144–151.

·12·

Effect of Dietary Fish Protein and Fish Oil on Azoxymethane-induced Aberrant Crypt Foci in A/J Mice

Kenji Fukunaga, Ryota Hosomi, and Munehiro Yoshida
Department of Life Science and Biotechnology, Faculty of Chemistry, Materials and Bioengineering, Kansai University, Suita, Osaka, Japan

Introduction

Recently, not only in developed countries but also world-wide, colon cancer has become a serious health problem. Epidemiologic and experimental reports have also shown a relationship between diet and in the etiology of colorectal cancers (Campos et al., 2005; Bonovas et al., 2008; Santarelli et al., 2008). Metabolic phenotypes and Western-style diets containing low dietary fiber and high fat levels are both risk factors for the development of colon cancer (Thomson et al., 2003). Conversely, fish-rich diets are considered beneficial, and several epidemiologic studies in Alaskan and Greenland Inuit revealed a low incidence of colon cancer compared with Western populations, which was linked to their high dietary consumption level of fish products (Friborg & Melbye, 2008; Calviello et al., 2007). In fact, fish is a good source of many important nutrients, such as n-3 polyunsaturated fatty acids (PUFAs), protein, vitamin A, vitamin D, vitamin B_{12}, calcium selenium, iron, and the other various microelements. Several studies document the beneficial effect on health of including fish in the diet. Increasing interest in the health benefits obtained by regular fish intake has been emphasized in developed Western countries. Fish products contain abundant n-3 PUFAs, namely icosapentaenoic acid (IPA) and docosahexaenoic acid (DHA), and a very small amount of n-6 PUFAs, such as linoleic acid. In addition, patients with chronic intestinal disorders, such as inflammatory bowel disease, had lower plasma levels of n-3 PUFAs than normal subjects (Belluzzi et al., 2000), thus supporting the protective role of the dietary intake of n-3 PUFAs (Calder et al., 2008).

Experimental animal reports have shown that n-3 PUFA-rich fish oil suppresses chemically induced colon cancer in rats (Good et al., 1998; Kohno et al., 2000; Rao et al., 2001; Corpet & Pierre, 2003; Reddy et al., 2005a; Kenar et al., 2008). Furthermore, in human studies, n-3 PUFAs have also been reported to suppress rectal cell proliferation (Hurlstone et al., 2005; Calviello et al., 2007). Thus, n-3 PUFAs in fish products can be significant suppression factors in colon carcinogenesis.

In terms of daily dietary requirements, humans do not need to eat fish protein or oil. Therefore, many investigators have considered healthy functional development from an intake of fish and the fish oil to be the same because of the strong bioactive effect of IPA and DHA included in fish oil. Previous studies were not designed as fish diets (i.e., to elucidate the possible effects of dietary intake of fish protein). However, no information is currently available on the effect of fish protein on the colon, colon tumors, or aberrant crypt foci (ACF) incidence, and multiplicity.

ACF have been reported to be putative preneoplastic lesions of colon cancer in both rodents (Stevens et al., 2007) and humans (Roncucci et al., 2000). To examine the anticarcinogenic activity of dietary constituents, 1,2-dimethylhydrazine (DMH) or azoxymethane (AOM) have been frequently used as carcinogens. These agents quickly and easily induce the formation of ACF and subsequent development of colon adenomas and adenocarcinomas. Investigators have utilized ACF number and type, tumor number and type, or the combination of these as endpoints to examine the effects of fish oils on colon cancer. Furthermore, ACF with a large numbers of crypts (i.e., four or more per focus) have been proposed as intermediate biomarkers for colon carcinogenesis (Mori et al., 2004; Takayama et al., 2005; Kim et al., 2008). In other words, ACF assay is a useful and suitable index for screening preventive agents for colon cancer. Thus far, IPA, DHA, and fish oil has been reported to suppress the formation and growth of ACF, as well as colon tumors, in rodents (Good et al., 1998; Dommels et al., 2003a). In the present study, male A/J mice were fed dietary fish protein and fish oil to examine the inhibitory effect of fish protein and its relationship with fish oil in the development of ACF induced by AOM in mice.

IPA and DHA, which have anti-inflammatory properties, are associated with the well-known ability of n-6 PUFAs to inhibit the production of various proinflammatory mediators, including eicosanoids, such as leukotriene B_4 or prostaglandin E_2 (PGE_2), and cytokines (Calder, 2002). An n-3 PUFA-rich diet suppresses the excessive production of PGE_2 in colon tumors, which may be accompanied with neoplastic formation, by competitively inhibiting cyclooxygenase activity (Mahmood et al., 2006). Therefore, this study was conducted to investigate the role of PGE_2 in fish oil and fish-protein-mediated effects on AOM induced colon carcinogenesis.

Furthermore, this study also examined the effect of bile acid, which is a colon carcinogenesis promoter. Bile acids have been implicated as important etiological factors in colon cancer (Debruyne et al., 2002; Shiraki et al., 2005). Damage to the intestinal tract mucous layer may render the underlying cells susceptible to intraluminal toxins or carcinogens. The cytotoxic concentrations of bile acids induce mucin release, presumably due to detergent effects. Fish oil, which contains IPA and DHA, has been suggested to reduce the plasma cholesterol level in rats by increasing the transfer of cholesterol into bile without an increase in bile acid secretion (Morgado et al., 2005), which may affect bile acid metabolism. On the other hand, there is a report that the amount of bile acids in the feces could provide information about the possible effects of fish protein on cholesterol excretion via bile acids (Shukla et al.,

2006). Therefore, the influence of dietary fish protein and fish oil on bile acid in the feces was also evaluated.

Materials and Methods

Chemicals and Experimental Diets

AOM was purchased from Wako Pure Chemical Co. Ltd. (Osaka, Japan). The vitamin mixture, mineral mixture, and other materials of the American Institute of Nutrition (AIN-93G) (Reeves et al., 1993) recommended diets were purchased from Oriental Yeast Co. Ltd. (Osaka, Japan). Fish oil was supplied by Nippon Chemical Feed Co. Ltd. (Hokkaido, Japan). Fish protein (Alaskan Pollock) was supplied by Suzuhiro Co. Ltd. (Kanagawa, Japan).

The semi-synthetic diet was prepared once per week in accordance with the standard formula of the AIN-93G diet. The fish protein and casein were almost lipid free (<0.05% w/w). The protein contents of the diets were adjusted to be isonitrogenous. At the expense of carbohydrates, all diets were prepared isoenergetically. Experimental diets were based on the AIN93G formula and contained fish oil (1.0% or 2.5%) and fish protein (4.0% or 10.0%). The compositions of experimental diets are summarized in Table 12.A, and the fatty acid and amino acid compositions of the

Table 12.A. Composition of Experimental Diets (g/kg).

Component	Experimental groups						
	Control	FPL	FPH	FOL	FOH	FPOL	FPOH
	(g/kg diet)						
Pregelatinized cornstarch	132.0	132.0	132.0	132.0	132.0	132.0	132.0
Corn starch	397.5	397.5	397.5	397.5	397.5	397.5	397.5
Casein	200.0	160.0	100.0	200.0	200.0	160.0	100.0
Fish protein	—	40.0	100.0	—	—	40.0	100.0
Sucrose	100.0	100.0	100.0	100.0	100.0	100.0	100.0
Cellulose powder	50.0	50.0	50.0	50.0	50.0	50.0	50.0
Mineral mixture*	35.0	35.0	35.0	35.0	35.0	35.0	35.0
Vitamin mixture*	10.0	10.0	10.0	10.0	10.0	10.0	10.0
L-cysteine	3.0	3.0	3.0	3.0	3.0	3.0	3.0
Choline bitartrate	2.5	2.5	2.5	2.5	2.5	2.5	2.5
Soybean oil	70.0	70.0	70.0	60.0	45.0	60.0	45.0
Fish oil	—	—	—	10.0	25.0	10.0	25.0

Each experimental diet was prepared once per week in accordance with the standard formula of the AIN-93G diet.

The prepared diets were subdivided and packed in plastic bags with deoxygenating materials and nitrogen gas and stored in the dark at –30ºC until use.

FPL, fish protein 2.0%; FPH, fish protein 5.0%; FOL, fish oil 1.0%; FOH, fish oil 2.5%; FPOL, fish protein 2.0% + fish oil 1.0%; FPOH, fish protein 5.0% + fish oil 2.5%.

Table 12.B. Experimental Protein and Lipid Profiles.

Amino acid composition of protein (%w/w)			Fatty acid composition of lipid (%w/w)			
Amino acids	Casein	Fish protein*	Fatty acids		Soy oil	Fish oil**
Alanine	2.6	6.0	14:0		0.2	6.7
Arginine	3.6	5.9	16:0		10.6	21.6
Aspartic acid	6.4	10.2	16:1		—	6.1
Cysteine	1.9	1.8	16:2		—	0.8
Glutamic acid	20.3	13.7	18:0		4.1	4.2
Glycine	1.8	3.2	18:1		24.1	18.6
Histidine	2.6	2.1	18:2	n-6	53.1	1.8
Isoleucine	4.7	5.1	18:3	n-3	6.9	0.7
Leucine	8.6	9.1	20:1		0.3	3.4
Lysine	7.4	10.6	20:4	n-6	—	0.9
Methionine	2.6	3.3	20:5	n-3	—	10.6
Phenylalanine	5	4.3	22:1		—	4.8
Proline	11.4	3.1	22:4	n-3	0.4	0.3
Serine	5.3	4.9	22:5	n-3	—	2.3
Threonine	3.7	5.6	22:5	n-6	—	0.3
Tryptophan	1.3	1.0	22:6	n-3	—	16.7
Tyrosine	5.3	4.1	24:0		—	0.8
Valine	5.6	6.7	others		0.3	1.3

*Fish protein was provided from Suzuhiro Co., Ltd., Kanagawa, Japan; **Fish oil was provided from Nippon Chemical Feed Co., Ltd. Hokkaido, Japan.; (–), non-detected.

experimental lipids and protein are summarized in Table 12.B. The prepared diets were subdivided and packed in plastic bags with deoxygenating materials and nitrogen gas substitution, and stored in the dark at $-30^{\circ}C$ until use.

Animal Care

At 3 weeks of age, male A/J mice were obtained from Shimizu Laboratory Supplies Co., Ltd (Kyoto, Japan) and were housed in polycarbonate cages maintained in a temperature and humidity controlled animal unit with a daily light:dark cycle of 12 h. The animals were allowed free access to drinking water and diet. After the first AOM treatment, mice were immediately divided between the respective experimental diets. The experimental protocol was reviewed and approved by the Animal Ethics Committee of Kansai Medical University and followed the "Guide for the Care and Use of Experimental Animals" of the Prime Minister's Office of Japan.

Study Design

The experimental design is shown in Fig. 12.1. Groups of male A/J mice were fed the control diet (AIN93G). At the start of the experiments, when the animals were 6 weeks old, a total of 210 mice were divided into 21 groups of 10 mice each. At 6 weeks of age 14 groups of mice intended for carcinogen treatment were injected intraperitoneally with AOM dissolved in 0.1 ml saline solution once a week at a dose of 7.5 mg/kg body weight, and to the remaining 7 groups of mice similar injections were given of an equal volume of saline as a vehicle control. After the first AOM or vehicle control treatment, mice were maintained continuously on each of the experimental diets shown in Table 12.A. Diet consumption was measured daily, and weight gain was recorded weekly. In the last week, feces were collected for fecal bile acid analysis, and its weight was recorded daily.

ACF Assay

Ten AOM-treated mice in each group were used to examine ACF, and those of the remaining ten mice were used for the assays of PGE_2 and fatty acid composition. All mice were sacrificed at 10 weeks after the first administration. The vehicle control groups of mice were treated in the same manner, but the assays for PGE_2 and fatty acid composition were omitted. The colons of the 10 mice in each group were removed, and the colon contents were removed by flushing from the fecal end with ~20 ml of phosphate-buffered saline (PBS) using a syringe. Then, the colon was opened longitudinally from the cecum to anus, placed between two pieces of filter

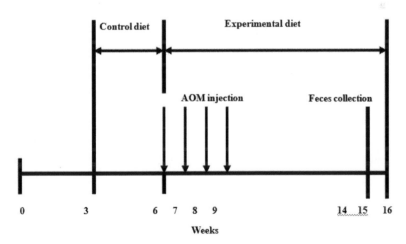

Fig. 12.1. Groups of male A/J mice were fed control diet (AIN9G). At the start of the experiments, when animals were 6 weeks old, azoxymethane (AOM) was administered intraperitoneally at a dose of 7.5 mg/ kg body weight. After the first AOM treatment, the mice were immediately divided into the respective experimental diets shown in Table 12.A. In the last week, feces were collected for fecal bile acid analysis.

paper, and fixed in neutral-buffered 10% formalin solution for 24 h. The colons were stained with 0.2% methylene blue in saline in PBS for 5–7 min or until the tissue had a uniform blue appearance (Paulsen et al., 2006). Tissues were rinsed with PBS for 1 min and stored in 0.4% formalin-PBS at 4°C. The colon samples were then placed mucosal side up on a microscope slide and observed; namely, the number of ACF/colon and the number of aberrant crypts in each focus were counted microscopically using side illumination. The criteria used to identify ACF topographically were crypts of increased size with a thicker and deeply stained epithelial lining, and an increased pericryptal zone compared with normal crypts. ACF were categorized according to crypt complexity (1, 2, 3, or more than 4 crypts per ACF). All colons were scored by one observer without revealing the identity of the agents under study; scores were checked at random by a second observer. Colons that failed to produce useable data due to poor fixation and/or staining were excluded from statistical analysis.

Prostaglandin E_2 Assay

For the determination of PGE_2 levels in the colonic mucosa, the colons of the 10 mice in each group were slit longitudinally, placed between two pieces of slide glass, and immediately frozen in liquid nitrogen. The mucosal layer, which remained attached to one piece of the slide glass, was separated from the muscular and serosal layers and was scraped off using a razor. Homogenates were prepared by homogenizing the mucosa in ice-cold PBS containing 1 mM EDTA, 0.1 mM indomethacin, and 100 U/ml aprotinin using a glass-Teflon homogenizer. The cell debris and nuclei were removed by centrifugation at 2000 g for 10 min at 4°C. PGE_2 levels were determined using an enzyme immunoassay determination (Amersham, Buckinghamshire, UK), according to the instructions provided by the manufacturer. These analyses were performed under low light to avoid photooxidation. The PGE_2 levels of colon mucosa were expressed as nanograms of per milligram of protein.

Fatty Acid Composition

The fatty acids of dietary lipid used in the experimental diets and colon mucosa lipids were extracted by the method of Bligh and Dyer (Bligh & Dyer, 1959). The colon mucosa was obtained from the same sample as for PGE_2 analysis. Colon phospholipids were separated from total lipid by silica gel cartridge column chromatography using a Supelco solid phase extraction system with chloroform and methanol as the eluent. Pentadecanoic acid was added to each lipid fractions as an internal standard. The samples were subjected to methanolysis in 10% HCl in methanol at 80°C for 1 h under N_2. Fatty acid methyl esters were extracted using n-hexane and analyzed by a G14B gas-chromatography (Shimadzu, Kyoto, Japan) with a HR-SS-10 column (Shimadzu, Kyoto, Japan) with N_2 carrier gas. The oven temperature was programmed from 140°C (2.0 min) to 220°C at 5°C/min and then maintained at 220°C. The detector temperature was set at 280°C and the injector port temperature at 250°C. Fatty acid methyl esters were detected using a flame ionization detector. The identification and quantitation of each fatty acid was carried out with commercially available authentic standard mixtures using a CR-6A Chromatopac integrator (Shimadzu, Kyoto, Japan).

Amino Acid Composition

The amino acids in dietary protein used in the experimental diets were hydrolyzed using 6 M HCl. The hydrolyzed sample solutions were automatically derivatized using o-phthalaldehyde with a robotic autosampler. After derivatization, the samples were separated on a reverse phased column (GL Sciences, Inertsil ODS-3, 5 µm, 250 × 4.6 mm) at 40°C. The eluate was monitored by fluorescence detection using an excitation wavelength of 340 nm and an emission wavelength of 450 nm. The separation was obtained at a flow rate of 2.0 mL/min with a gradient program with 50 mM sodium phosphate buffer (pH 7.5) and acetonitrile. The identification and quantitation of each amino acid were carried out with authentic standard mixtures available commercially using a CR-6A Chromatopac integrator (Shimadzu, Kyoto, Japan).

Fecal Bile Acid Assay

In the last week of the study period, feces were collected, then lyophilized, pulverized using a pestle and mortar, and weighed. Dry feces were hydrolyzed using KOH at 95°C for 5 h. Cooled hydrolysates were extracted with diethyl ether to remove non-saponifiable components and acidified to pH 2 using HCl. Pooled total diethyl ether extracts were evaporated using a rotary evaporator and redissolved in ethanol, and total fecal bile acids were determined enzymatically using 3-hydroxysteroid dehydrogenase by the assay method of Sheltawy and Losowsky (1975). The fecal bile acid content was expressed as micrograms per gram of dry feces.

Statistical Analysis

Data are expressed as means ± SD. Statistical differences between multiple groups were determined by ANOVA. Statistical comparisons were made using the Tukey-Kramer test. The difference was considered significant at p values <0.05 and <0.01. The analyses were performed using StatView-J version 5.0 software (Abacus Concepts, Berkeley, CA, USA).

Results

Body Weight and Diet Intake

Body weight and diet intake are shown in Table 12.C. There were no significant differences in average diet intake among all groups during the experimental period. At the end of the experimental period, there were no significant differences in body weight among all groups. This result indicated that the consumption of fish protein

Table 12.C. Body Weight and Average Diet Intake of Mice.

	Experimental groups						
	Control	FPL	FPH	FOL	FOH	FPOL	FPOH
Initial body weight (g)	17.7±0.9	19.7±0.9	17.7±1.0	17.6±0.9	17.6±0.8	17.5±1.0	17.6±0.9
Final body weight (g)	28.5±1.6	28.1±2.1	28.3±1.4	28.3±0.9	27.9.±1.8	28.1±2.0	27.6±2.3
Average diet intake (g/ mouse/day)	4.3±0.2	4.2±0.3	4.3±0.3	4.4±0.4	4.2.±0.2	4.4±0.3	4.3±0.2

FPL, fish protein 2.0%; FPH, fish protein 5.0%; FOL, fish oil 1.0%; FOH, fish oil 2.5%; FPOL, fish protein 2.0% + fish oil 1.0%; FPOH, fish protein 5.0% + fish oil 2.5%.

Value are expressed as mean ±SD, n=20.

and fish oil did not affect normal growth and appetite compared with the control casein group.

ACF Assay

The ACF number/colon and aberrant crypts/focus, which indicates crypt multiplicity for ACF, were determined by the methylene blue staining method. ACF showed dilated irregular luminal openings, thicker epithelial linings, and protrusions toward the lumen. No ACF developed in the colons of mice in vehicle control groups (data not shown). In contrast, the mice treated with AOM showed a 100% incidence of ACF. Subcutaneous injection of AOM resulted in the formation of ACF in all mice. The total number of ACF, those classified as 1, 2, 3, and 4+ crypts per focus in AOM-treated mice are shown in Table 12.D. AOM-treated mice fed with fish protein, fish oil, and fish protein + fish oil diets had significantly fewer ACF per colon compared with rats fed on the control diet. Both fish oil and fish protein feeding led to lower numbers of ACF with 4+ crypts. Fish-oil-fed mice tended to have lower numbers of ACF/colon and lower crypt multiplicity compared with the fish protein group, and this effect was more pronounced in the group fed on fish oil + fish protein compared with each individually. The suppressive effect of fish oil and fish protein on the total number of ACF/colon and crypt multiplicity was dose-dependent, and an additive effect was found. The inhibitory effect of ACF on multiplicity was clear; as a result there was a higher number of AC1 in the fish protein and fish oil groups.

PGE$_2$ Content in Colon Mucosa

The PGE$_2$ content in the colon mucosa of mice fed each experimental diet is shown in Table 12.E. A high PGE$_2$ level in the colon was observed in the control group. The PGE$_2$ level in the fish oil group was reduced significantly, in a dose-dependent manner, to approximately half, or 20% that of the control group. In addition, the PGE$_2$

Table 12.D. Effect of Fish Protein and Fish Oil on AOM-Induced ACF Formation in the Colon of A/J Mice.

Experimental groups	No. of mice with ACF	Crypt multiplicity for ACF											Total No. of ACF/ colon		
		AC1			AC2			AC3			AC4+					
Control	10/10	12.3	±	3.9	19.7	±	4.5	12.4	±	1.7	14.9	±	3.4	59.4	±	8.6
FPL	10/10	11.5	±	4.1	13.9	±	3.2*	10.5	±	2.4	10.2	±	2.7	46.1	±	5.2*
FPH	10/10	15.7	±	2.2	9.8	±	2.1**	7.9	±	0.9	8.6	±	1.9*	37.2	±	4.1*
FOL	10/10	11.7	±	2.9	7.5	±	3.6**	4.7	±	0.7*	7.3	±	1.7**	31.2	±	9.2**
FOH	10/10	12.6	±	3.4	6.4	±	1.0**	3.4	±	1.2**	3.3	±	0.7**	25.6	±	3.1**
FPOL	10/10	13.8	±	2.6	7.8	±	1.7**	4.3	±	1.3*	4.6	±	0.9**	30.5	±	7.0**
FPOH	10/10	15.8	±	2.1	5.4	±	0.6**	2.1	±	0.5**	2.4	±	0.4**	25.8	±	5.3**

AOM, Azoxymethane; ACF, aberrant crypt foci; AC, aberrant crypt; FPL, fish protein 2.0%; FPH, fish protein 5.0%; FOL, fish oil 1.0%; FOH, fish oil 1.0%; FPOL, fish protein 2.0% + fish oil 1.0%; FPOH, fish protein 5.0% + fish oil 2.5%; AC4 + four or more aberrant crypts. Value are expressed as mean ±SD, n=10. Statistically significant difference from the control group by Tukey-Kramer test; * p<0.05; ** p<0.01. No ACF were present in the vehicle injected mice (data not shown).

Table 12.E. Effect of Fish Protein and Fish Oil on PGE$_2$ Content in Colon Mucosa of AOM Treated Mice.

Experimental groups	PGE$_2$ (ng/mg protein)		
Control	12.3	±	2.1
FPL	11.5	±	1.2
FPH	10.9	±	1.7
FOL	6.3	±	1.4**
FOH	2.5	±	0.3**
FPOL	5.2	±	1.2**
FPOH	2.0	±	0.8**

AOM, Azoxymethane; PGE$_2$, Prostaglandin E$_2$; FPL, fish protein 2.0%; FPH, fish protein 5.0%; FOL, fish oil 1.0%; FOH, fish oil 2.5%; FPOL, fish protein 2.0% + fish oil 1.0%; FPOH, fish protein 5.0% + fish oil 2.5%.

Value are expressed as mean ±SD, n=10. Statistically significant difference from control group by Tukey-Kramer test; ** $p < 0.01$.

level in the fish protein group was slightly lower in the control group, although the difference was not statistically significant. Combined feeding with fish protein and fish oil further down-regulated PGE$_2$ levels in the colon in a dose-dependent manner compared with the each treatment individually.

Colon Mucosa Phospholipids Fatty Acid Composition

The colon mucosa phospholipid fatty acid composition of mice fed each experimental diet is shown in Table 12.F. The overall composition of fatty acids reflected essentially the fatty acid composition of each diet. The contents of total n-6 PUFAs (mostly linoleic acid and arachidonic acid) in the colon were significantly lower in the fish oil diet group than in the control and fish protein diet groups. IPA and DHA were the predominant colon n-3 PUFAs in the fish oil group, and the colon levels of IPA plus DHA were higher in the fish oil group than in the control and fish protein groups. The ratio of n-6/n-3 fatty acids in the colon was 4.7-fold higher in the control and 12.4-fold higher in the fish protein diet group compared with the control diet group. Dietary intake of fish protein did not significantly alter the colon fatty acid composition.

Fecal Output and Bile Acid Content

Fecal output and bile acid content are shown in Table 12.G. The fecal output was not affected by the experimental diets. On the other hand, the content of fecal bile acid was significantly higher in rats fed the fish protein diet in a dose dependent manner. The fecal bile acid concentration was not affected by the difference in dietary lipids. The fecal bile acid content from the combined feeding with fish protein and fish oil was almost the same as in fish protein feeding group.

Table 12.F. Fatty Acid Composition of Colon Mucosa Phospholipids Fatty Acid Composition (%w/w) in AOM Treated Mice.

Fatty acid	Control	FPL	FPH	FOL	FOH	FPOL	FPOH
14:0	1.3 ± 0.2	1.2 ± 0.1	1.4 ± 0.2	1.2 ± 0.1	1.3 ± 0.3	1.0 ± 0.1	1.1 ± 0.1
16:0	15.4 ± 2.2	14.9 ± 1.9	15.9 ± 0.9	16.2 ± 1.1	15.7 ± 1.8	15.9 ± 1.1	16.3 ± 1.2
16:1 n-7	2.3 ± 0.2	2.1 ± 0.3	2.1 ± 0.2	2.3 ± 0.3	2.2 ± 0.3	2.1 ± 0.3	2.4 ± 0.2
16:2	0.2 ± 0.04	0.3 ± 0.03	0.4 ± 0.03	0.2 ± 0.04	0.5 ± 0.04	0.3 ± 0.04	0.4 ± 0.03
18:0	20.3 ± 2.3	21.1 ± 2.1	20.9 ± 3.3	16.9 ± 3.1	14.6 ± 2.6**	15.6 ± 3.1	14.9 ± 2.3
18:1 n-9	12.6 ± 1.1	11.9 ± 0.9	13.1 ± 0.7	18.2 ± 0.6**	17.4 ± 0.5**	18.1 ± 0.6**	17.9 ± 0.4**
18:2 n-6	19.2 ± 0.6	18.7 ± 0.7	18.2 ± 0.9	10.6 ± 0.8**	8.6 ± 0.6**	12.9 ± 0.8**	8.3 ± 0.3**
18:3 n-3	1.2 ± 0.1	1.4 ± 0.1	0.9 ± 0.1	1.6 ± 0.2	1.9 ± 0.2	1.8 ± 0.2	1.4 ± 0.3
20:1	0.3 ± 0.01	0.2 ± 0.03	0.4 ± 0.03	0.3 ± 0.04	0.2 ± 0.04	0.5 ± 0.04	0.3 ± 0.02
20:4 n-6	17.5 ± 1.0	18.1 ± 0.9	17.2 ± 1.2	6.3 ± 1.6**	5.4 ± 0.9	6.8 ± 1.6**	4.8 ± 0.7**
20:5 n-3	0.6 ± 0.02	0.5 ± 0.03	0.4 ± 0.02	9.4 ± 1.2**	14.9 ± 3.1**	8.7 ± 1.2**	15.4 ± 0.9**
22:1	0.4 ± 0.01	0.5 ± 0.02	0.3 ± 0.02	0.4 ± 0.04	0.6 ± 0.01	0.3 ± 0.04	0.4 ± 0.03
22:4 n-3	3.6 ± 0.2	4.2 ± 0.2	3.1 ± 0.5	2.1 ± 0.4*	1.3 ± 0.7**	2.0 ± 0.4*	1.5 ± 0.3**
22:5 n-3	0.7 ± 0.01	0.8 ± 0.03	0.5 ± 0.02	2.4 ± 0.03**	2.3 ± 0.02**	2.6 ± 0.03**	2.1 ± 0.04**
22:5 n-6	0.2 ± 0.02	0.2 ± 0.04	0.2 ± 0.01	0.1 ± 0.03	0.1 ± 0.04	0.2 ± 0.03**	0.3 ± 0.01**
22:6 n-3	1.2 ± 0.2	1.3 ± 0.3	1.1 ± 0.4	9.3 ± 0.6**	10.9 ± 0.9**	8.9 ± 0.9**	10.5 ± 1.3**
others	3.2 ± 0.4	2.9 ± 0.6	3.9 ± 0.8	2.6 ± 0.3	2.1 ± 0.5	2.3 ± 0.3	2.0 ± 0.6
n-6/n-3	5.2 ± 0.9	4.5 ± 1.3	5.9 ± 1.2	1.1 ± 0.2	0.42 ± 0.6	0.8 ± 0.3	0.39 ± 0.5

AOM, Azoxymethane; ACF, aberrant crypt foci; FPL, fish protein 2.0%; FPH, fish protein 5.0%; FOL, fish oil 1.0%; FOH, fish oil 2.5%; FPOL, fish protein 2.0% + fish oil 1.0%; FPOH, fish protein 5.0% + fish oil 2.5%; AC4 + four or more aberrant crypts.

Value are expressed as mean ±SD, n=10. Statistically significant difference from the control group by Tukey-Kramer test; * $p < 0.05$; ** $p < 0.01$.

Table 12.G. Effect of Fish Oil and Fish Protein on Fecal Output and Fecal Bile Acid Content in Mice Treated with AOM.

Experimental groups	Fecal output (dry weight g/day)			Fecal bile acid (μg/g dry feces)		
Control	1.9	±	0.3	1029	±	126
FPL	1.7	±	0.2	1523	±	236*
FPH	1.8	±	0.3	2344	±	298**
FOL	1.9	±	0.4	1123	±	154
FOH	2.0	±	0.5	987	±	213
FPOL	1.9	±	0.3	1489	±	187*
FPOH	1.8	±	0.4	2496	±	255**

AOM, Azoxymethane; ACF, aberrant crypt foci; FPL, fish protein 2.0%; FPH, fish protein 5.0%; FOL, fish oil 1.0%; FOH, fish oil 2.5%; FPOL, fish protein 2.0% + fish oil 1.0%; FPOH, fish protein 5.0% + fish oil 2.5%.

Value are expressed as mean ±SD, n=10. Statistically significant difference from the control group by Tukey-Kramer test; * $p < 0.05$; ** $p < 0.01$.

Discussion

The fish oil diet, with its high n-3 PUFA content, had a significantly stronger protective effect against ACF formation than the control diet. The inhibitory effect of fish oil against colon carcinogenesis is supported by several studies. Epidemiologic studies have provided strong evidence for the role of fish oil in colon cancer prevention. There is, for example, an inverse correlation between fish oil consumption and colon cancer risk identified from data from a 22-year prospective study (Hall et al., 2008).

The anticancer effects of dietary n-3 PUFAs, such as IPA and DHA in fish products, have attracted great interest (Berquin et al., 2008). Epidemiologic studies have presented a low mortality level from colon cancer in areas where a large quantity of fish or its products are consumed (Calviello et al., 2007; Friborg & Melbye, 2008). It is thought that n-3 PUFAs have inhibitory effects on tumor cell growth (Good et al., 1998; Rao et al., 2001). In most in vitro studies, the n-3 fatty acids has antitumor effects through the inhibition of cell proliferation (Hurlstone et al., 2005; Calviello et al., 2007) or induction of apoptosis (Hong et al., 2003), whereas several in vivo studies hypothesized that corn oil, with its high content of n-6 PUFAs such as linoleic acid, might enhance colorectal carcinogenesis via stimulation of colonic cell proliferation (Dommels et al., 2003b).

In daily dietary habits we usually eat fish itself or various types of processed fish products rather than fish oil extracts, and it is clear that fish protein plays an important role as a protein source in the Japanese diet. The beneficial health effects of fish proteins, which have been clarified by various epidemiologic studies about the effects of the consumption of fish, have shown that fish proteins, as well as lipids, are important. There are a considerable number of reports on the importance of the nutritional component of fish protein to lipid metabolism (Boukortt et al., 2006; Demonty et al., 2003; Gunnarsdottir et al., 2008). However, no information is currently available on the effect of fish protein on colon cancer. The effect of protein sources such as soya, whey, and casein on promotion of colorectal neoplasia by carcinogen such as DMH or AOM in a rat model was reported previously (Badger et al., 2005; Xiao et al., 2005). Furthermore, DMH treatment of mice has been examined, and it was reported that lean meat promotes colon cancer more than other sources of protein (McIntosh & Le, 2001; Pence et al., 2003). Further studies have been required to clarify the effect of fish protein and fish lipid on the development of colonic cancer. Therefore, the present study was designed to analyze fish diets for the suppressive effect of fish protein and fish oil on colon carcinogenesis in an AOM-induced mouse colon cancer model.

The previous observation of body weight loss in rats on a fish oil or n-3 PUFA rich diet might, in part, have contributed to the reduction of colon tumor incidence (Reddy et al., 1986; Reddy et al., 1987). The nutritional effect associated with the change of triglycelyride metabolism caused by the IPA diet may affect colon carcinogenesis. The ability of calorie restriction to inhibit or delay cancer incidence and progression is mediated, in part, by changes in energy balance, body mass, and/or body composition rather than calorie intake, suggesting that excess calorie retention, rather than consumption, confers cancer risk (Huffman et al., 2007). In the present study, there was no significant difference in food consumption or body weight gain throughout the experimental period among the dietary groups. Therefore, the present results have shown that the retardation of cancer initiation and progression in the mouse colon is not related to the amount of calories consumed.

AOM stimulates both proliferation and apoptosis in the colonic mucosa (Kawamori et al., 2003). ACF are thought to be the earliest observable pre-neoplastic lesions to arise in this tissue (Roncucci et al., 2000; Alrawi et al., 2006) and have also been confirmed to be present in the human colon (Takayama et al., 2005). ACF are widely regarded as early markers of incipient neoplasticity (Cheng & Lai, 2003; Orlando et al., 2008). However, the validity of ACF as an intermediate biomarker for colon cancer is controversial. Larger multi-cryptal ACF are generally correlated with tumor incidence in rodents and humans (Roncucci et al., 2000; Mori et al., 2004; Stevens et al., 2007; Kim et al., 2008). In the present study, ACF was employed as an index of early neoplastic induction. The number of crypts/focus was shown to increase with time after carcinogen treatment, and ACF demonstrated increased cell proliferation in rodents (Mori et al., 2002; Selvam et al., 2008). In fact, the development of ACF has also been reported to be enhanced by supplementation with n-6 PUFAs as compared to supplementation with fish oil, which is rich in n-3 PUFAs (Rao et al., 2001; Dupertuis et al., 2007).

In this study, fish oil suppressed the development of AOM-induced ACF in rat colons in a dose-dependent manner, suggesting that dietary n-3 PUFA was effective in preventing neoplastic changes in the initiation phase of colon carcinogenesis. Furthermore, fish protein also reduced the ACF number and multiplicity.

Dietary fish oil is known to induce a selective incorporation of n-3 PUFA and a competitive exclusion of n-6 PUFA in the membrane phospholipid fraction. In this study, alterations in fatty acid composition in the colon mucosa were observed in the colonic mucosa of rats fed fish oil group, whereas fish protein did not affect the fatty acid composition of the colon mucosa. Changes in the ratio of n-3 to n-6 PUFA in the membrane can affect the function of the mucosa membrane fluidity, permeability, and membrane bound receptors, such as epidermal growth factor receptor (EGFR) (Dougherty et al., 2008).

A possible mechanism to explain the anticancer role of n-3 PUFAs could be their ability to suppress inflammation by down-regulating the cyclooxygenase (COX)-2 enzyme, which is involved in the synthesis of pro-inflammatory prostaglandins. COX-2 activity is enhanced in experimental rat studies in response to feeding large amounts of n-6 PUFAs (Kim et al., 2008). COX-2 has also been detected in human colon tumors and in chemically induced colon tumors in rats (Reddy et al., 2005b). Because PGE_2 has been shown to be elevated in rat colon tumors (Kawamori et al., 2003), the mechanism responsible for the inhibitory effects of n-3 PUFAs on the formation of ACF and colorectal tumors can be related in part to the inhibition of PGE_2 synthesis from AA and reduction the level of AA itself. An association between the down-regulation of COX-2 expression and PGs, particularly the type-2 series, is believed to be closely involved in colon carcinogenesis, as elevated PG levels have been found in colon cancer tissues (Fosslien et al., 2000). Furthermore, PGE_2 induces hyperproliferation in the colonic mucosa (Gostner et al., 2000), and inhibitors of PG synthesis, such as indomethacin, inhibit colon carcinogenesis in rats. It has been reported that n-3 PUFAs inhibit the production of type-2 series of eicosanoids, including PGE_2, from arachidonic acid (Tapiero et al., 2002; Das, 2005). The overexpression of COX-2 may contribute to ACF growth and sequential tumor growth. In the present study, after 10 weeks of feeding fish oil, the fatty acid compositions of mucosal membrane phospholipids were changed, and there were significantly reduced PGE_2 levels in the colonic mucosa, whereas a fish protein diet did not significantly affect the PGE_2 content of the colon mucosa or change the fatty acid composition. These findings suggest that fish protein can play an important role in inhibiting the incidence of ACF in the mouse colon without passing through arachidonic acid metabolism.

Bacteria in the intestine turn bile acids into cancer-promoting substances called secondary bile acids. High dietary fat increases the secretion of secondary bile acids (Debruyne et al., 2001; Tong et al., 2008). Linoleic acid increases the excretion of fecal secondary bile acids, resulting in the promotion of colon carcinogenesis (Rao et al., 2001; Juste, 2005), whereas fish oil has been suggested to reduce the plasma cholesterol level in rats by increasing the transfer of cholesterol into bile without an increase of bile acid secretion (Bartram et al., 1998), which may affect bile acid metabolism. Furthermore, a detailed experiment by Narisawa et al. (1994) showed that the amount and concentration of total and secondary bile acids in the feces were not different between rats fed on fish oil (n-3 PUFAs) and fed on safflower oil (n-6 PUFAs). Therefore, it seems unlikely that the inhibitory effect of fish oil can be attributed to a change in fecal bile acids. In the present study, the fecal excretion of bile acid was greater in rats fed on fish protein than in the rats fed on casein or fish oil. Hence, dietary fish protein effectively inhibited the absorption of cholesterol and bile acid in the small intestine. These results suggest that the suppression of cholesterol absorption by the micellar solubility of cholesterol and the binding capacity of cholate of digested fish protein in the jejunal epithelia. Furthermore, the digestion of fish protein may also inhibit the re-absorption of bile acids in the ileum. These effects might also contribute to protection against colon carcinogenesis. Another possible mechanism, which may explain the protection against colon cancer during the initiation phase, is the postulated ability of n-3 PUFA or fish protein to increase the rate of detoxification of AOM in the liver.

The protein-dependent difference in ACF suppression effect may be attributed in part to the difference in the amino acid composition. Therefore, the present study evaluated whether the amino acid composition of fish protein inhibited the development of the ACF in another experiment. The equivalent amino acid composition was prepared as a model of the fish protein component diet by adding each amino acid to casein, and then the mice were fed this experimental diet. The ACF suppression effect of this diet was examined, but it was found that the ACF number/colon and crypts/focus were not affected (data not shown). This result suggests that the fish protein digest products cause suppression of ACF through detoxification and bile acid metabolism in the liver or colon, whereas the amino acid composition in each diet did not contribute to the ACF suppression effect.

Conclusion

The present results suggest that fish protein, in combination with fish oil, may contribute to the beneficial health effects of fish consumption in protection against ACF formation. Daily supplementation of fish or fish products has a beneficial effect against chemically induced colonic preneoplastic progression in mice induced by AOM, which provides an effective dietary chemopreventive approach to disease management. However, other definitive bioassays, including protein and mRNA expression, are now in progress to establish surrogate end-point biomarkers in dietary fish-protein- and oil-product-mediated cancer chemoprevention. Currently, many reports are available on the nutritional properties of fish protein, including alteration of plasma cholesterol levels (Jacques et al., 1995; Boukortt et al., 2006; Gunnarsdottir et al., 2008), lipoprotein metabolism (Jacques, 1990; Wergedahl et al., 2004), bile acid metabolism (Shukla et al., 2006; Matsumoto et al., 2007), and antioxidative effects (Yahia et al., 2003; Boukortt et al., 2004). These attributes could be related to the mechanism of the chemopreventive effect of fish protein. Although clinical applications and the safety of daily long term administration still require further evaluation, fish protein is established as an important food factor for health promotion and is considered to be a promising candidate as a chemopreventive agent against colon cancer or inflammatory diseases such as ulcerative colitis and Crohn's disease. In the future, these findings from fish protein may aid the creation of a new field of functional food materials. In addition, they may contribute to the development of the improved understanding of food chemistry by clarifying the health functionality of fish proteins.

Acknowledgments

The authors would like to thank Professor T. Nishiyama, Department of Public Health, Kansai Medical University for helpful discussions and assistance in the project.

References

Alrawi, S.J.; Schiff, M.; Carroll, R.E.; Dayton, M.; Gibbs, J.F.; Kulavlat, M.; Tan, D.; Berman, K.; Stoler, D.L.; Anderson, G.R. Aberrant crypt foci. Anticancer Res. 2006, 26(1A), 107–119.

Badger, T.M.; Ronis, M.J.; Simmen, R.C.; Simmen, F.A. Soy protein isolate and protection against cancer. J. Am. Coll. Nutr. 2005, 24(2), 146S–149S.

Bartram, H.P.; Gostner, A.; Kelber, E.; Dusel, G.; Scheppach, W.; Kasper, H. Effect of dietary fish oil on fecal bile acid and neutral sterol excretion in healthy volunteers. Z. Ernahrungswiss. 1998, 37(Suppl 1), 139–141.

Belluzzi, A.; Boschi, S.; Brignola, C.; Munarini, A.; Cariani, G.; Miglio, F. Polyunsaturated fatty acids and inflammatory bowel disease. Am. J. Clin. Nutr. 2000, 71(Suppl 1), 339S–342S.

Berquin, I.M.; Edwards, I.J.; Chen, Y.Q. Multi-targeted therapy of cancer by omega-3 fatty acids. Cancer Lett. 2008, 269(2), 363–377.

Bligh, E.G.; Dyer, W.J. A rapid method of total lipid extraction and purification. Can. J. Biochem. Physiol. 1959, 37(8), 911–917.

Bonovas, S.; Tsantes, A.; Drosos, T.; Sitaras, N.M. Cancer chemoprevention: a summary of the current evidence. Anticancer Res. 2008, 28(3B), 1857–1866.

Boukortt, F.O.; Girard, A.; Prost, J.L.; Ait-Yahia, D.; Bouchenak, M. Belleville Fish protein improves the total antioxidant status of streptozotocin-induced diabetes in spontaneously hypertensive rat. J.Med. Sci. Monit. 2004, 10(11) , 397–404.

Boukortt, F.O.; Girard, A.; Prost, J.; Belleville, J.; Bouchenak, M. Fish proteins moderate triacylglycerols, activities of hepatic triacylglycerol lipase and tissue lipoprotein lipases in hypertensive and diabetic rats. Arch. Mal. Coeur. Vaiss. 2006. 99(7–8), 727–731.

Calder, P.C. Dietary modification of inflammation with lipids. Proc. Nutr. Soc. 2002, 61(3), 345–358.

Calder, P.C. Polyunsaturated fatty acids, inflammatory processes and inflammatory bowel diseases. Mol. Nutr. Food Res. 2008, 52(8), 885–897.

Calviello, G.; Serini, S.; Piccioni, E. n-3 polyunsaturated fatty acids and the prevention of colorectal cancer: molecular mechanisms involved. Curr. Med. Chem. 2007, 14(29), 3059–3069.

Campos, F.G.; Logullo Waitzberg, A.G.; Kiss, D.R.; Waitzberg, D.L.; Habr-Gama, A.; Gama-Rodrigues, J. Diet and colorectal cancer: current evidence for etiology and prevention. Nutr. Hosp. 2005, 20(1), 18–25.

Cheng, L.; Lai, M.D. Aberrant crypt foci as microscopic precursors of colorectal cancer. World J. Gastroenterol. 2003, 9(12), 2642–2649.

Corpet, D.E.; Pierre, F. Point: From animal models to prevention of colon cancer. Systematic review of chemoprevention in min mice and choice of the model system. Cancer Epidemiol. Biomarkers Prev. 2003, 12(5), 391–400.

Das, U.N. COX-2 inhibitors and metabolism of essential fatty acids. Med. Sci. Monit. 2005, 11(7), 233–237.

Debruyne, P.R.; Bruyneel, E.A.; Li, X.; Zimber, A.; Gespach, C.; Mareel, M.M. The role of bile acids in carcinogenesis. Mutat. Res. 2001, 480–481, 359–369.

Debruyne, P.R.; Bruyneel, E.A.; Karaguni, I.M.; Li, X.; Flatau, G.; Miller, O.; Zimber, A.; Gespach, C.; Mareel, M.M. Bile acids stimulate invasion and haptotaxis in human colorectal cancer cells through activation of multiple oncogenic signaling pathways. Oncogene. 2002, 21(44), 6740–6750.

Demonty, I.; Deshaies, Y.; Lamarche, B.; Jacques, H. Cod protein lowers the hepatic triglyceride secretion rate in rats. J Nutr. 2003, 133(5), 1398–1402.

Dommels, Y.E.; Haring, M.M.; Keestra, N.G.; Alink, G.M.; van Bladeren, P.J.; van Ommen, B. The role of cyclooxygenase in n-6 and n-3 polyunsaturated fatty acid mediated effects on cell proliferation, PGE(2) synthesis and cytotoxicity in human colorectal carcinoma cell lines. Carcinogenesis. 2003a, 24(3), 385–392.

Dommels, Y.E.; Heemskerk, S.; van den Berg, H.; Alink, G.M.; van Bladeren, P.J.; van Ommen, B. Effects of high fat fish oil and high fat corn oil diets on initiation of AOM-induced colonic aberrant crypt foci in male F344 rats. Food Chem. Toxicol. 2003b, 41(12), 1739–1747.

Dougherty, U.; Sehdev, A.; Cerda, S.; Mustafi, R.; Little, N.; Yuan, W.; Jagadeeswaran, S.; Chumsangsri, A.; Delgado, J.; Tretiakova, M.; et al. Epidermal growth factor receptor controls flat dysplastic aberrant crypt foci development and colon cancer progression in the rat azoxymethane model. Clin. Cancer Res. 2008, 14(8), 2253–2262.

Dupertuis, Y.M.; Meguid, M.M.; Pichard, C. Colon cancer therapy: new perspectives of nutritional manipulations using polyunsaturated fatty acids. Curr. Opin. Clin. Nutr. Metab. Care. 2007, 10(4), 427–432.

Fosslien, E. Molecular pathology of cyclooxygenase-2 in neoplasia. Ann. Clin. Lab. Sci. 2000, 30(1), 3–21.

Friborg, J.T.; Melbye, M. Cancer patterns in Inuit populations. Lancet Oncol. 2008, 9(9), 892–900.

Good, C.K.; Lasko, C.M.; Adam, J.; Bird, R.P. Diverse effect of fish oil on the growth of aberrant crypt foci and tumor multiplicity in F344 rats. Nutr. Cancer 1998,31(3), 204–211.

Gostner, A.; Dusel, G.; Kelber, E.; Scheppach, W.; Bartram, H.P. Comparisons of the antiproliferative effects of butyrate and aspirin on human colonic mucosa in vitro. Eur. J. Cancer Prev. 2000, 9(3), 205–211.

Gunnarsdottir, I.; Tomasson, H.; Kiely, M.; Martinez, J.A.; Bandarra, N.M.; Morais, M.G.; Thorsdottir, I. Inclusion of fish or fish oil in weight-loss diets for young adults: effects on blood lipids. Int. J. Obes. (Lond.) 2008, 32(7), 1105–1112.

Hall, M.N.; Chavarro, J.E.; Lee, I.M.; Willett, W.C.; Ma, J. A 22-year prospective study of fish, n-3 fatty acid intake, and colorectal cancer risk in men. Cancer Epidemiol. Biomarkers Prev. 2008, 17(5), 1136–1143.

Hong, M.Y.; Chapkin, R.S.; Davidson, L.A.; Turner, N.D.; Morris, J.S.; Carroll, R.J.; Lupton, J.R. Fish oil enhances targeted apoptosis during colon tumor initiation in part by downregulating Bcl-2. Nutr. Cancer. 2003, 46(1), 44–51.

Huffman, D.M.; Johnson, M.S.; Watts, A.; Elgavish, A.; Eltoum, I.A.; Nagy, T.R. Cancer progression in the transgenic adenocarcinoma of mouse prostate mouse is related to energy balance, body mass, and body composition, but not food intake. Cancer Res. 2007, 67(1), 417–424.

Hurlstone, D.P.; Cross, S.S. Role of aberrant crypt foci detected using high-magnification-chromoscopic colonoscopy in human colorectal carcinogenesis. J. Gastroenterol. Hepatol. 2005, 20(2), 173–181.

Jacques, H. Effects of dietary fish protein on plasma cholesterol and lipoproteins in animal modes and in humans. Monogr. Atheroscler. 1990, 16, 59–70.

Jacques, H.; Gascon, A.; Bergeron, N.; Lavigne, C.; Hurley, C.; Deshaies, Y.; Moorjani, S.; Julien, P. Cardiol. Role of dietary fish protein in the regulation of plasma lipids. Can. J. Cardiol. 1995, 11(Suppl G), 63G–71G.

Juste, C. Dietary fatty acids, intestinal microbiota and cancer. Bull. Cancer. 2005, 92(7), 708–721.

Kawamori, T.; Uchiya, N.; Sugimura, T.; Wakabayashi, K. Enhancement of colon carcinogenesis by prostaglandin E2 administration. Carcinogenesis. 2003, 24(5), 985–90.

Kenar, L.; Kenar, L.; Karayilanoglu, T.; Aydin, A.; Serdar, M.; Kose, S.; Erbil, M.K. Protective effects of diets supplemented with omega-3 polyunsaturated fatty acids and calcium against colorectal tumor formation. Dig. Dis. Sci. 2008, 53(8), 2177–2182.

Kim, J.; Ng, J.; Arozulllah, A.; Ewing, R.; Llor, X.; Carroll, R.E.; Benya, R.V. Aberrant crypt focus size predicts distal polyp histopathology. Cancer Epidemiol. Biomarkers Prev. 2008, 17(5), 1155–1162.

Kohno, H.; Yamaguchi, N.; Ohdoi, C.; Nakajima, S.; Odashima, S.; Tanaka, T. Modifying effect of tuna orbital oil rich in docosahexaenoic acid and vitamin D3 on azoxymethane-induced colonic aberrant crypt foci in rats. Oncol. Rep. 2000, 7(5), 1069–1074.

Mahmood, B.; Higgs, N.; Howl, L.; Warhurst, G. Colonic secretion studied in vitro in rats fed polyunsaturated fatty acids. Bangladesh Med. Res. Counc. Bull. 2006, 32(3), 72–77.

Matsumoto, J.; Enami, K.; Doi, M.; Kishida, T.; Ebihara, K. Hypocholesterolemic effect of katsuobushi, smoke-dried bonito, prevents ovarian hormone deficiency-induced hypercholesterolemia. J. Nutr. Sci. Vitaminol. 2007, 53(3), 225–231.

McIntosh, G.H.; Le Leu, R.K. The influence of dietary proteins on colon cancer risk. Nutr. Res. 2001, 21(7), 1053–1066.

Morgado, N.; Rigotti, A.; Valenzuela, A. Comparative effect of fish oil feeding and other dietary fatty acids on plasma lipoproteins, biliary lipids, and hepatic expression of proteins involved in reverse cholesterol transport in the rat. Ann. Nutr. Metab. 2005, 49(6), 397–406.

Mori, H.; Yamada, Y.; Hirose, Y.; Kuno, T.; Katayama, M.; Sakata, K.; Yoshida, K.; Sugie, S.; Hara, A.; Yoshimi, N. Chemoprevention of large bowel carcinogenesis; the role of control of cell proliferation and significance of beta-catenin-accumulated crypts as a new biomarker. Eur. J. Cancer Prev. 2002, 11(Suppl 2), S71–75.

Mori, H.; Yamada, Y.; Kuno, T.; Hirose, Y. Aberrant crypt foci and beta-catenin accumulated crypts; significance and roles for colorectal carcinogenesis. Mutat. Res. 2004, 566(3), 191–208.

Narisawa, T.; Fukaura, Y.; Yazawa, K.; Ishikawa, C.; Isoda, Y.; Nishizawa, Y. Colon cancer prevention with a small amount of dietary perilla oil high in alpha-linolenic acid in an animal model. Cancer. 1994, 73(8), 2069–2075.

Orlando, F.A.; Tan, D.; Baltodano, J.D.; Khoury, T.; Gibbs, J.F.; Hassid, V.J.; Ahmed, B.H.; Alrawi, S.J. J Aberrant crypt foci as precursors in colorectal cancer progression. Surg. Oncol. 2008, 98(3), 207–213.

Paulsen, J.E.; Knutsen, H.; Ølstørn, H.B.; Løberg, E.M.; Alexander, J. Identification of flat dysplastic aberrant crypt foci in the colon of azoxymethane-treated A/J mice. Int. J. Cancer. 2006, 118(3), 540–546.

Pence, B.C.; Landers, M.; Dunn, D.M.; Shen, C.L.; Miller, M.F. Feeding of a well-cooked beef diet containing a high heterocyclic amine content enhances colon and stomach carcinogenesis in 1,2-dimethylhydrazine-treated rats. Nutr. Cancer. 1998, 30(3), 220–226.

Rao, C.V.; Hirose, Y.; Indranie, C.; Reddy, B.S. Modulation of experimental colon tumorigenesis by types and amounts of dietary fatty acids. Cancer Res. 2001, 61(5), 1927–1933.

Reddy, B.S.; Maruyama, H. Effect of dietary fish oil on azoxymethane-induced colon carcinogenesis in male F344 rats. Cancer Res., 1986, 46(7), 3367–3370.

Reddy, B.S.; Patlolla, J.M.; Simi, B.; Wang, S.H.; Rao, C.V. Prevention of colon cancer by low doses of celecoxib, a cyclooxygenase inhibitor, administered in diet rich in omega-3 polyunsaturated fatty acids. Cancer Res. 2005a, 65(17), 8022–8027.

Reddy, B.S.; Rao, C.V. Chemoprophylaxis of colon cancer. Curr. Gastroenterol. Rep. 2005b, 7(5), 389–395.

Reddy, B.S.; Wang, C.X.; Maruyama, H. Effect of restricted caloric intake on azoxymethane-induced colon tumor incidence in male F344 rats. Cancer Res. 1987, 47(5), 1226–1228.

Reeves, P.G.; Nielsen, F.H.; Fahey, G.C. Jr. AIN-93 purified diets for laboratory rodents: final report of the American Institute of Nutrition ad hoc writing committee on the reformulation of the AIN-76A rodent diet. J. Nutr. 1993, 123(11), 1939–1951.

Roncucci, L.; Pedroni, M.; Vaccina, F.; Benatti, P.; Marzona, L.; De Pol, A. Aberrant crypt foci in colorectal carcinogenesis. Cell and crypt dynamics. Cell Prolif. 2000, 33(1), 1–18.

Santarelli, R.L.; Pierre, F.; Corpet, D.E. Processed meat and colorectal cancer: a review of epidemiologic and experimental evidence. Nutr. Cancer. 2008, 60(2), 131–144.

Selvam, J.P.; Aranganathan, S.; Nalini, N. Aberrant crypt foci and AgNORs as putative biomarkers to evaluate the chemopreventive efficacy of pronyl-lysine in rat colon carcinogenesis. Invest. New Drugs 2008, 26(6), 531–540.

Sheltawy, M.J.; Losowsky, M.S. Determination of faecal bile acids by an enzymic method. Clin. Chim. Acta. 1975, 64(2), 127–132.

Shiraki, K.; Ito, T.; Sugimoto, K.; Fuke, H.; Inoue, T.; Miyashita, K.; Yamanaka, T.; Suzuki, M.; Nabeshima, K.; Nakano, T.; Takase, K. Different effects of bile acids, ursodeoxycholic acid and deoxycholic acid, on cell growth and cell death in human colonic adenocarcinoma cells. Int. J. Mol. Med. 2005, 16(4), 729–733.

Shukla, A.; Bettzieche, A.; Hirche, F.; Brandsch, C.; Stangl, G.I.; Eder, K. Dietary fish protein alters blood lipid concentrations and hepatic genes involved in cholesterol homeostasis in the rat model. Br. J. Nutr. 2006, 96(4), 674–682.

Stevens, R.G.; Swede, H.; Rosenberg, D.W. Epidemiology of colonic aberrant crypt foci: review and analysis of existing studies. Cancer Lett. 2007, 252(2), 171–183.

Takayama, T.; Miyanishi, K.; Hayashi, T.; Kukitsu, T.; Takanashi, K.; Ishiwatari, H.; Kogawa, T.; Abe, T.; Niitsu, Y. Aberrant crypt foci: detection, gene abnormalities, and clinical usefulness. Clin. Gastroenterol. Hepatol. 2005, 7(Suppl 1), S42–45.

Tapiero, H.; Ba, G.N.; Couvreur, P.; Tew, K.D. Polyunsaturated fatty acids (PUFA) and eicosanoids in human health and pathologies. Biomed. Pharmacother. 2002, 56(5), 215-222.

Thomson, C.A.; LeWinn, K.; Newton, T.R.; Alberts, D.S.; Martinez, M.E. Nutrition and diet in the development of gastrointestinal cancer. Curr. Oncol. Rep. 2003, 5(3), 192–202.

Tong, J.L.; Ran, Z.H.; Shen, J.; Fan, G.Q.; Xiao, S.D. Association between fecal bile acids and colorectal cancer: a meta-analysis of observational studies. Yonsei. Med J. 2008, 49(5), 792–803.

Wergedahl, H.; Liaset, B.; Gudbrandsen, O.A.; Lied, E.; Espe, M.; Muna, Z.; Mørk, S.; Berge, R.K. Fish protein hydrolysate reduces plasma total cholesterol, increases the proportion of HDL cholesterol, and lowers acyl-CoA:cholesterol acyltransferase activity in liver of Zucker rats. J. Nutr. 2004, 134(6), 1320–1327.

Xiao, R.; Badger, T.M.; Simmen, F.A. Dietary exposure to soy or whey proteins alters colonic global gene expression profiles during rat colon tumorigenesis. Mol. Cancer. 2005, 4(1), 1–17

Yahia, D.A.; Madani, S.; Prost, E.; Prost, J.; Bouchenak, M.; Belleville, J. Tissue antioxidant status differs in spontaneously hypertensive rats fed fish protein or casein. J. Nutr. 2003, 133(2), 479–482.

Index

Printed and bound by CPI Group (UK) Ltd, Croydon, CR0 4YY

11/05/2025

01866606-0001